POLYMER SCIENCE AND TECHNOLOGY SERIES

ADVANCES IN POLYMER LATEX TECHNOLOGY

Polymer Science and Technology Series

Oligomeric State of Substances
S.M. Mezhikovskii, A.E. Arinstein, and R.Ya. Deberdeev
2009 ISBN: 978-1-60741-344-8

Polycyclic Aromatic Hydrocarbons: Pollution, Health Effects and Chemistry
Pierre A. Haines and Milton D. Hendrickson
2009 ISBN: 978-1-60741-462-9

Advances in Polymer Latex Technology
Vikas Mittal
2009 ISBN: 978-1-60741-170-3

POLYMER SCIENCE AND TECHNOLOGY SERIES

ADVANCES IN POLYMER LATEX TECHNOLOGY

VIKAS MITTAL

Nova Science Publishers, Inc.
New York

For permission to use material from this book please contact us:
Telephone 631-231-7269; Fax 631-231-8175
Web Site: http://www.novapublishers.com

LIBRARY OF CONGRESS CATALOGING-IN-PUBLICATION DATA

Mittal, Vikas.
Advances in polymer latex technology / Vikas Mittal.
p. cm. -- (Polymer science and technology)
Includes bibliographical references and index.
ISBN 978-1-60741-170-3 (hardcover : alk. paper)
1. Polymerization. 2. Latex. I. Title.
TP156.P6M48 2009
668.9'2--dc22

2009004634

Published by Nova Science Publishers, Inc. ✛ New York

CONTENTS

PREFACE

Polymer latex technology by emulsion polymerization offers significant advantages compared with bulk and solution polymerization technologies owing to the better control of heat and viscosity of the medium along with the possibility of increasing the molecular weight of the polymer chains without affecting the rate of polymerization. These emulsion polymerization methods have undergone significant advances in recent years and, at present, synthesis of structured latexes with well-defined morphologies and properties is possible. The development of miniemulsion polymerization has helped to polymerize very hydrophobic and very hydrophilic monomers, which otherwise are difficult to polymerize with conventional emulsion polymerization. The combination of controlled living polymerization methods, like nitroxide mediated polymerization, atom transfer radical polymerization as well as reversible addition fragmentation chain transfer with the emulsion and miniemulsion polymerization methods has resulted in the synthesis of functional block copolymer chains, graft copolymer chains and star or multi-arm copolymer chains in the polymer particles. Apart from that, specific surface modifications on the surface of the emulsion particles have been achieved, which have expanded the area of application of such polymer particles. The kinetics and mechanism of particle generation with living polymerization in emulsion and miniemulsion are very complex; however, recent advances have helped us to understand these systems more clearly, thus allowing us to steer them advantageously. Therefore, these recent advances in latex technology need to be gathered into a cumulative text in order to provide insight into the various possibilities of achieving the optimum latexes.

The first three chapters of the text deal with the basic mechanisms and techniques of generating polymers, and focus on the kinetics of radical polymerization. A detailed elaboration of the concept of emulsion polymerization is provided to make the reader aware of the background of this technology, before stepping towards the detailed documentation of the various advances achieved in these technologies. Chapter 4 compiles the various advances in polymerization methods by using more controlled means of polymerization, such as nitroxide mediated polymerization, atom transfer radical polymerization and reversible addition fragmentation chain transfer. The application of these methods for heterogeneous polymerization is also been detailed. The miniemulsion polymerization is described in a separate chapter (Chapter 5), owing to its significant potential, and various controlled radical polymerization techniques are also described in the context of miniemulsion polymerization. Chapters 6 and 7 focus on the generation of core shell morphologies or structured latexes, where the synthesis and characterization of such latexes are described both by using the conventional non-living polymerization as well as more controlled living polymerization methods. The role of microscopy in the characterization of the size as well as surface morphology of the particles is underlined in Chapter 8. Various microscopy techniques, including SEM, AFM, TEM, EELS, Cryo-SEM, etc. are discussed for their importance in the characterization of polymer latexes. Advances in areas of more commercial significance, such as latex stabilization, high solids and scale up, are discussed in Chapter 9. A special study on the generation of thermally reversible polymer particles is presented in Chapter 10. Generation of such thermally reversible effects has been studied a great deal which leads to the application of these functional particles in a large number of applications.

ACKNOWLEDGMENTS

The author would like to take this opportunity to acknowledge the contributions of those who helped to bring this effort to reality. I would like to thank Nova Science Publishers for their kind offer to publish this work. I am indebted to Professor Massimo Morbidelli at the Swiss Federal Institute of Technology, Zurich, Switzerland, in whose guidance I acquired the know-how of latex technology. I would like to express my heartfelt thanks to my family, and especially to my mother, whose support was absolutely vital for the realization of this work.

I would like to dedicate this book to my dear wife, Preeti, for correcting the manuscript, and for her patience throughout the duration of the writing of this book.

Chapter 1

POLYMERS AND POLYMERIZATION

A. INTRODUCTION

What are polymers? Since when have they been known? Natural polymers have surely been in use for centuries owing to their beneficial mechanical and chemical properties. Scientific developments in recent years have tremendously helped in understanding the structure property relationships of various natural polymers, and subsequently led to the generation of a number of tailored synthetic polymers to suit today's requirements. If we were to look around, we would find more and more products made of polymers and, because of their tailorable properties, polymers are also replacing other conventional materials such as glass or steel in a number of applications, including packaging and automobiles. Because of their high thermal and mechanical properties, the polymers of today find applications in almost every spectrum of products, whether the tip of a spacecraft, a bulletpoof jacket, food packaging, the dashboard of an automobile or even an artificial organ transplant.

Polymers are defined simply as the long-chain macromolecules formed by the joining of the repeating units, or monomers. Polymers are versatile materials and, depending upon the route of their synthesis and nature of the starting materials, can form numerous materials with properties varying greatly, having traits like branching, crosslinking, crystallinity, atacticity, isotacticity, syndiotacticity, hydrophobicity, hydrophilicity, etc. Apart from that, the nature of polymers can also be changed by the number of monomer units attached in the chains, i.e., the degree of polymerization. Lesser numbers of these repeating units in the polymer chains (or lower degree of polymerization) would create more tacky polymers, whereas the same polymer chains would become completely unsticky after crossing a threshold degree of polymerization (generally 10,000). Over the years, a number of methods have been generated to achieve polymers, including conventional methods like free radical polymerization, ionic polymerization, condensation polymerization, redox polymerization, and coordination polymerization, as well as more advanced living or controlled polymerizations like nitroxide mediated polymerization, atom transfer radical polymerization, radical addition chain fragmentation transfer polymerization, reverse atom transfer radical polymerization, activator generated by electron transfer for atom transfer radical polymerization, etc. In addition, remarkable developments in polymerization techniques have also been achieved with polymerization methods of bulk, solution, emulsion, suspension, miniemulsion, micro-emulsion and reverse emulsion polymerization, etc. which are now understood in more detail.

B. Classification of Polymers

Based on the characteristics mentioned above, polymers can be classified into different categories in many different ways. From the composition or structural viewpoint, polymers are classified as condensation and addition polymers [1]. Condensation polymers are defined as the polymers generated by the chemical reactions that lead to the generation of small byproduct molecules, mostly water. Common examples of polymers synthesized by using such a methodology are polyesters (reaction of bifunctional or multifunctional acids with bifunctional or multifunctional alcohols), polyamides (reactions of bifunctional or multifunctional acids with bifunctional or multifunctional amines), etc. An example of a condensation reaction is shown in equation 1 for the synthesis of polyesters, where m = 2n-1.

$$\mathrm{n\ HOOC{-}R{-}COOH} + \mathrm{n\ HO{-}R'{-}OH} \longrightarrow \mathrm{HO}\left[\mathrm{OC{-}R{-}\overset{\displaystyle O}{\overset{\|}{C}}{-}O{-}R'{-}O}\right]_n\mathrm{H} + \mathrm{m\ H_2O} \tag{1}$$

Addition polymers were classified by Carothers as polymers during the synthesis of which no byproduct or small molecule is generated; thus, the monomer unit in full forms the repeat unit of the polymer chain. Examples of addition polymers are polyethylene, polypropylene, polystyrene, poly(methyl methacrylate), etc. Equation 2 is an example of the addition polymerization of methyl methacrylate to form poly(methyl methacrylate).

$$\mathrm{n}\ \begin{matrix} \mathrm{H} & \mathrm{CH_3} \\ | & | \\ \mathrm{C} = & \mathrm{C} \\ | & | \\ \mathrm{H} & \mathrm{COOCH_3} \end{matrix} \longrightarrow \left[\begin{matrix} \mathrm{H} & \mathrm{CH_3} \\ | & | \\ \mathrm{C} - & \mathrm{C} \\ | & | \\ \mathrm{H} & \mathrm{COOCH_3} \end{matrix}\right]_n \tag{2}$$

However, as mentioned by Odian, this led to the classification of polyurethanes under the category of addition polymers, even though structurally the nature of these polymers resembles those of condensation polymers [1]. It was further correctly defined that condensation polymers are the polymers in which the repeating units are joined together by one functional group or the other. On the other hand, polymers synthesized by addition polymerization would either have no functional group in the chain or, in the case that they are present, they are present as the pendant groups, as in polystyrene, poly(methyl methacrylate), etc.

There have been further attempts to classify the polymers based on the synthesis mechanisms as step and chain polymers. Step polymers represent the synthesis of the polymers by the reactions of the functional groups, and during these reactions, generally small molecules are generated as byproducts. Removal of these molecules during the reaction process shifts the reaction equilibrium towards the generation of products. Step polymerization proceeds literally in steps of monomers reacting with each other to form dimers, dimers reacting with each other to form tetramers, monomers reacting with dimers to form trimers, and so on. Polycondensation, polyaddition, amidation, esterification and ring opening reactions are some of the most important reaction types in this category of step synthesis of polymers. The following steps describe the process of step polymerization:

$$M + M \longrightarrow M{-}M$$

$$M{-}M + M \longrightarrow M{-}M{-}M$$

$$M{-}M + M{-}M \longrightarrow M{-}M{-}M{-}M$$

$$M{-}M{-}M + M \longrightarrow M{-}M{-}M{-}M$$

An example of such a reaction is the polyamidation reaction between the diacids and diamines to form polyamides, as shown in equation 3, where m = 2n-1.

$$n\,H_2N{-}(CH_2)_6{-}NH_2 + n\,HOOC{-}(CH_2)_4{-}COOH \longrightarrow H{-}\left[NH{-}(CH_2)_6{-}NHCO{-}(CH_2)_4{-}CO\right]_n{-}OH + m\,H_2O \quad (3)$$

Chain polymerization follows completely different routes compared with the step polymerization. In this process, external initiators are used to react with the monomer molecules, generating an active centre with either a cation, anion or free radical depending on the type of initiator used. This then reacts with further monomer molecules, forming long chains. The special feature of chain polymerization is that the monomers do not react with each other because they do not have any reactivity toward each other. The high molecular weight is generated at earlier stages of polymerization in the case of chain polymerization, but, owing to the different route of polymerization, step polymerization generates a high molecular weight only at the high monomer conversion. Equation 4 is an example of chain polymerization to polymerize styrene, where R* can represent an anion, a cation or a free radical.

$$CH_2{=}CH(C_6H_5) \xrightarrow{R^*} R{-}CH_2{-}\overset{H}{C^*}(C_6H_5) \xrightarrow{CH_2=CH(C_6H_5)} R{-}CH_2{-}\overset{H}{C}(C_6H_5){-}CH_2{-}\overset{H}{C^*}(C_6H_5) \quad (4)$$

There are also other numerous classifications of polymers based on the structure and properties of polymers, as explained in Figure 1. It is worth noting that the whole spectrum of polymer classification is very wide and is also representative of the potential applications of polymers in numerous fields. Based on the structural differences, polymers are classified as thermoplastics and thermosets. Thermoplastics are defined as the materials that can be formed into any required shape by the use of heat. Common thermoplastics are polyethylene, polypropylene, polystyrene, poly(methyl methacrylate), etc. and can be recycled and reformed into different articles. These polymers have different glass transition temperatures, and at room temperature these polymers are either below or above the glass transition temperatures. In both cases, the chains are present as an entangled mass and differ only in the extent of their segmental mobility. By heating the chains to their melting point, they achieve a large segmental mobility and lose their earlier form. Thermosets are materials that cannot be reformed into other shapes by the use of temperature. Typical examples of these materials are epoxies, polyurethane, unsaturated polyesters, etc. Their setting into one shape also depends on the degree of crosslinking. Once crosslinked to a certain value, their shape is fixed forever, and the use of high temperature only destroys the material rather than changing the shape.

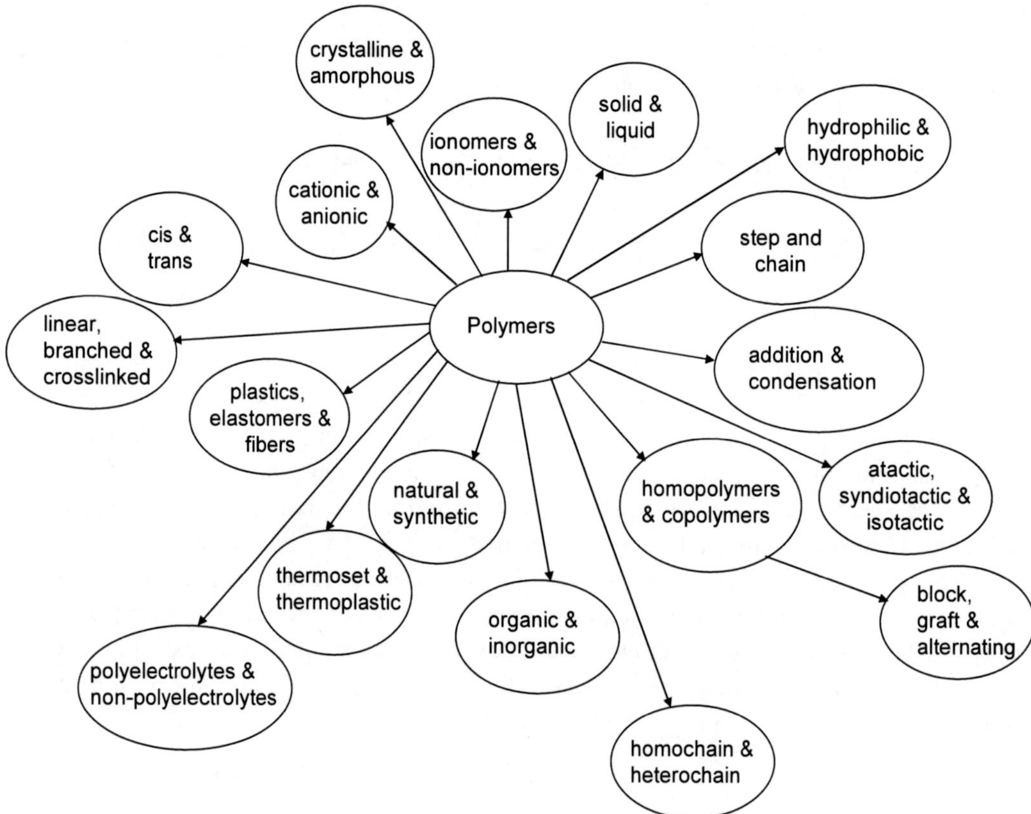

Figure 1. Classification of polymers based on structure and properties of polymers.

Based on the elements present in the backbone of the chains, polymers are also classified as organic and inorganic polymers. Organic polymers have their backbones essentially made of carbon and the side groups as well as atoms attached to these carbon atoms can contain further carbon atoms or oxygen, hydrogen and nitrogen atoms. The majority of the synthetic polymers are, in general, organic polymers. Inorganic polymers generally do not contain any carbon atom in the backbone, e.g., silicone polymer with chains containing silicon and oxygen atoms in the backbone. Based on the one or two kinds of atoms present in the polymer chain, the polymers can be classified into homochain and heterochain polymers [2].

Homochain polymers are the materials with only one kind of atoms present in the backbone, as an example, polyethylene with only carbon atoms in the backbone. Heterochain polymers contain two or more kinds of atoms present in the backbone. Silicones are an example of heterochain polymers with oxygen and silicon atoms in the backbone.

Based on the origin, the polymers are also defined as natural and synthetic polymers. Natural polymers are derived from natural sources, e.g., natural rubber, silk, wool, etc., whereas the polymers synthesized from natural raw materials are called synthetic polymers, e.g., polyethylene made from ethylene, polypropylene made from propylene, etc. Based on the physical form, the polymers are also divided into elastomers, plastics and fibers.

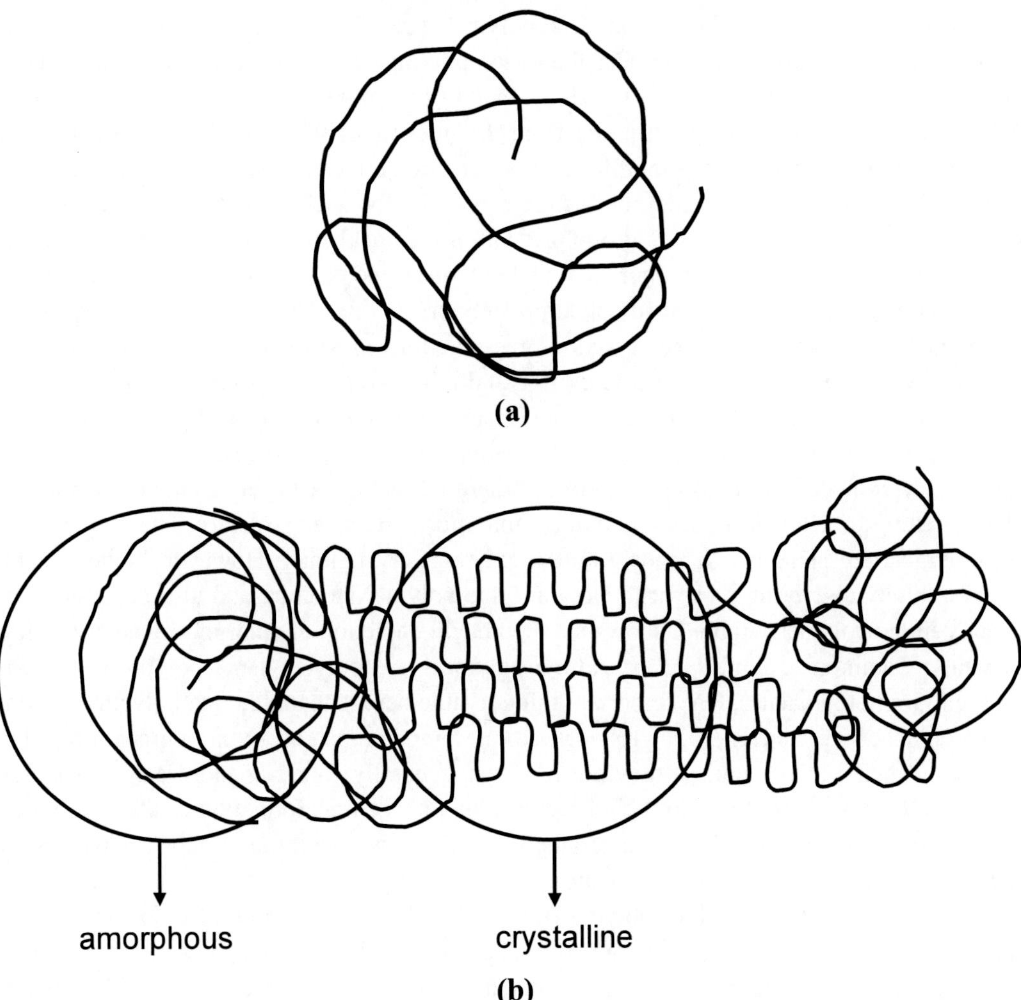

Figure 2. (a) Totally amorphous polymer chains and (b) a mixture of crystalline and amorphous domains in the polymer matrix.

Elastomers are the materials which generally have rubbery nature with larger extensibility and sometime also foamy nature. These polymers generally are above their glass transition temperature at the room temperature and thus have higher extents of segmental mobility in the structure. General examples are natural and synthetic rubbers, etc. Plastics are more compact materials in which the chains are more compactly arranged by the use of pressure and temperature. The common examples are polyethylene, polypropylene, polystyrene, etc. The majority of these plastics are below their glass transition temperature at room temperature leading to more restricted segmental mobility in the chains (only exceptions are polyethylene and polypropylene). When the polymers are unilaterally stretched, it results into polymer fibers, e.g., nylon fibers. Similarly, the physical form also allows one to classify polymers into liquids and solids. Majority of polymers are solids at high molecular weights, e.g., polyethylene, polystyrene, etc.

Depending on the nature of chain packing, the polymers are classified as crystalline and amorphous polymers. Crystalline polymers are the polymers in which the chains are packed into crystalline domains, thus generating polymer specific crystals with specific crystal dimensions. The folded chain lamella theory of polymer crystallization defines lamella as polymer single crystals growing in the shape of thin two dimensional platelets. Spherulite is defined as a three-dimensional structure from the nucleus of which multilayers of lamellae grow outwards to form fibrils. Amorphous polymers are defined as the polymers that lack this regular chain packing. There may be different reasons for such a behavior, most common being the presence of a bulky side group present on the backbone which hinders the packing of the chains in an organized manner. Completely crystalline polymers are rarely found as the crystalline polymers themselves are not completely crystalline. There is always a presence of some amorphous part around the crystalline regions, thus most commonly used terminology is semi-crystalline polymers. The extent of crystallinity can also be controlled in the manner the polymers are cooled from melt. If cooled slowly, higher extent of crystallinity can be achieved as it provides chains ample time to organize into crystallites. However, if the chains are quickly cooled, it leads to only a partial ordered structure with lot of entangled polymer chains getting frozen before they could move into more organized structure. The properties of the polymers are markedly dependant on the extent of crystallinity. However, higher extents of crystallinity also bring more brittleness into the polymer structure and also generate more opacity in the polymer. Figure 2 shows the theoretical depiction of amorphous and crystalline domains. Common examples of crystalline polymers are polyethylene, polypropylene and poly(ethyleneterepthalate). One important thing to note here is that although polyethylene and polypropylene are above their glass transition temperatures at room temperature, thus signifying their rubbery nature, they are however generally found as more compact solids. Higher extents of crystallinity in their structures lead to the transformation of these polymers from rubbery to more plastic like materials. Common examples of amorphous polymers are polystyrene and poly(methyl methacrylate).

Depending on the method of polymerization as well as the location of monomer units in the polymer chains, the polymers are also classified as linear, branched and crosslinked polymers. Linear polymers are the polymers in which the monomer units only form the backbone of the polymer chains leading to linear chains of large number of monomer units attached to each other. Branched polymers are the polymers in which the monomer units are also attached to the main chain as branches leading to the attachment of these monomer units to some of the monomer units in the main chain. This branching can occur due to many reasons like proton abstraction from the main chain and subsequent generation of active centre which can then attach monomer units as branches. Branching leads to reduction in the molecular weight as well as extent of crystallinity. Crosslinked polymers can be formed by two ways. The branches formed from one main chain in the case of branched polymers can attach to the other main chain thus cross linking the two chains. Similarly, as the polymerization proceeds, more and more links can be generated to achieve crosslinked polymers. Crosslinked polymers can also be achieved by using multifunctional monomer units especially when polymers are formed by step polymerization, e.g., using multifunctional amines and multifunctional alcohols to form crosslinked polyamides. Figure 3 shows the structures of these linear, branched and crosslinked polymers. Polyethylene structures for linear, branched and crosslinked chains are shown for clarity. Figure 4 also shows the mechanism of crosslinking in the case of polyurethane (using isocyanate and alcohols) and epoxy (using alcohol and DGEBA) polymers.

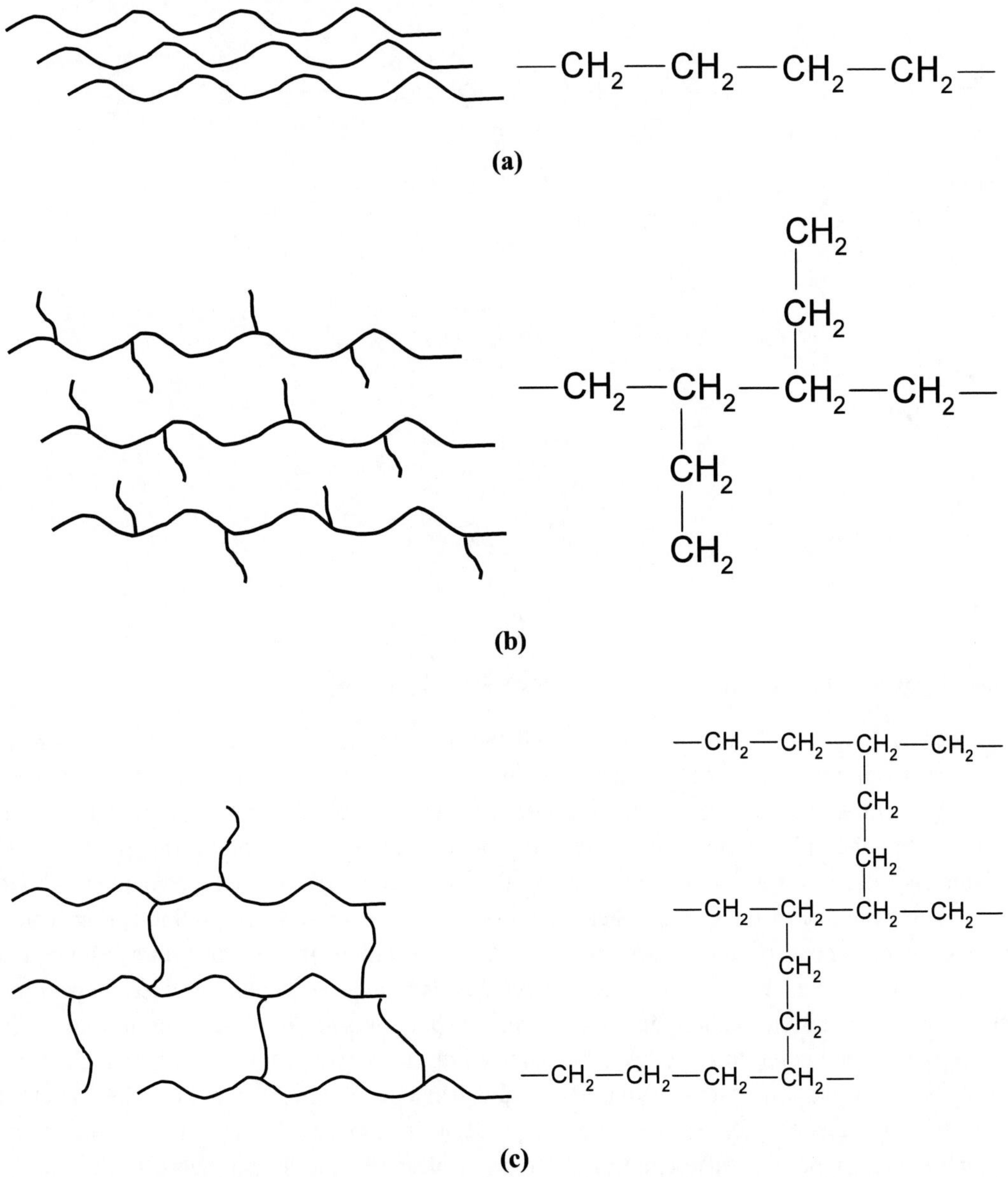
—CH2—CH2—CH2—CH2—
(a)
CH2
CH2
—CH2—CH2—CH2—CH2—
CH2
CH2
(b)
—CH2—CH2—CH2—CH2—
CH2
CH2
—CH2—CH2—CH2—CH2—
CH2
CH2
—CH2—CH2—CH2—CH2—
(c)

Figure 3. (a) Linear, (b) branched and (c) crosslinked polymer chains in general (on the left) and polyethylene structures in specific (on the right).

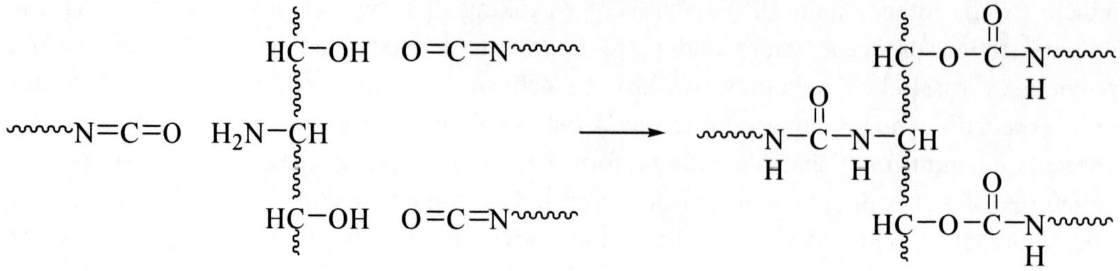

(a)

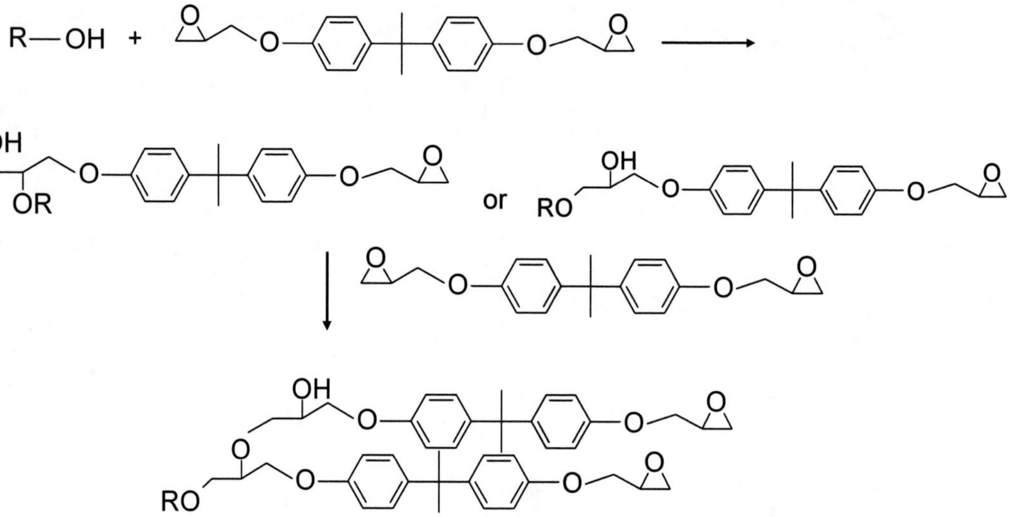

(b)

Figure 4. Mechanism of crosslinking of (a) polyurethane and (b) epoxy.

Polymers can also be classified based on isomerism. When optical isomerism is considered, polymers are generally classified into atactic, isotactic and syndiotactic polymers. These classifications are based on the conformations of the side groups around the main chain. When the side groups are present on one side of the plane of the main chain, the polymer is called as isotactic polymer. When the side groups are present alternatively above and below the plane of the carbon backbone, the polymer is termed as syndiotactic polymer. In case, there is no particular order present in the positioning of the side groups above and below the plane of the main chain, the polymer then is termed as atactic polymer. Depending on the polymerization techniques, one form of polymers chains can be preferentially synthesized as compared to the others thus controlling the properties of the polymer. Figure 5 shows the examples of atactic, syndiotactic and isotactic polymers. On the other hand, if geometric isomerism is considered, the polymers can also be divided into cis and trans polymers. One important difference in this type of classification as compared to the optical isomerism based classification is that in the case of geometrical isomerism, only double bonds are involved and it is based on the configurations of the side groups around the double bond of the chain. One-well known example of polymer showing such a behavior is polybutadiene in which butadiene with two double bonds in the chemical structure is polymerized into polybutadiene thus resulting in one double bond, as shown in equation 5.

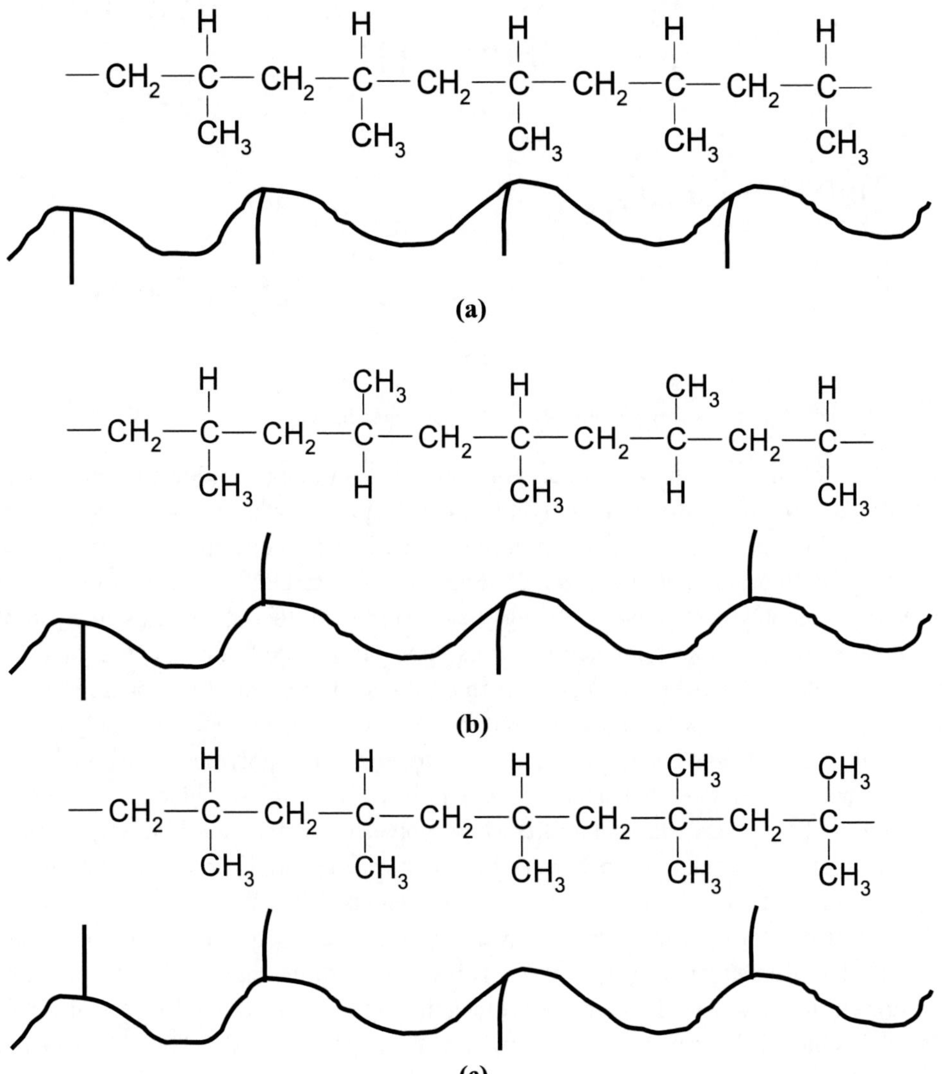

Figure 5. Optical isomerism of polymers. (a) Isotactic, (b) syndiotactic and (c) atactic conformations of polypropylene in specific and polymer chains in general.

$$CH_2{=}CH{-}CH{=}CH_2 \longrightarrow \left[CH_2{-}CH{=}CH{-}CH_2 \right]_n \qquad (5)$$

The position of the CH_2 groups around the double bonds can generate optical isomers in the polymer chains. If both the CH_2 groups are present on one side of the chain, the arrangement is defined as cis-configuration, whereas if the CH_2 groups are present on the opposite sides of the double bond, the configuration is termed as trans. Figure 6 shows the cis and trans configurations for polybutadiene.

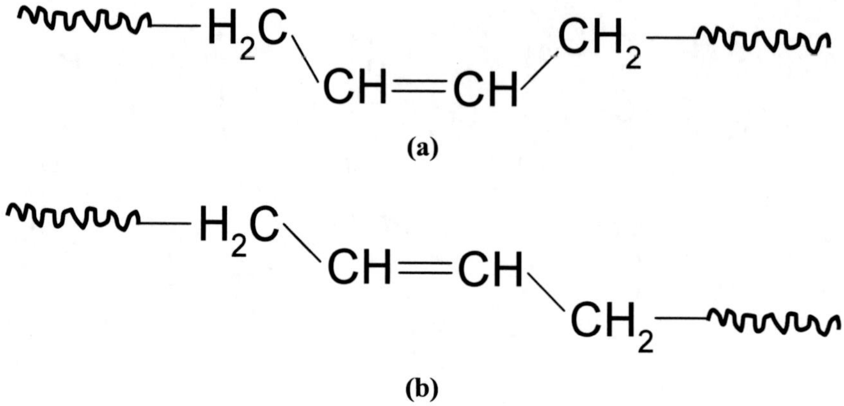

$$\sim\!\sim\!\sim - H_2C\diagdown CH{=}CH \diagup CH_2 - \sim\!\sim\!\sim$$

(a)

$$\sim\!\sim\!\sim - H_2C\diagdown CH{=}CH \diagdown CH_2 - \sim\!\sim\!\sim$$

(b)

Figure 6. (a) Cis and (b) trans conformations of polybutadiene chains.

Based on the interactions with water, polymers are also categorized into hydrophobic and hydrophilic polymers. Hydrophilic polymers have affinity towards water and can either dissolve fully in water or can become highly swollen in water due to the sorption of water molecules inside the polymer structure. Water molecules are also known to form hydrogen bonds with these polymers thus increasing their swelling further. Examples of hydrophilic polymers are polyurethanes, polyacrylic acids, polyacrylamides, etc. On the other hand, hydrophobic polymers are the polymers, which do not tend to swell or solubilize in water. In fact, these properties are extremely well used in the suspension and emulsion polymerization methods of polymer synthesis. Examples of hydrophobic polymers are polyethylene, polypropylene, poly(methyl methacrylate) and polystyrene, etc. Based on the interactions with water, polymers are also classified into polyelectrolytes and non-polyelectrolytes. Polyelectrolytes are the charged polymers in which the repeating unit bears an electrolyte group. Due to this, these polymers are also termed as polysalts. These polymers when dissolved in water form hydronium ions with water molecules by giving protons from the polymer chains (first equation of Figure 7). This results into the negatively charged polymer chains present in water. As the chains have negative charges all around it, because of their repulsion it can no longer coil randomly and thus stretches out. This phenomenon is commercially exploited into a number of applications like rheology modifiers to increase viscosity. Their stretched out conformation is easily destroyed if salt like NaCl is added as sodium ions balance the negative charges present on the chains and the chains return to more coiled forms. Most common examples of such polyelectrolytes are polyacrylic acids and Figure 7 details the behavior of these polymers in water. Polymers are also classified as ionomers and non-ionomers. Ionomers are a special form of polyelectrolytes, but they are copolymers with polymer chains containing ionic and non-ionic parts. An example of ionomers is the copolymer of ethylene with acrylic acid to form poly(ethylene-co-acrylic acid). The polar ionic groups have attraction towards each other and such groups present on different chains form bonds with each other forming structures similar to crosslinked polymers. However, the crosslinked structure in ionomers is non-permanent. Heating the polymer causes the bonds between the polar components to break and the chains achieve full segmental mobility. Thus these polymers have properties of elastomer and processability like thermoplastics. Therefore, these ionomers are also termed as thermoplastic elastomers. Figure 8 is the depiction of these ionomers and their behavior.

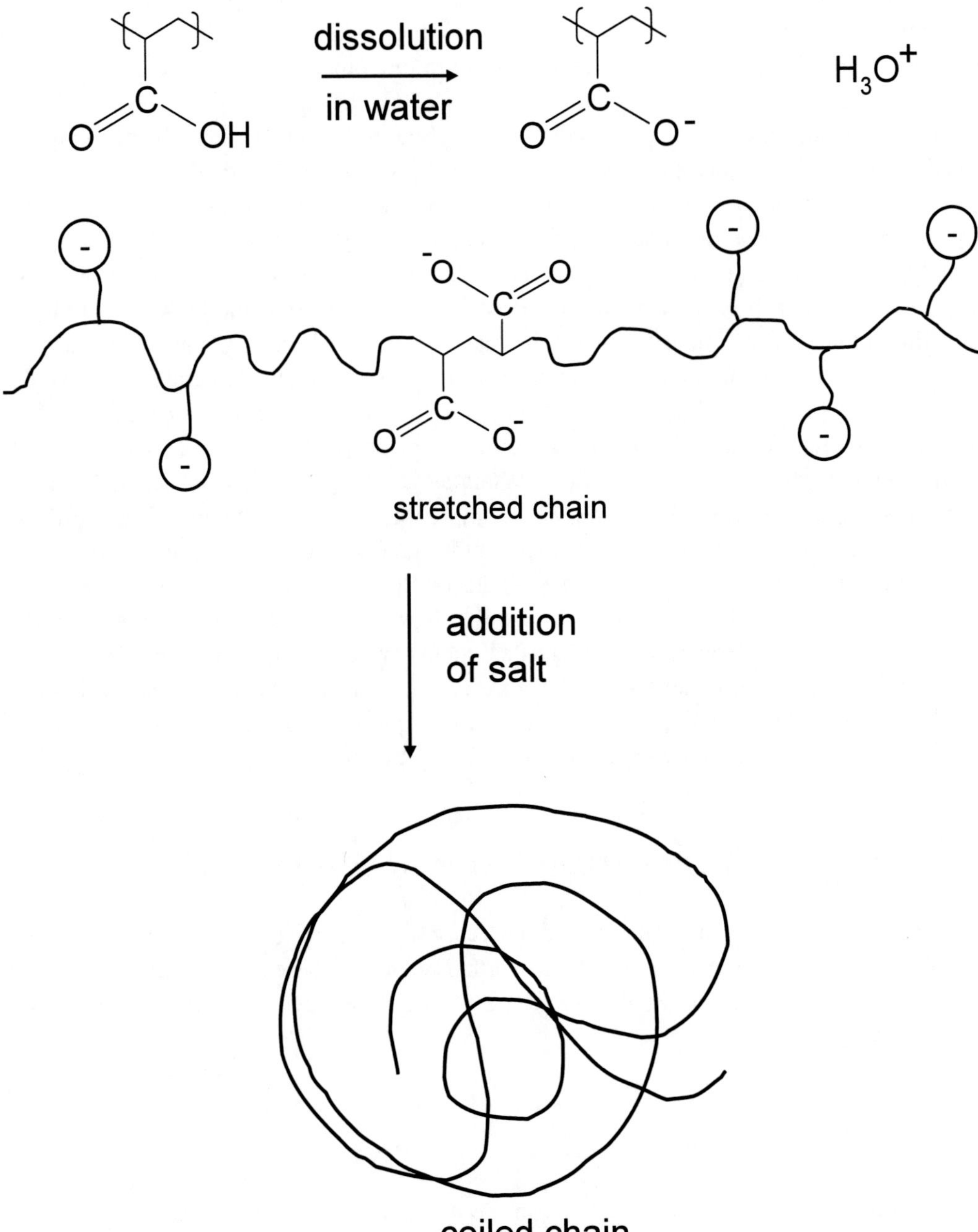

Figure 7. Behavior of polyelectrolytes in water and after addition of salt.

C. Homopolymers and Copolymers

As mentioned above, copolymers are very special form of polymers that combine the properties of two or more polymers and depending on the type of copolymer can have very different applications. Thus, polymers are can also be classified as homopolymers and copolymers. Homopolymers are polymers in which the repeating units in the backbone are from one monomer unit. Whereas when two or more monomers are polymerized together, a copolymer is formed in which the monomer units of all the monomers are present. When these repeating units are distributed randomly throughout the polymer chain, the copolymer is termed as a random copolymer. Figure 9a shows the example of a random copolymer of ethylene and styrene. When the repeating units are present alternatively throughout the polymer chain, the copolymer is termed as alternating copolymer. Common example of such a copolymer is styrene-co-methyl methacrylate copolymer shown in Figure 9b. In this case, the styrene units attach only to methyl methacrylate units owing to the higher reactivity of this addition than the styrene monomer units reacting with each other. Same is true also for the methyl methacrylate monomer units thus leading to an alternating structure. There is also a possibility of generating block of one polymer followed by addition of another block of some other polymer to the main chain to generate linear polymer chains containing blocks of polymer units. Controlled polymerization techniques are very effective in generating these kinds of structures. Figure 9c shows an example of block copolymer of acrylonitrile, butadiene and styrene (ABS). There is also another category of copolymers which can be generated by grafting molecules or polymer chains on the main chain. Thus these copolymers are termed as graft copolymers. Common example of such a system is polyethylene grafted with maleic anhydride. However, if the grafted chains are also very long, these can also be visualized as blocks of polymers and the nature of polymer becomes more near to that of a block copolymer. One such example is the polypropylene grafted with long chains of polypropylene glycol monobutyl ether, as shown in Figure 10.

D. Polymerization Methods

Various polymerization methods and the classification of polymers based on these methods were mentioned in section B. Some further details of these methods are described in this section. Figure 11 also shows the whole spectrum of these methods which can be used to synthesize polymers.

D.1. Chain Polymerization

As mentioned earlier, the double bond in the monomer unit is polymerized generally by using chain polymerization. This method is based on the addition of monomer units to the polymer chains without the generation of any byproduct. The polymerization is initiated by using an initiator. The whole polymerization process is divided into three steps of initiation, propagation and termination. The initiator reacts with the pi electrons in the double bond of the monomer units to form covalent bonds with one of the carbon atoms of the double bond

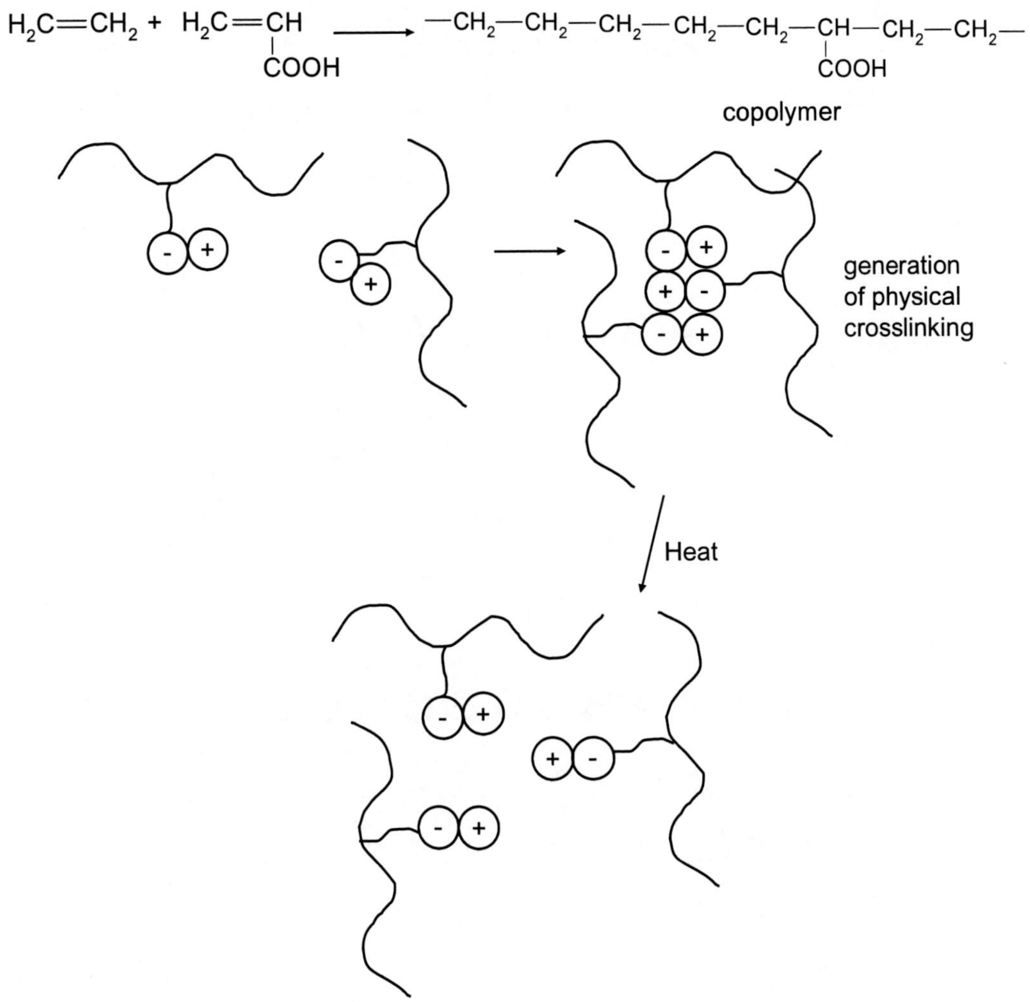

$$H_2C{=}CH_2 + H_2C{=}\underset{\displaystyle COOH}{CH} \longrightarrow -CH_2-CH_2-CH_2-CH_2-CH_2-\underset{\displaystyle COOH}{CH}-CH_2-CH_2-$$

copolymer

generation of physical crosslinking

Heat

Examples:

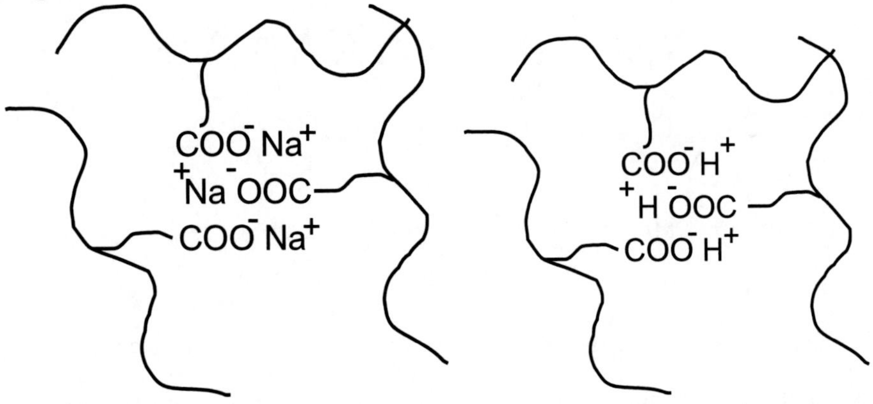

$COO^- Na^+$
$^+Na\ ^-OOC$
$COO^- Na^+$

$COO^- H^+$
$^+H\ ^-OOC$
$COO^- H^+$

Figure 8. Ionomers with COOH groups neutralized to generate ionic charges on the polymer chains.

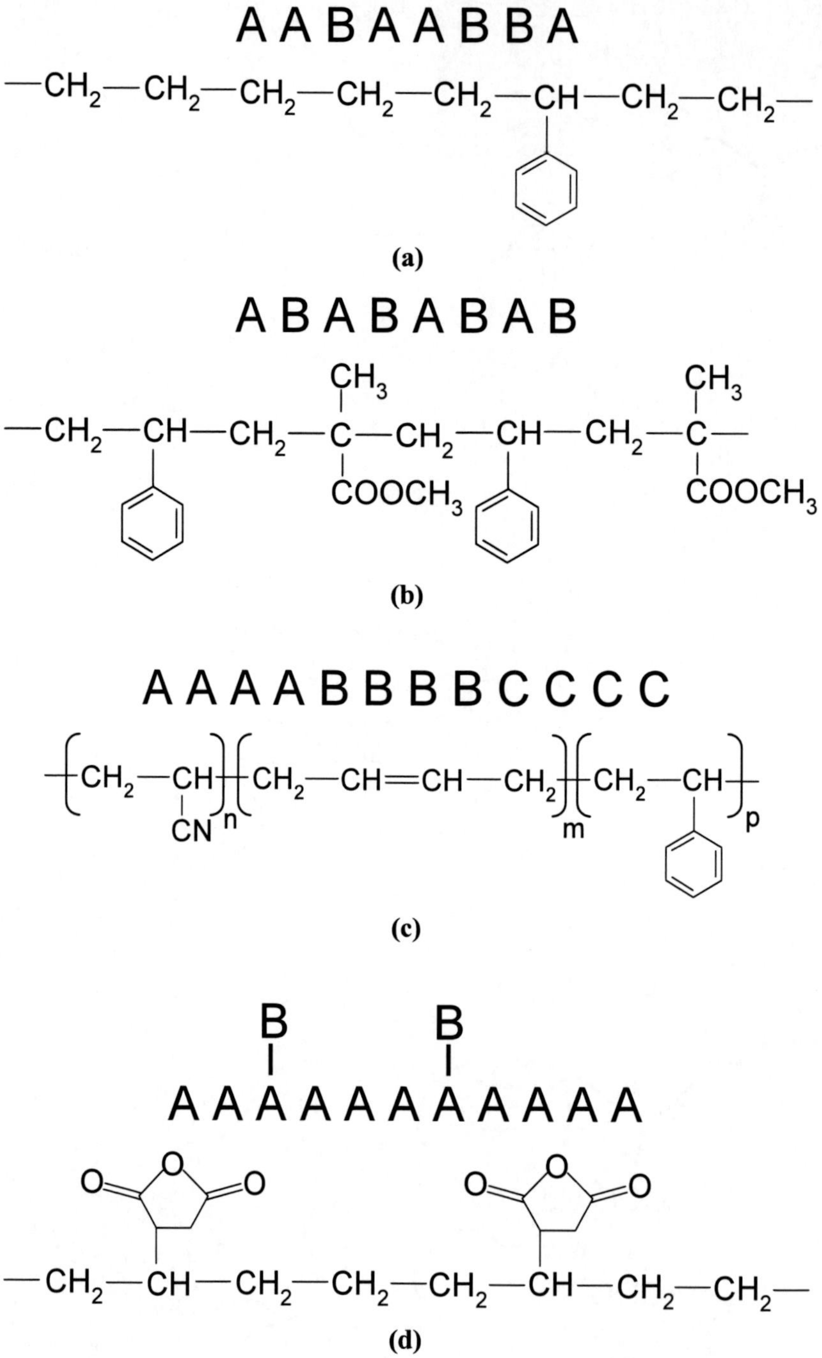

A A B A A B B A

$-CH_2-CH_2-CH_2-CH_2-CH_2-CH(C_6H_5)-CH_2-CH_2-$

(a)

A B A B A B A B

$-CH_2-CH(C_6H_5)-CH_2-C(CH_3)(COOCH_3)-CH_2-CH(C_6H_5)-CH_2-C(CH_3)(COOCH_3)-$

(b)

A A A A B B B B C C C C

$\left[CH_2-CH(CN)\right]_n\left[CH_2-CH{=}CH-CH_2\right]_m\left[CH_2-CH(C_6H_5)\right]_p$

(c)

B B

A A A A A A A A A A A

$-CH_2-CH-CH_2-CH_2-CH_2-CH-CH_2-CH_2-$

(d)

Figure 9. Examples of (a) random, (b) alternating, (c) block and (d) graft copolymers.

using one of the pi electron, whereas the remaining pi electron from the double bond is pushed to the other carbon atom forming again an entity that is reactive and attacks further monomer units. The process of addition of the monomer units is termed propagation. The chains come to termination if the two growing chains react with each other. Chain polymerization can further be subdivided into free radical, cationic and anionic polymerization.

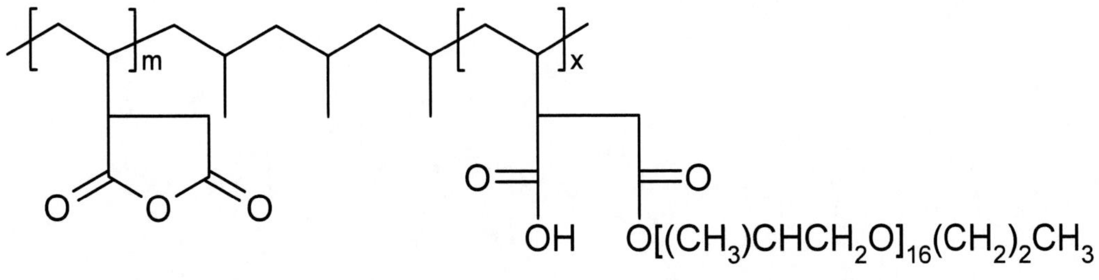

Figure 10. Copolymer of polypropylene with polypropylene glycol monobutyl ether.

In free radical polymerization, the initiator is decomposed into two free radicals by the use of temperature owing to the thermal unstability of the initiators. These free radicals subsequently attack the double bond in the monomer units to start a polymer chain. A number of azo compounds, peroxides and sulphates are used as initiators. Equations 6–8 show the chemical structure of the common initiators azo-bis isobutyro nitrile, benzoyl peroxide and potassium persulphate and the corresponding radical generated upon their thermal decomposition.

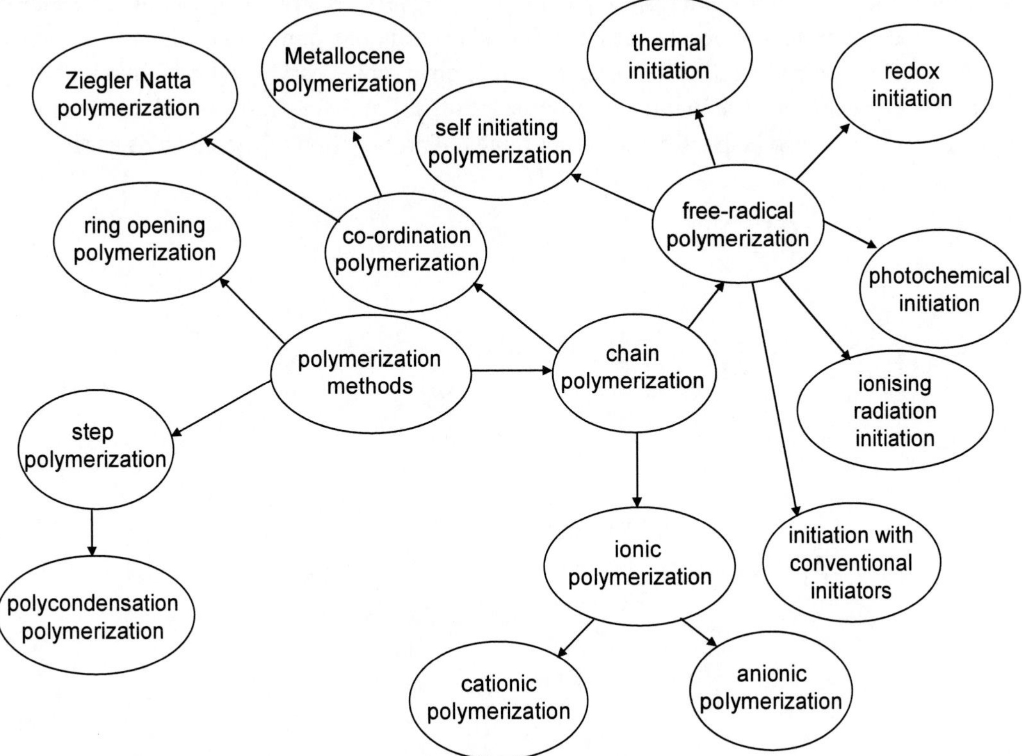

Figure 11. The wide spectrum of polymerization methods to synthesize polymers.

$$(CH_3)_2C(CN)\text{—}N\text{=}N\text{—}C(CN)(CH_3)_2 \longrightarrow 2\ (CH_3)_2\dot{C}(CN) + N_2 \qquad (6)$$

$$C_6H_5\text{—}C(\text{=}O)\text{—}O\text{—}O\text{—}C(\text{=}O)\text{—}C_6H_5 \longrightarrow 2\ C_6H_5\text{—}C(\text{=}O)\text{—}O^{\bullet} \qquad (7)$$

$$^{+}K\ O_3S\text{—}O\text{—}O\text{—}SO_3\ K^{+} \longrightarrow 2\ SO_4^{\bullet -} \qquad (8)$$

These initiators decompose at different temperatures based on their own chemical structures and thus take different amounts of time to achieve half time, i.e., time required to attain decomposition of 50% of the initiator molecules. The free radicals generated from the initiator are very reactive, and to achieve stability attack the pi electron of the double bond in the monomer unit. As a result, the radical forms a bond with one of the double bonds carrying carbon, leading to the breaking of the double bond and leaving the other carbon of the double bond with a lone electron. The resulting entity is again very reactive owing to the presence of an unpaired electron, and reacts further with other monomer units and this way the chain propagates by adding together thousands of monomer units very quickly. Equations 9–12 show these processes of initiation and propagation for the free radical polymerization of styrene. There can also be possible different routes of propagation of these chains as shown in

$$I-I \longrightarrow 2\,I^{\bullet} \qquad (9)$$

$$I + CH_2\text{=}CH(C_6H_5) \longrightarrow I\text{—}CH_2\text{—}\dot{C}H(C_6H_5) \qquad (10)$$

$$I\text{—}CH_2\text{—}\dot{C}H(C_6H_5) + CH_2\text{=}CH(C_6H_5) \longrightarrow I\text{—}CH_2\text{—}CH(C_6H_5)\text{—}CH_2\text{—}\dot{C}H(C_6H_5) \qquad (11)$$

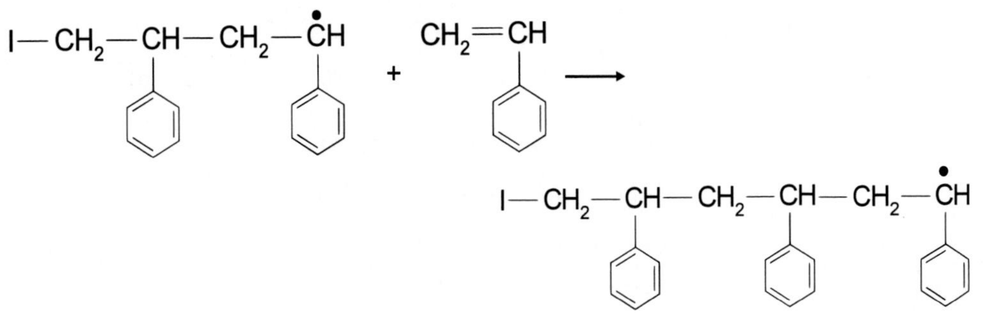

$$I-CH_2-CH-CH_2-\dot{C}H + CH_2{=}CH \longrightarrow I-CH_2-CH-CH_2-CH-CH_2-\dot{C}H \quad (12)$$

equations 13–15, in which either the head-to-head or tail-to-tail, head-to-tail or tail-to-head propagation arrangements are shown.

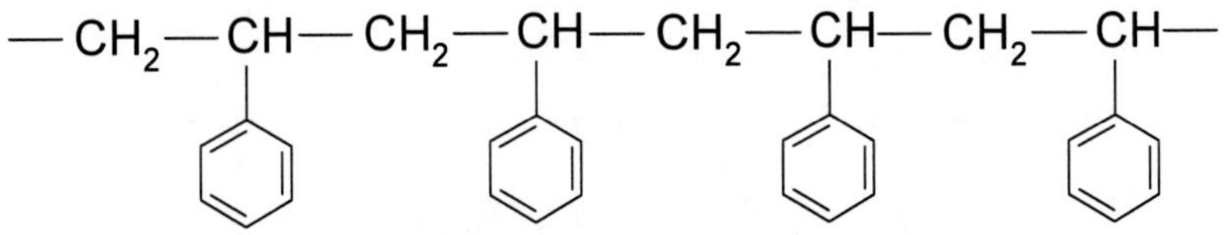

$$-CH_2-CH-CH_2-CH-CH_2-CH-CH_2-CH- \quad (13)$$

head to tail or tail to head

$$-CH-CH_2-CH_2-CH- \quad (14)$$

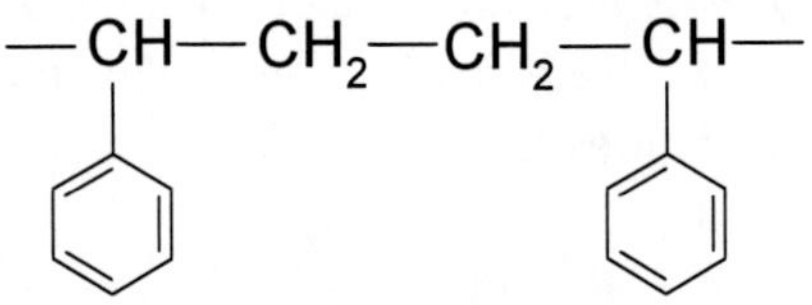

tail to tail

$$-CH_2-CH-CH-CH_2- \quad (15)$$

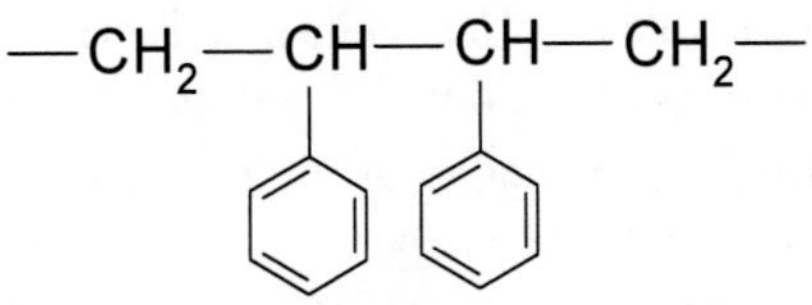

head to head

During the course of propagation, the growing chains come in contact with each other and terminate themselves. There are two common routes of termination, viz. termination by coupling and termination by disproportionation. When the two propagating chains couple with each other to form a single bond at the point of contact, the process is called termination by coupling. However, it is also possible that one of the chains abstracts a proton from the other chain and the other chain stabilizes itself by forming the double bond with the two free radicals in the chain resulting in dead chains. This process is called termination by disproportionation. Equations 16 and 17 represent these two phenomena.

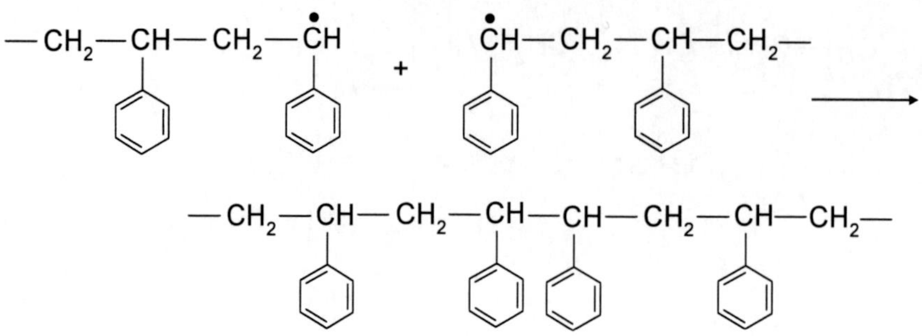

$$-CH_2-CH-CH_2-\dot{C}H \quad + \quad \dot{C}H-CH_2-CH-CH_2- \longrightarrow$$

$$-CH_2-CH-CH_2-CH-CH-CH_2-CH-CH_2- \qquad (16)$$

$$-CH_2-CH-CH_2-\dot{C}H \quad + \quad \dot{C}H-CH_2-CH-CH_2- \longrightarrow$$

$$-CH_2-CH-CH_2-CH_2 \quad + \quad CH{=}CH-CH-CH_2- \qquad (17)$$

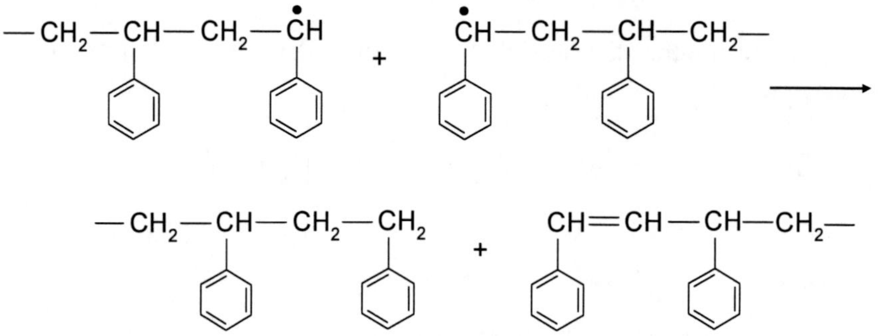

Apart from initiators decomposing to initiate the polymerization, many other routes of initiation have been developed. The most common of them are redox initiation (equation 18), photoinitiaton, self initiation of monomers at high temperatures, etc.

$$H_2O_2 + Fe^{2+} \longrightarrow HO^- + HO^{\bullet} + Fe^{3+} \qquad (18)$$

Ionic polymerization can also be used in a way similar to free radical polymerization. However, ionic polymerization is more sensitive to reaction conditions as well as selective to monomer functional groups; thus, many monomers do not polymerize using these techniques. In cationic polymerization, Lewis acids such as BF_3 are used as a catalyst and water is used as a co-catalyst. Together these form an ion pair, as shown in equation 19, where H^+ is the initiating cationic species and $BF3OH^-$ is the stabilizing counter ion. Equations 20–22 show the initiation and propagation steps for the cationic polymerization of styrene. The two growing chains can react and terminate each other. In anionic polymerization, organometallic compounds like butyl lithium C_4H_9Li are used as catalysts. In reality, the anionic polymerization is a living polymerization and the chains propagate again as soon as a new batch of monomer is added. Equations 23 to 25 show the anionic polymerization of styrene.

$$BF_3 + OH_2 \longrightarrow [BF_3OH]^{\ominus} \cdots\cdots {}^{\oplus}H \qquad (19)$$

$$[BF_3OH]^{-}\cdots H^{+} \quad + \quad CH_2{=}CH \longrightarrow CH_3{-}CH^{+}\,{}^{-}BF_3OH \tag{20}$$

$$CH_3{-}CH^{+}\,{}^{-}BF_3OH \quad + \quad CH_2{=}CH \longrightarrow CH_3{-}CH{-}CH_2{-}CH^{+}\,{}^{-}BF_3OH \tag{21}$$

$$CH_3{-}CH{-}CH_2{-}CH^{+}\,{}^{-}BF_3OH \quad + \quad CH_2{=}CH \longrightarrow CH_3{-}CH{-}CH_2{-}CH{-}CH_2{-}CH^{+}\,{}^{-}BF_3OH \tag{22}$$

$$C_4H_9Li \quad + \quad CH_2{=}CH \longrightarrow C_4H_9{-}CH_2{-}\overset{H}{\overset{|}{C}}{:}^{-}\;Li^{+} \tag{23}$$

$$C_4H_9{-}CH_2{-}\overset{H}{\overset{|}{C}}{:}^{-}\;Li^{+} \quad + \quad CH_2{=}CH \longrightarrow C_4H_9{-}CH_2{-}\overset{H}{\overset{|}{C}}{-}CH_2{-}\overset{H}{\overset{|}{C}}{:}^{-}\;Li^{+} \tag{24}$$

$$C_4H_9{-}CH_2{-}\overset{H}{\overset{|}{C}}{-}CH_2{-}\overset{H}{\overset{|}{C}}{:}^{-}\;Li^{+} \quad + \quad CH_2{=}CH \longrightarrow C_4H_9{-}CH_2{-}\overset{H}{\overset{|}{C}}\left[CH_2{-}\overset{H}{\overset{|}{C}}\right]_n CH_2{-}\overset{H}{\overset{|}{C}}{:}^{-}\;Li^{+} \tag{25}$$

D.2. Step Polymerization

As mentioned earlier, step polymerization differs significantly from chain polymerization. At any point during step polymerization, the reaction mixture contains dimers, trimers and higher molecules. However, only monomers and high molecular weight chains would be present in the chain polymerization. High molecular weight is generated at once in chain polymerization, and during increasing conversion only the number of polymer chains increases and the molecular weight remains unaffected. However, high molecular weight is not achieved until high conversion is obtained in the case of step polymerization. Figure 12 shows the behavior of molecular weight evolution as a function of the conversion of the monomer in both chain and step polymerization methodologies. Step polymerization reactions of polyamides, polyesters, polyurethanes and epoxies have been mentioned earlier. The following equations (26–28) show the synthesis of polycarbonates (equation 26) and phenolic resins (equations 27–28).

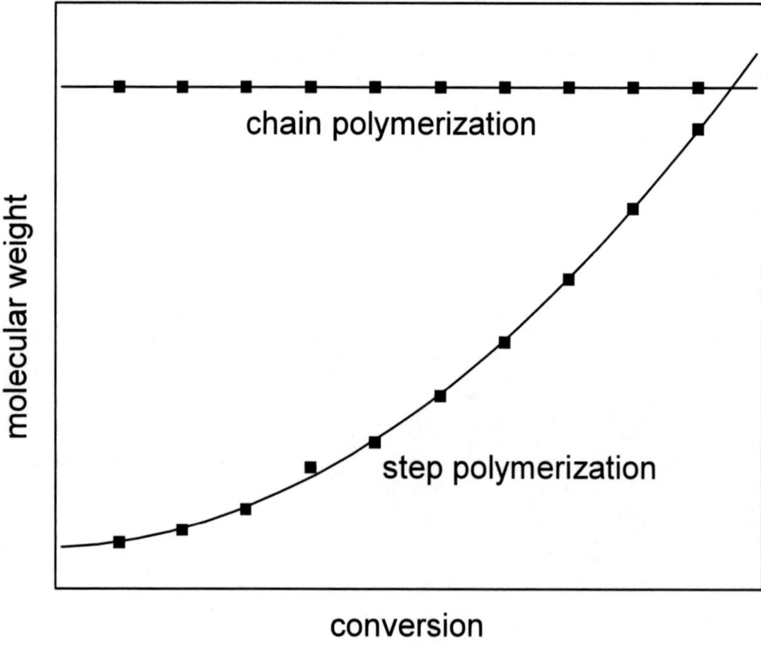

Figure 12. Comparison of the molecular weight evolution in step and chain polymerization reactions as a function of conversion.

$$HO-C_6H_4-C(CH_3)_2-C_6H_4-OH \xrightarrow{Cl-CO-Cl} \left[O-C_6H_4-C(CH_3)_2-C_6H_4-O-CO \right]_n \quad (26)$$

$$C_6H_5OH + HCHO \longrightarrow o\text{-}HOC_6H_4CH_2OH + p\text{-}HOC_6H_4CH_2OH \quad (27)$$

$$o\text{-}HOC_6H_4CH_2OH + C_6H_5OH \longrightarrow HOC_6H_4\text{-}CH_2\text{-}C_6H_4OH \quad (28)$$

D.3. Ring Opening Polymerization

Ring opening polymerization is another type of addition polymerization in which the cyclic monomers are polymerized into long polymer chains by opening their rings. The most common example is that of the synthesis of nylon 6 from carpolactam (equation 29). The initiators are the same types as those used for ionic polymerization, and polymerization is achieved by the generation of oxonium ions or anionic centers.

$$\text{caprolactam} \xrightarrow{ROP} \left[NHCOCH_2CH_2CH_2CH_2CH_2 \right]_n \quad (29)$$

REFERENCES

[1] G. Odian (2004). *Principles of Polymerization, Fourth Edition*, John Wiley & Sons, Inc., New Jersey.

[2] V. R. Gowariker, N. V. Viswanathan and J. Sreedhar (1986). *Polymer Science*, John Wiley & Sons, Wiley Eastern Limited, New Delhi.

Chapter 2

KINETICS OF POLYMERIZATION

A. INTRODUCTION

The qualitative details of various polymerization reactions in the synthesis of a large spectrum of polymers, as well as the classification of these polymers into a number of categories based on structure, morphology or application, are important to know in order to obtain insight into the full potential of polymer science. However, quantification of these polymerization reactions in terms of reaction kinetics is necessary to determine the exact reaction conditions that lead to efficient polymerization reactions. The determination of polymerization kinetics is important, as it affects the final polymer properties in a given set of reaction conditions. The kinetics differ for different polymerization methods, as described in the following paragraphs.

B. CHAIN POLYMERIZATION

B.1. Free Radical Chain Polymerization

As the initiator decomposes into two radicals to initiate the free radical chain polymerization, therefore, the rate of initiator decomposition R_d can be written as shown below in equations 1 to 2.

$$I \xrightarrow{k_d} 2R^{\bullet} \tag{1}$$

$$R_d = \frac{-d[I]}{dt} = 2fk_d[I] \tag{2}$$

where [I] is the initiator concentration, f is the initiator efficiency and k_d is the rate constant of decomposition reaction. Typical values of f are in the range of 0.5 to 0.8 depending on the viscosity of the reaction system [1]. The value of k_d varies in the range of 10^{-4} to 10^{-9} s^{-1} depending on the reaction temperature and type of initiator [2]. As the radicals are very reactive in nature, they start attacking the monomer molecules in the vicinity, and the initiation process is started. The rate of initiation R_i is thus defined kinetically by the following equations 3 to 4.

$$R^{\bullet} + M \xrightarrow{k_i} RM^{\bullet} \tag{3}$$

$$R_i = \frac{-d\,[M]}{dt} = k_i\,[R^{\bullet}]\,[M] \tag{4}$$

where [M] is the monomer concentration, $[R^{\bullet}]$ is the free radical concentration and k_i is the rate constant for the initiation reaction. As the radicals react with the monomer molecules very rapidly, therefore, it is generally assumed that the rate of generation of free radicals is equal to the rate of their consumption. Also, in most of the polymerization reactions, the second step is faster than the first step, thus making the first step a rate-determining step. Therefore, the rate of initiation R_i can also be written according to equation 5.

$$R_i = 2\,f\,k_d\,[I] \tag{5}$$

Propagation follows by the consumption of monomer and formation of polymer chains by the addition of large numbers of monomer molecules. The propagation process and the rate of propagation can be defined according to the following equations 6 to 8.

$$RM^{\bullet} + M \xrightarrow{k_p} RMM^{\bullet} \tag{6}$$

$$RMM^{\bullet} + M \xrightarrow{k_p} RMMM^{\bullet} \tag{7}$$

$$R_p = \frac{-d\,[M]}{dt} = k_p\,[M^{\bullet}]\,[M] \tag{8}$$

where $[M^{\bullet}]$ is the concentration of growing polymer chains and k_d is the propagation rate constant. Chains terminate when the two growing chains come in contact with each other. The termination can occur either by coupling or by disproportionation leading to dead chains in both the cases. The termination process as well as its rate R_t can be described by the following equations 9 to 11.

$$M_n^{\bullet} + M_m^{\bullet} \xrightarrow{k_t} M_{n+m} \tag{9}$$

$$M_n^{\bullet} + M_m^{\bullet} \xrightarrow{k_t} M_n + M_m \tag{10}$$

$$R_t = \frac{-d\,[M\,]}{dt} = 2\,k_t\,[M^{\bullet}]^2 \tag{11}$$

where k_t is the termination rate constant. The use of factor 2 in the rate equation is based on the phenomenon of radicals pairs being consumed during the termination reaction. It is often assumed in the free radical reactions that the steady state reaches quite early i.e. the rate of change of initiator concentration become zero. Thus, at steady state R_i can be equated to R_t according to the equation 12 as

$$R_i = R_t \tag{12}$$

and the above defined equations can thus be compared to each other to generate an equation to calculate $[M^{\bullet}]$. Substituting equation 8 for R_p into the equation 14 leads to general rate of polymerization or rate of propagation equations 15 and 16 as follows

$$2\,f\,k_d\,[I] = 2\,k_t\,[M^{\bullet}]^2 \tag{13}$$

$$[M^{\bullet}] = \left(\frac{k_d\,f\,[I]}{k_t}\right)^{1/2} \tag{14}$$

$$R_p = k_p\,M\left(\frac{R_i}{2\,k_t}\right)^{1/2} \tag{15}$$

$$R_p = k_p\,M\left(\frac{k_d\,f\,[I]}{k_t}\right)^{1/2} \tag{16}$$

Also important is to know about the half life of initiators $t_{1/2}$ which is defined as the time taken by the initiator molecules to decompose to half of its original concentration and it is defined by the following equation

$$t_{1/2} = \frac{0.693}{k_d} \tag{17}$$

It has also been observed that the termination can also shift from bimolecular termination to primary termination, a process in which the growing polymer chains react with the primary radicals as shown below in equation 18 and the rate of polymerization is then defined by the equation 19

$$M^{\bullet} + M^{\bullet} \xrightarrow{k_t} M - M \tag{18}$$

$$R_p = \frac{k_p k_i [M]^2}{k_t} \tag{19}$$

Kinetic chain length is defined as the average number of monomer molecules consumed by each free radical which was successful in initiating a chain. This quantity is defined as the ratio of polymerization rate to the initiation or termination rate (as at steady state both the initiation and termination rates are identical). Therefore, the kinetic chain length υ is defined by the following equations 20 to 21 as

$$v = \frac{R_p}{R_i} \text{ or } \frac{R_p}{R_t} \tag{20}$$

$$v = \frac{k_p [M]}{2 (f k_d k_t [I])^{1/2}} \tag{21}$$

Degree of polymerization, D_p or X_m is defined as average number of monomer molecules in a polymer chains and it is related to average kinetic chain length by the following relations:

$D_p = 2$ υ, when the polymer chains terminate by coupling route
$D_p =$ υ, when the polymer chains terminate by disproportionation route

Chain transfer is also a very common phenomenon in free radical polymerization. In this process, a radical species reacts with a non-radical species and as a result, the radical is transferred to the non-radical species and the original radical species results into dead polymer chains. The chain termination reaction can be represented by the following equation

$$\sim\sim M^{\bullet} + RH \xrightarrow{k_{tr}} \sim\sim MH + R^{\bullet} \quad (22)$$

where k_{tr} is the chain transfer coefficient and RH can either be a chain transfer agent or a variety of other species like initiator, monomer, solvent or polymer. These transfer reactions lead to lower molecular weights than expected and thus need to be controlled to achieve higher molecular weight. In some cases, however, the chain transfer agents are deliberately added to stop the chains from growing further. Equations 23 and 24 show the typical chain transfer reactions with common chain transfer agents like mercaptans and carbon tetrachloride

$$\sim\sim CH_2-\dot{C}H(C_6H_5) + RSH \xrightarrow{k_{tr,CT}} \sim\sim CH_2-CH_2(C_6H_5) + RS^{\bullet} \quad (23)$$

$$\sim\sim CH_2-\dot{C}H(C_6H_5) + CCl_4 \xrightarrow{k_{tr,CT}} \sim\sim CH_2-CHCl(C_6H_5) + \dot{C}Cl_3 \quad (24)$$

where k_{tr} is the rate coefficient for the chain transfer reaction and the rate of polymerization is thus defined as

$$R_p = k_{tr}[RH][M^{\bullet}] \quad (25)$$

where [RH] is the concentration of chain transfer agent and $[M^{\bullet}]$ is the concentration of polymer radicals.

As the concentration of initiator in the system is quite low, therefore, the chain transfer to initiator is not of much significance. However, it can be detrimental in a few cases owing to disturbance in the control of polymerization reaction. Equation 26 shows one example of chain transfer to benzoyl peroxide during polymerization of acetonitrile.

$$\sim\!\!\sim CH_2{-}\dot{C}H(CN) + C_6H_5{-}C(=O){-}O{-}O{-}C(=O){-}C_6H_5 \xrightarrow{k_{tr,I}} \sim\!\!\sim CH_2{-}CH(CN){-}O{-}C(=O){-}C_6H_5 + C_6H_5{-}C(=O){-}\dot{O} \quad (26)$$

where $k_{tr,I}$ is the reaction rate coefficient for the chain transfer to initiator reaction.

Chain transfer to monomer can be quite significant as the monomer concentration is quite high especially in the beginning of polymerization reaction. However, the rate constants of the chain transfer to the monomer are quite low, in the range of 1×10^{-5} to 15×10^{-5} L $mol^{-1}s^{-1}$ [2], thus the final effect of monomer chain transfer can be quite less. Equation 27 is an example of chain transfer to monomer process.

$$\sim\!\!\sim CH_2{-}\dot{C}H(C_6H_5) + CH_2{=}CH(C_6H_5) \xrightarrow{k_{tr,M}} \sim\!\!\sim CH_2{-}CH_2(C_6H_5) + CH_2{=}\dot{C}(C_6H_5) \quad (27)$$

The chain transfer to solvent is a very common phenomenon. Many times solvents are also added as chain transfer agents to control the molecular weight of polymer chains. The solvent in this case is called a telogen and this process is called telomerization. This process leads to mostly the synthesis of liquid polymers. Equation 28 is an example of such a process of bromotrifluoro ethylene using bromotrifluoro methane as telogen.

$$\sim\!\!\sim CF_2{-}\dot{C}F(Br) + CF_3Br \xrightarrow{k_{tr,S}} \sim\!\!\sim CF_2{-}CF(Br)_2 + \dot{C}F_3 \quad (28)$$

It has also been observed that chain transfer can also occur to already dead polymer chains by the abstraction of hydrogen atom from the polymer backbone by a growing polymer chain. This results into the dead chain formed from initially growing chain and a living chain from initially dead polymer chain. The generated radical can then start adding monomer molecules in the backbone of polymer chain. Equation 29 is an example of this process.

$$\sim\!\sim CH_2-\underset{X}{\overset{H}{C}}\sim\!\sim \;+\; \sim\!\sim CH_2-\underset{X}{\overset{H}{C^{\bullet}}} \xrightarrow{k_{tr,P}}$$

$$\sim\!\sim CH_2-\underset{X}{CH_2} \;+\; \sim\!\sim CH_2-\underset{X}{\overset{\bullet}{C}}\sim\!\sim \longrightarrow$$

$$\sim\!\sim CH_2-\underset{X}{\overset{\sim\sim}{C}}\sim\!\sim \qquad (29)$$

Mayo developed kinetic equations to represent degree of polymerization when such chain transfer reactions take place [2]. The degree of polymerization was defined by the following equation

$$\bar{X}_m = \frac{R_p}{(R_t/2) + k_{tr,M}[M^{\bullet}][M] + k_{tr,S}[M^{\bullet}][S] + k_{tr,I}[M^{\bullet}][I] + k_{tr,RH}[M^{\bullet}][RH]} \qquad (30)$$

where [RH] is the concentration of chain transfer agent, [S] is the solvent concentration and [I] is the initiator concentration. To simplify the equation further chain transfer constant, C was defined which is the ratio of chain transfer coefficient to propagation rate coefficient. Thus the chain transfer constants for chain transfer to monomer, initiator, solvent and chain transfer agents can be depicted as

$$C_M = \frac{k_{tr,M}}{k_p}$$

$$C_I = \frac{k_{tr,I}}{k_p}$$

$$C_S = \frac{k_{tr,S}}{k_p}$$

$$C_{RH} = \frac{k_{tr,RH}}{k_p} \qquad (31)$$

When equation 30 was substituted with above defined chain transfer constants, it takes a simple form as shown below, which is also termed as Mayo equation [2,3].

$$\frac{1}{\overline{X}_m} = \frac{k_t R_p}{k_p^2 [M]^2} + C_M + C_S \frac{[S]}{[M]} + C_I \frac{[I]}{[M]} + C_{RH} \frac{[RH]}{[M]} \tag{32}$$

It has to be noted that in this equation, it has been assumed that the termination takes place by coupling, thus requiring modifications in the equation when the termination is observed to take place by disproportionation. The Mayo equation is very useful in determination of the effect of termination and transfer reactions on the final molecular weight of the polymer chains and thus allows one to tune the reaction parameters to obtain the required molecular weight.

Chain transfer to polymer sometimes does not involve a dead and a growing polymer chain, rather the growing polymer chain itself starts to abstract hydrogen atoms from the back bone of same polymer chain. This process is called backbiting reaction and is sometimes a common mechanism to generate branching and molecular weight shortening in polyethylene [2]. The typical backbiting reaction involves abstraction of hydrogen atom from fifth, sixth or seventh methylene group from the radical end as shown in the scheme below (equation 33).

Inhibitors are also added to the polymerization reactions to control the molecular weight or conversion. The inhibitors react with the propagating polymer chains to convert them into either dead chains or stable radicals with low reactivity. The polymerization thus is completely stopped and starts only when the whole of inhibitor has been consumed and this lag in the polymerization time is termed as inhibition period. The inhibitors are added in the monomers to eliminate their premature polymerization and thus the polymerization does not start until all the inhibitor has been consumed. Commonly used inhibitors are hydroquinone, benzoquinone, oxygen, polyalkyl ring substituted phenols and nitroxides, etc. [2]. A common example of nitroxide is

$$\sim CH_2-CH_2-CH_2-CH_2-CH_2-CH_2-CH_2-\dot{C}H_2$$

$$\downarrow$$

$$\sim CH_2-\dot{C}H-CH_2-CH_2-CH_2-CH_2-CH_2-CH_3$$

$$\sim CH_2-CH_2-CH_2-\dot{C}H-CH_2-CH_2-CH_2-CH_3$$

$$\sim CH_2-CH_2-\dot{C}H-CH_2-CH_2-CH_2-CH_2-CH_3 \tag{33}$$

2,2,6,6-teramethylpiperidine-1-oxyl or TEMPO, which is very efficient in scavenging the radical species and is very stable to generation of radicals on its own. Equations 34 to 37 show the inhibition reactions of polyalkyl ring substituted phenols (equation 34), benzoquinone (equations 35–36) and nitroxide (equation 37) inhibitors.

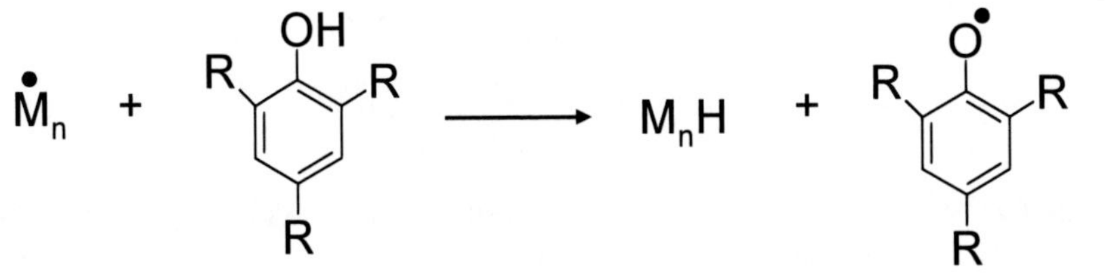

(34)

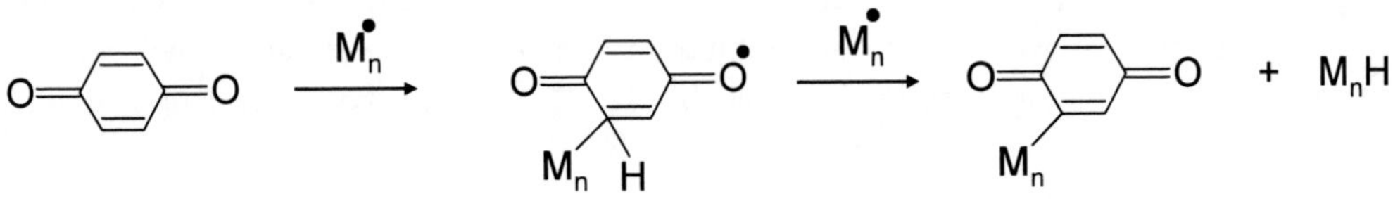

(35)

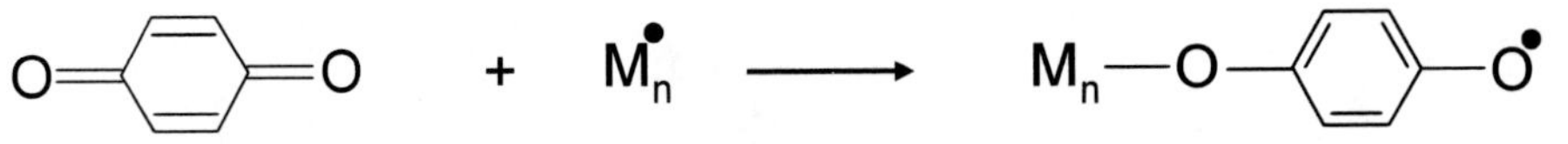

(36)

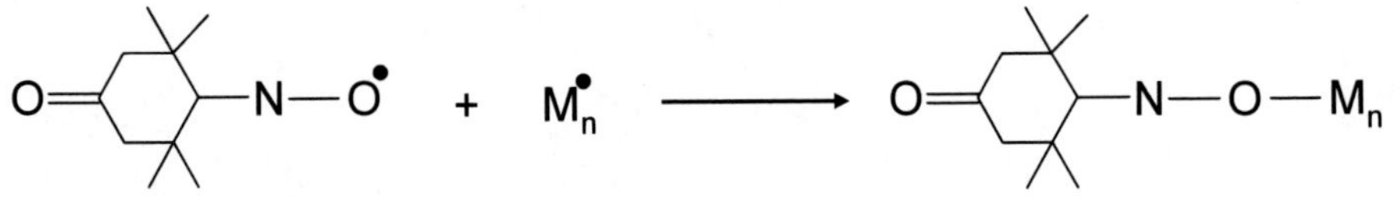

(37)

Retarders are also similar type of species; however, they do not completely quench the polymerization, but slow it down by consuming a part of radicals. Thus by adding retarder to the polymerization system, inhibition period is never experienced, only a decreased propagation rate occurs. Nitrobenzene is a very common retarder and works as explained in the following equations in which either the benzyl ring or the NO_2 group is involved [2].

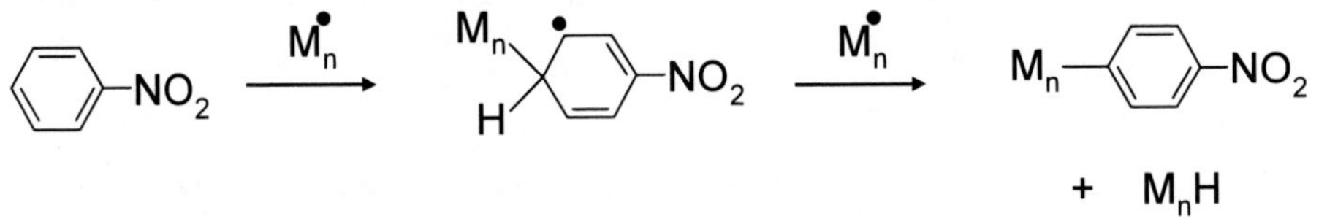

(38)

$$C_6H_5NO_2 \xrightarrow{M_n^\bullet} C_6H_5N(OM_n)O^\bullet \xrightarrow{M_n^\bullet} C_6H_5N{=}O + M_n{-}O{-}M_n$$

$$C_6H_5N(OM_n)O^\bullet \xrightarrow{M_n^\bullet} C_6H_5N(OM_n)_2 \quad (39)$$

Apart from that, there are also a number of other compounds which can act both as inhibitor as well as retarder. Nitrosobenzene is a common example; it acts initially as inhibitor however later on during the polymerization, its behavior changes to retarder. Figure 1 is a depiction of these phenomena, where the effect of these species on conversion as a function of time has been shown. Kinetics of inhibition and retardation reactions had been widely studied and the rate of polymerization in such cases can be described by equation 40 as

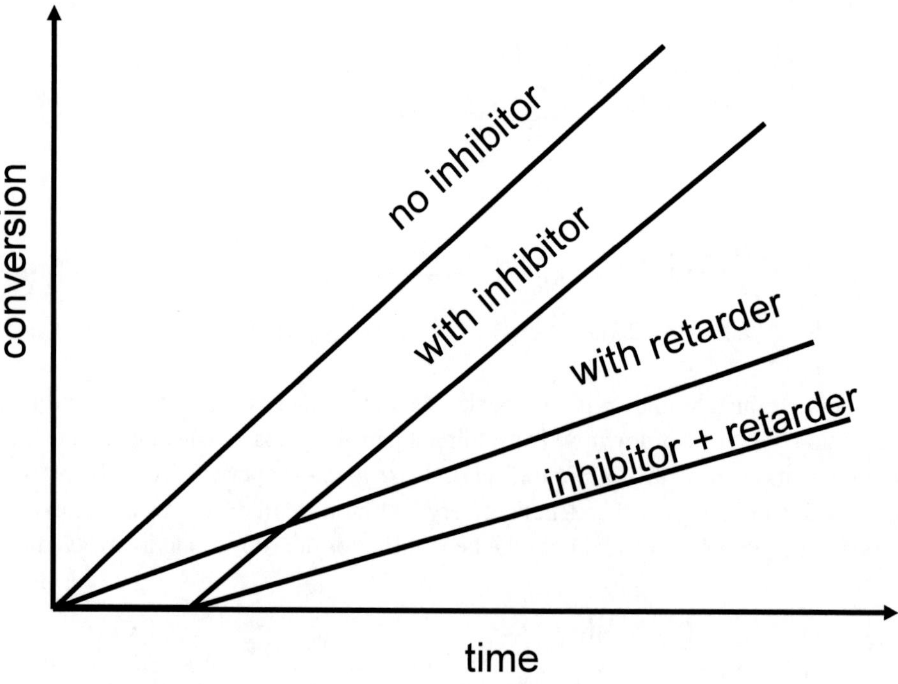

Figure 1. The conversion vs. time behavior of inhibitors and retarders in the polymerization reaction.

$$R_p = \frac{k_p\,[M]\,R_i}{k_Q\,[Q]} \quad (40)$$

where [Q] is the concentration of inhibitor or retarder and k_Q is the reaction rate coefficient. Figure 2 is a summary of various types of reactions taking place in the free radical polymerization.

Initiation

$$I \xrightarrow{k_d} R^{\bullet}$$

$$R^{\bullet} + M \xrightarrow{k_i} M^{\bullet}$$

Propagation

$$M^{\bullet} + M \xrightarrow{k_p} MM^{\bullet}$$

$$M_n^{\bullet} + M \xrightarrow{k_p} M_nM^{\bullet}$$

$$M_nM^{\bullet} + M \xrightarrow{k_p} M_{n+1}M^{\bullet}$$

Termination

$$M_mM^{\bullet} + M_nM^{\bullet} \xrightarrow{k_{tc}} M_{m+n+2}$$

$$M_mM^{\bullet} + M_nM^{\bullet} \xrightarrow{k_{td}} M_{m+1} + M_{n+1}$$

Chain Transfer

$$\sim\!\sim\!\sim M^{\bullet} + RH \xrightarrow{k_{tr}} \sim\!\sim\!\sim MH + R^{\bullet}$$

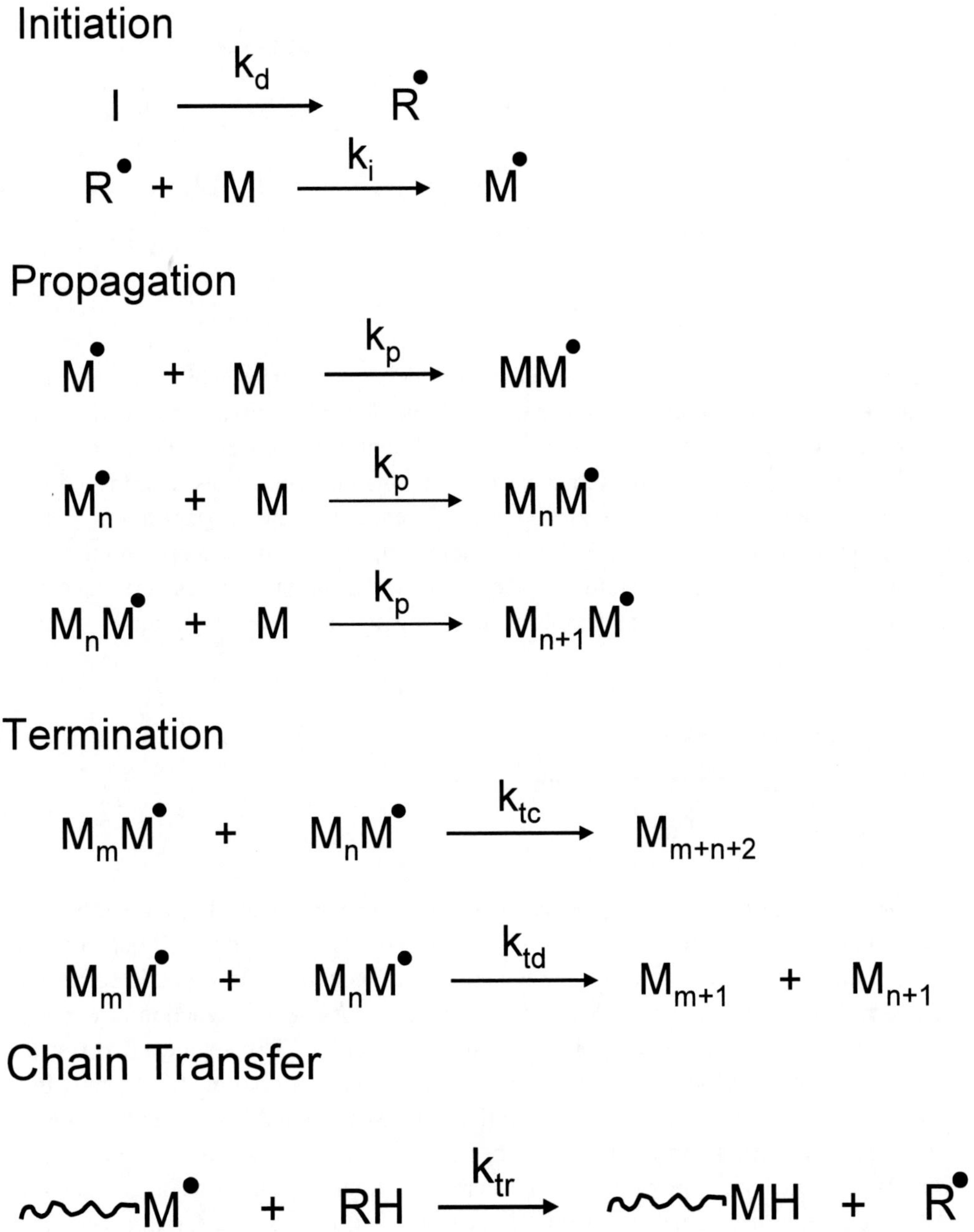

Figure 2. The summary of various reactions taking place during the free-radical polymerization reactions.

Inhibition occurs not only by the addition of inhibitors, but also by autoinhibition reactions in some of the monomers. Allylic monomers are common example where, owing to inhibition, the polymerization rate is very slow and the degree of polymerizations is significantly smaller. The common inhibition reaction is the chain transfer to the monomer, as shown in equation 41.

$$\sim\!\sim CH_2-\underset{CH_2Y}{\overset{H}{C}}\bullet \;+\; CH_2=\underset{CH_2Y}{\overset{H}{C}} \longrightarrow \sim\!\sim CH_2-\underset{CH_2Y}{CH_2} \;+\; CH_2=\underset{\underset{\bullet}{CHY}}{\overset{H}{C}} \tag{41}$$

There is an also very important phenomenon of depolymeriztaion which occurs at higher temperature. The propagation reaction can be again described by the following reaction (equation 42), where k_p is the propagation rate coefficient. However, once the depropagation or depolymerization occurs, the reaction becomes reversible with k_{dp} as depropagation rate coefficient. Initially by increasing the reaction temperature, the rate of propagation increases. However, the reaction temperature is further increased, the rate of depropagation becomes significant and finally a temperature is reached where the two rates are equal. This temperature is termed as ceiling temperature and at this temperature the net rate or amount of polymer production is zero.

$$M^{\bullet} + M \underset{k_{dp}}{\overset{k_p}{\rightleftharpoons}} MM^{\bullet} \tag{42}$$

Autoaccelaration is also an important kinetic phenomenon which is observed when the reaction mixtures are particularly viscous. The rate of polymerization generally decreases as a function of time or conversion owing to the depletion of monomer. However, in some special cases, the rate of polymerization is actually observed to increase as a function of conversion and such an effect is termed as gel or Trommsdorff effect. The segmental diffusion is increased with conversion owing to the less dissolution of formed polymer in the polymerization medium due to high concentration. The translation diffusion on the other hand is decreased with conversion as the reaction medium becomes viscous as more and more polymer is generated. At a certain point, the decrease in translational diffusion is much higher than the increase in segmental diffusion and at this point, the autoaccelaration occurs. The diffusion controlled reduction in termination leads to high rate of polymerization, however, the rate of polymerization is also affected due to diffusion, but this effect is insignificant as compared to the reduced termination.

Molecular weight distribution or polydispersity in the formed polymer chains can also be derived by using following equations [2] for number average and weight average degree of polymerization

$$\bar{X}_n = \frac{2}{1-p} \tag{43}$$

$$\bar{X}_w = \frac{2+p}{1-p} \tag{44}$$

where p is defined as the probability that a propagating radical would continue to propagate without termination and thus the distribution or dispersity index can be defined by the following equation

$$\frac{\bar{X}_w}{\bar{X}_n} = \frac{2+p}{p} \tag{45}$$

B.2. Cationic Chain Polymerization

Lewis acids are very common initiators in the cationic polymerization of the carbon carbon double bond and they work much better in the presence of a proton donor such as water. The water in this case is termed as initiator and the Lewis acid is called coinitiator. The kinetic equations of initiation in cationic polymerization process along with rate of initiation can thus be described as follows (equations 46-48).

$$A + BH \rightleftharpoons AB^{\ominus} H^{\oplus} \tag{46}$$

$$AB^{\ominus} H^{\oplus} + M \xrightarrow{k_i} HM^{\oplus} BA^{\ominus} \tag{47}$$

$$R_i = k_i [M] [H^{\oplus}] \tag{48}$$

where A is coinitiator like boron trifluoride, BH is initiator like water. These form an initiator-coinitiator complex, which then reacts with the double bond of the monomer molecules and initiates the polymer chains. The propagation follows by the addition of a large number of monomer molecules to the growing chains and thus can be described as follows

$$HM^{\oplus} BA^{\ominus} + M \xrightarrow{k_p} HMM^{\oplus} BA^{\ominus} \tag{49}$$

$$\mathrm{HMM^{\oplus}BA^{\ominus} + nM \xrightarrow{k_p} HM_{n+1}M^{\oplus}BA^{\ominus}} \qquad (50)$$

$$R_p = k_p [M] [M^{\oplus}] \qquad (51)$$

The termination of the chains occurs by a number of different ways. Chain transfer to monomer involves transfer of a proton to monomer from the growing chain resulting in a dead polymer chain with terminal unsaturation as shown in equations 52 and 53 and the rate of termination thus can be described as in equation 54.

$$\mathrm{CH_3{-}CH(C_6H_5){-}[CH_2{-}CH(C_6H_5)]_n{-}CH_2{-}\overset{\oplus}{C}H(C_6H_5)\ {}^{\ominus}BF_3OH + CH_2{=}CH(C_6H_5) \longrightarrow CH_3{-}\overset{\oplus}{C}H(C_6H_5)\ {}^{\ominus}BF_3OH + CH_3{-}CH(C_6H_5){-}[CH_2{-}CH(C_6H_5)]_n{-}CH{=}CH(C_6H_5)} \qquad (52)$$

$$\mathrm{HM_nM^{\oplus}BA^{\ominus} \xrightarrow{k_{tr,M}} HM^{\oplus}BA^{\ominus} + HM_{n+1}} \qquad (53)$$

$$R_t = k_{tr,M} [M] [M^{\oplus}] \qquad (54)$$

Another mode of termination is spontaneous termination which involves the transfer of a hydrogen atom to counterion as shown below.

$$\mathrm{HM_nM^{\oplus}BA^{\ominus} \xrightarrow{k_{tS}} H^{\oplus}BA^{\ominus} + M_{n+1}} \qquad (55)$$

$$R_t = k_{tS} [M^{\oplus}] \qquad (56)$$

Combination with the counterion is also observed as a mode of termination and this process is described in the following equations 57-59.

$$HM_nM^{\oplus}BA^{\ominus} \xrightarrow{k_t} HM_nMBA \quad (57)$$

$$R_t = k_t[M^{\oplus}] \quad (58)$$

$$CH_3{-}\underset{C_6H_5}{CH}{-}\left[CH_2{-}\underset{C_6H_5}{CH}\right]_n{-}CH_2{-}\underset{C_6H_5}{\overset{\oplus}{C}H}\ \ {}^{\ominus}OCOCF_3 \longrightarrow CH_3{-}\underset{C_6H_5}{CH}{-}\left[CH_2{-}\underset{C_6H_5}{CH}\right]_n{-}CH_2{-}\underset{C_6H_5}{CH}{-}OCOCF_3 \quad (59)$$

Apart from that, chain transfer to solvents or deliberately added chain transfer agents can lead to termination of polymer chains as described below.

$$HM_nM^{\oplus}BA^{\ominus} + ZX \xrightarrow{k_{tr,S}} HM_nMX + Z^{\oplus}BA^{\ominus} \quad (60)$$

$$R_t = k_{tr,S}[M^{\oplus}][ZX] \quad (61)$$

where ZX is the chain transfer agent.

Assuming chain termination occurs either by spontaneous termination or by combination, the initiation and termination rates can be equated to each other at steady state to achieve an equation for rate of propagation or polymerization as shown in equations 62 to 64..

$$k_i[M][H^{\oplus}] = k_t[M^{\oplus}] \quad (62)$$

$$[M^{\oplus}] = \frac{k_i}{k_t}[M][H^{\oplus}] \quad (63)$$

$$R_p = k_p[M][M^{\oplus}] = \frac{k_i k_p}{k_t}[M]^2[H^{\oplus}] \tag{64}$$

The degree of polymerization can also be defined by the following equations by taking into account all modes of chain terminations.

$$\bar{X}_n = \frac{R_p}{R_t + R_{tS} + R_{tr,M} + R_{tr,ZX}} \tag{65}$$

$$\bar{X}_n = \frac{k_p[M]}{k_t + k_{tS} + k_{tr,M}[M] + k_{tr,ZX}[ZX]} \tag{66}$$

B.3. Anionic Chain Polymerization

Alkyl lithium initiators like butyl lithium are the most common initiators used for anionic chain polymerization (equation 67). Apart from that, tertiary amines are also used as initiators and in this case propagating species is a zwitterion (equation 68). Apart from that, a dianion system has also been developed in which the chain propagation takes place from both sides e.g. sodium naphthalene (equations 69-72).

$$C_4H_9Li + CH_2{=}CH(C_6H_5) \longrightarrow C_4H_9{-}CH_2{-}\overset{H}{\underset{C_6H_5}{C}}{:}^{\ominus}\ Li^{\oplus} \tag{67}$$

$$R_3N{:} + CH_2{=}CH(C_6H_5) \longrightarrow R_3\overset{\oplus}{N}{-}CH_2{-}\overset{H}{\underset{C_6H_5}{C}}{:}^{\ominus} \tag{68}$$

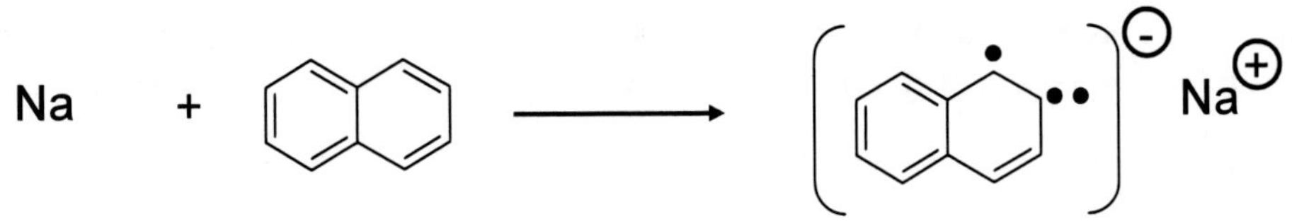

(69)

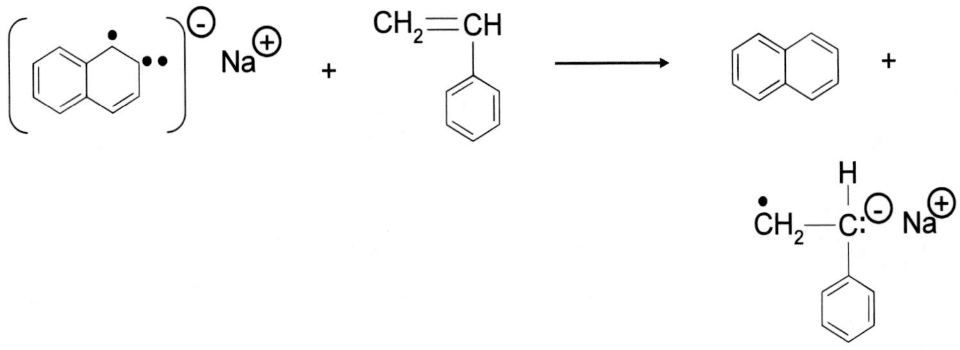

(70)

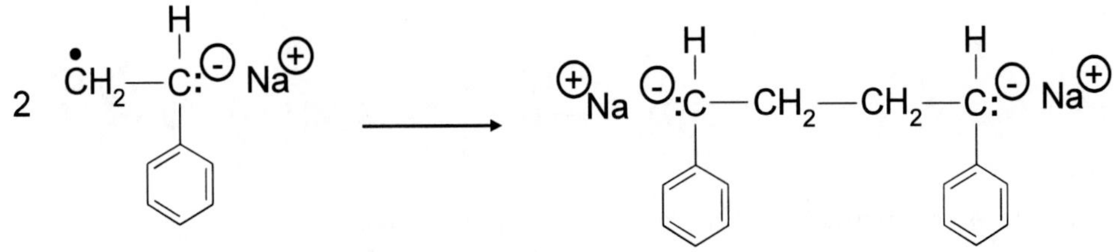

(71)

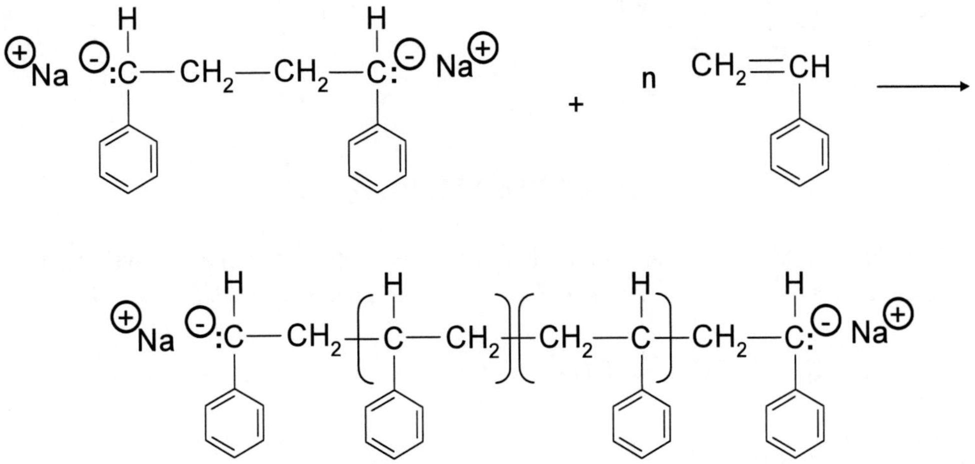

(72)

The kinetics of anionic polymerization can be descried by the following equations. The ion pair attacks the monomer molecules to initiate polymerization, the generated polymer radicals then add a large number of monomers to the chains thus propagating the polymerization reaction. The termination in the anionic polymerization system is rare, thus this polymerization technique is also called living polymerizations, the chains would propagate further as soon as fresh monomer is added to the system. However, as these polymerizations are very sensitive to any trace impurities, the termination does take place, e.g., with moisture.

$$I - X \longrightarrow I^{\oplus} X^{\ominus} \tag{73}$$

$$I^{\oplus} X^{\ominus} + M \xrightarrow{k_i} X - M^{\ominus} I^{\oplus} \tag{74}$$

$$X - M^{\ominus} I^{\oplus} + M \xrightarrow{k_p} X - MM^{\ominus} I^{\oplus} \tag{75}$$

$$X - MM^{\ominus} I^{\oplus} + nM \xrightarrow{k_p} X - M_{n+1}M^{\ominus} I^{\oplus} \tag{76}$$

$$X - M_{n+1}M^{\ominus} I^{\oplus} + H_2O \xrightarrow{k_t} X - M_nMH + OH^{\ominus} I^{\oplus} \tag{77}$$

The rate of polymerization reaction then can be deduced as

$$R_p = \frac{k_i k_p}{k_t} \frac{[X^{\ominus}] [M]^2}{[H_2O]} \tag{78}$$

C. Step Polymerization

The kinetics of step polymerization is quite different from the chain polymerization. As an example, is shown the common esterification reaction between a diacid and a diol.

$$HOOC - R - COOH + OH - R' - OH \longrightarrow HOOC - R - COO - R' - OH \tag{79}$$

The reaction can be achieved either in non-catalyzed conditions or acid-catalyzed conditions. In non-catalyzed conditions, the reactant acid itself acts as catalyst, therefore the rate of monomer consumption is written with the square of concentration of acid as

$$\frac{-d[OH]}{dt} = \frac{-d[COOH]}{dt} = k\,[COOH]^2\,[OH] \tag{80}$$

where k is the reaction rate coefficient. As the functional groups react in stoichiometric quantities, therefore, their concentrations can be equated and can be replaced as that of one reactant.

$$\frac{-d[M]}{dt} = k\,[M]^3 \tag{81}$$

Integration of the equation leads to the following equation, where $[M]_0$ is the initial concentration of reactants. To generate more information from the expression, a new term p is defined which is the fraction of functional groups which have reacted and can be given as

$$2\,k\,t = \frac{1}{[M]^2} - \frac{1}{[M]_0^2} \tag{82}$$

$$p = \frac{[M]_0 - [M]}{[M]} \tag{83}$$

[M] expression generated from equation 83 can be substituted in equation 82 to generate a more general equation of step polymerization as

$$2\,[M]_0^2\,k\,t + 1 = \frac{1}{(1-p)^2} \tag{84}$$

When the esterification is performed in the presence of a catalyst, the rate expression can be written as

$$\frac{-d[OH]}{dt} = \frac{-d[COOH]}{dt} = k\,[C]\,[COOH]\,[OH] \tag{85}$$

where [C] is the acid catalyst concentration. Integration of the rate equation leads to the similar expression as for the non-acid catalyzed scenario

$$k\,[C]\,[M]_0\,t \;=\; \frac{1}{(1-p)} - 1 \tag{86}$$

The number average and weight average molecular weight distribution of the formed polymer chains can be described by following equations:

$$N_n \;=\; N_0\,(1-p)^2\,p^{n-1} \tag{87}$$

$$W_n \;=\; n\,(1-p)^2\,p^{n-1} \tag{88}$$

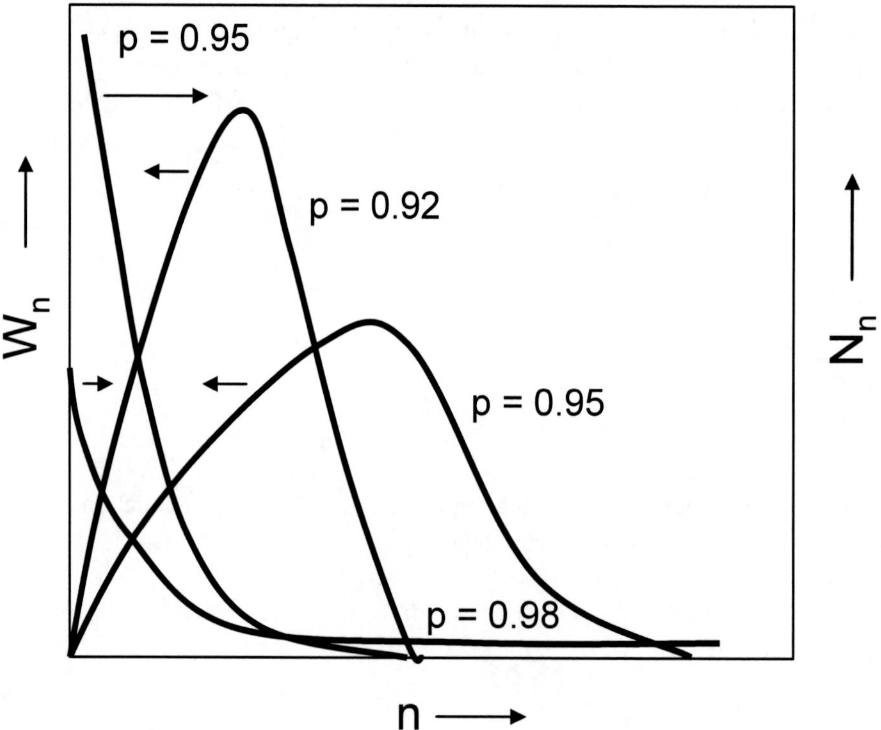

Figure 3. Number average distributions and weight average distributions of polymer chains at different p values.

where N_n is the number fraction of chains with n monomer molecules out of total polymer chains, N. N_o is the number of reactant molecules present initially, n is the number of repeat units, p is the extent of conversion and W_n is the weight fraction of chains with n monomer molecules out of total polymer chains. Figure 3 is the representation of weight and number fractions distribution of polymer chains using different p values [3].

D. Copolymerization

Copolymerization is route to generate a variety of commercially important polymers. As it involves two or more monomers, therefore, the kinetics of the copolymerization reactions is fairly complex. Following equations provide a glimpse of the complexity of the system and multitude of various reactions which take place during this process.

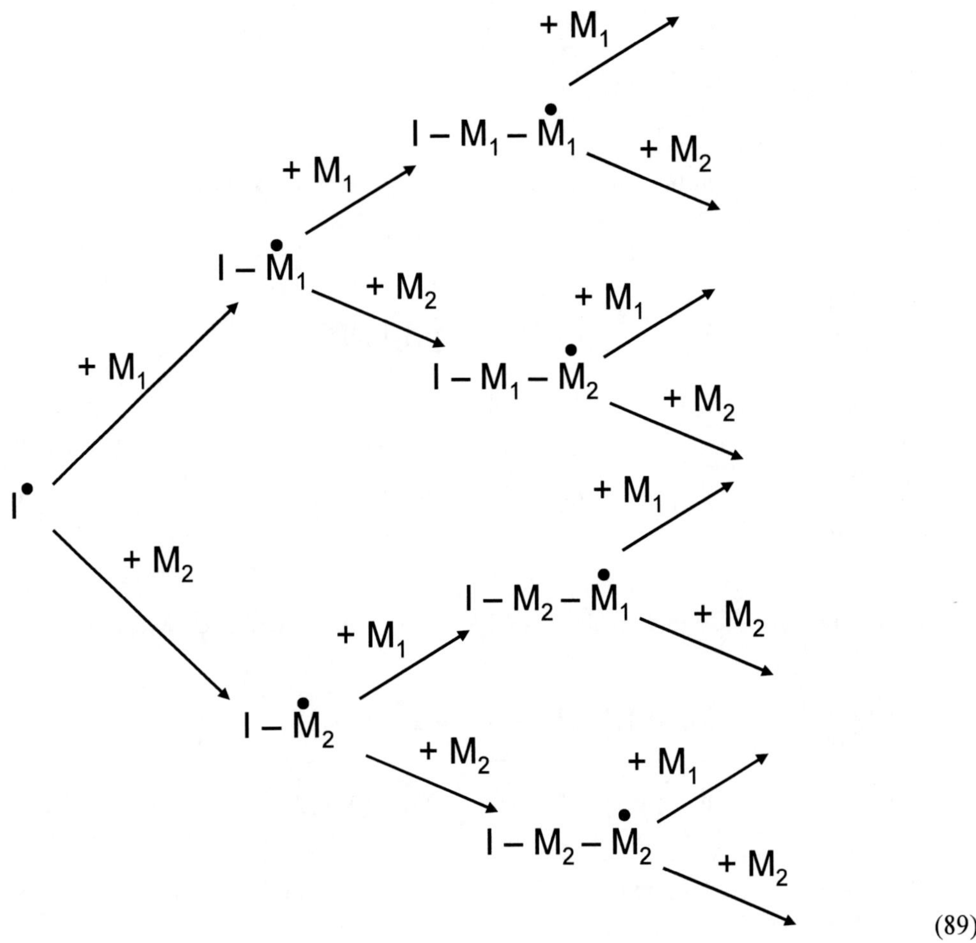

(89)

To address copolymerization of two monomers, the reaction paths can be broken down into following four possibilities, where k_{11} is the reaction rate coefficient for the reaction of radical with chain end consisting of M_1 with M_1 and so on.

$$M_1^{\bullet} + M_1 \xrightarrow{k_{11}} M_1^{\bullet} \tag{90}$$

$$M_1^{\bullet} + M_2 \xrightarrow{k_{12}} M_2^{\bullet} \tag{91}$$

$$M_2^{\bullet} + M_1 \xrightarrow{k_{21}} M_1^{\bullet} \tag{92}$$

$$M_2^{\bullet} + M_2 \xrightarrow{k_{22}} M_2^{\bullet} \tag{93}$$

The rates of consumption of monomers can thus be written according to the following two equations

$$\frac{-d[M_1]}{dt} = k_{11}[M_1^{\bullet}][M_1] + k_{21}[M_2^{\bullet}][M_1] \tag{94}$$

$$\frac{-d[M_2]}{dt} = k_{12}[M_1^{\bullet}][M_2] + k_{22}[M_2^{\bullet}][M_2] \tag{95}$$

and the division of these equations leads to the general copolymerization equation as

$$\frac{-d[M_1]}{-d[M_2]} = \frac{k_{11}[M_1^{\bullet}][M_1] + k_{21}[M_2^{\bullet}][M_1]}{k_{12}[M_1^{\bullet}][M_2] + k_{22}[M_2^{\bullet}][M_2]} \tag{96}$$

The reaction rate coefficients can be reorganized as r_1 and r_2 as shown below and the above equation can be transformed as

$$r_1 = \frac{k_{11}}{k_{12}} \tag{97}$$

$$r_2 = \frac{k_{22}}{k_{21}} \tag{98}$$

$$\frac{d[M_1]}{d[M_2]} = \frac{[M_1]\,(r_1[M_1] + [M_2])}{[M_2]\,([M_1] + r_2[M_2])} \tag{99}$$

The above equation is more general form of copolymerization equation and r_1 and r_2 are called monomer reactivity ratios [1,2]. Reactivity ratio is defined as the ratio of the rate constant for a reacting species adding the monomer of its own type to the rate constant for the reacting species adding the other monomer. It is also possible to convert the above equation in form of monomer mole ratios f_1 and f_2 in feed and $F_1$and F_2 in copolymer as

$$f_1 = 1 - f_2 = \frac{[M_1]}{[M_1] + [M_2]} \tag{100}$$

$$F_1 = 1 - F_2 = \frac{d[M_1]}{d[M_1] + d[M_2]} \tag{101}$$

$$F_1 = \frac{r_1 f_1^2 + f_1 f_2}{r_1 f_1^2 + 2 f_1 f_2 + r_2 f_2^2} \tag{102}$$

Monomer reactivity ratios are useful tools to ascertain the type of copolymer formed during the copolymerization reaction. If $r_1{*}r_2 = 1$ or $r_2 = 1/r_1$, then copolymer is termed as ideal copolymer and a random copolymer is formed. This means that the propagating species $[M_1^\bullet]$ and $[M_2^\bullet]$ do not have a preference of monomer, but can add any of the M_1 or M_2 monomer molecules in vicinity. F_1 can be defined in this case as

$$F_1 = \frac{r_1 f_1}{r_1 f_1 + f_2} \tag{103}$$

On the other hand, if $r_1 = r_2 = 0$, the copolymer formed is strictly alternating as in this case $k_{11} = k_{22} = 0$. It means that the propagating species $[M_1^\bullet]$ and $[M_2^\bullet]$ only can add monomer M_2 and M_1 respectively, not the opposite. A block copolymer is formed when r_1, $r_2 > 1$, indicating that $[M_1^\bullet]$ may add both M_1 and M_2 monomers to the chain, but the preference is for M_1.

The above treatment is valid for the two monomer system, however, many chemical reactions involve three or more monomers and the kinetic treatment of such systems is further complex owing to the large number of permutations of the different reactions between the monomers. The various possibilities of such chemical reactions are described as follows:

$$I-M_3-M_1^{\bullet} \qquad I-M_3-M_2^{\bullet} \qquad I-M_3-M_3^{\bullet}$$

$$I-M_3^{\bullet}$$

$$M_3$$

$$+$$

$$I-M_2-M_1^{\bullet} \qquad\qquad\qquad\qquad\qquad\qquad I-M_1-M_1^{\bullet}$$

$$I-M_2-M_2^{\bullet} \leftarrow I-M_2^{\bullet} \leftarrow M_2 + I^{\bullet} + M_1 \rightarrow I-M_1^{\bullet} \rightarrow I-M_1-M_2^{\bullet}$$

$$I-M_2-M_3^{\bullet} \qquad\qquad\qquad\qquad\qquad\qquad I-M_1-M_3^{\bullet} \tag{104}$$

The reactivity ratios and the rate equations for the reactions can then be written as

$$r_{12} = \frac{k_{11}}{k_{12}} \tag{105}$$

$$r_{13} = \frac{k_{11}}{k_{13}} \tag{106}$$

$$r_{21} = \frac{k_{22}}{k_{21}} \tag{107}$$

$$r_{23} = \frac{k_{22}}{k_{23}} \tag{108}$$

$$r_{31} = \frac{k_{33}}{k_{31}} \tag{109}$$

$$r_{32} = \frac{k_{33}}{k_{32}} \tag{110}$$

$$R_{11} = k_{11}[M_1^{\bullet}][M_1] \tag{111}$$

$$R_{12} = k_{12}[M_1^\bullet][M_2] \tag{112}$$

$$R_{13} = k_{13}[M_1^\bullet][M_3] \tag{113}$$

$$R_{21} = k_{21}[M_2^\bullet][M_1] \tag{114}$$

$$R_{22} = k_{22}[M_2^\bullet][M_2] \tag{115}$$

$$R_{23} = k_{23}[M_2^\bullet][M_3] \tag{116}$$

$$R_{31} = k_{31}[M_3^\bullet][M_1] \tag{117}$$

$$R_{32} = k_{32}[M_3^\bullet][M_2] \tag{118}$$

$$R_{33} = k_{33}[M_3^\bullet][M_3] \tag{119}$$

E. Polymerization Techniques

Polymerization techniques are equally important, as polymerization kinetics as different monomers require different polymerization conditions, e.g., some monomers polymerize in gas phase, others in a solution with solvent or in bulk or in emulsion or suspension. Such requirements have thus led to the development of a number of polymerization techniques as described below. Figure 4 also details the wide spectrum of homogenous and heterogeneous polymerization techniques.

E.1. Bulk Polymerization

Bulk polymerization comes under the category of homogenous polymerization. In this technique, liquid monomer is mixed with monomer soluble initiator and is polymerized. Though the reaction mixture is free from any unwanted impurities or contamination leading to clean polymer, however, it is very difficult to control such a process. As the polymerization proceeds, the viscosity of the system increases significantly due to the generation of polymer chains and thus the mixing of the system becomes extremely difficult leading to very broad molecular weight distributions in the polymer chains. Apart from that, the increased viscosity leads to slower termination as the polymer chains do no diffuse freely leading to the accumulation of radicals at particular sites where the polymerization rate increase exponentially, which is termed as autoaccelaration. This leads to runaway reactions which can also turn into explosions. Therefore, bulk polymerization is not used extensively for the commercial generation of polymers. Styrene and methyl methacrylate are polymerized by this method to some extent. A possible solution to circumvent the problem of increased viscosity and associated effects to carry out polymerizations at low conversions and using multiple reactors.

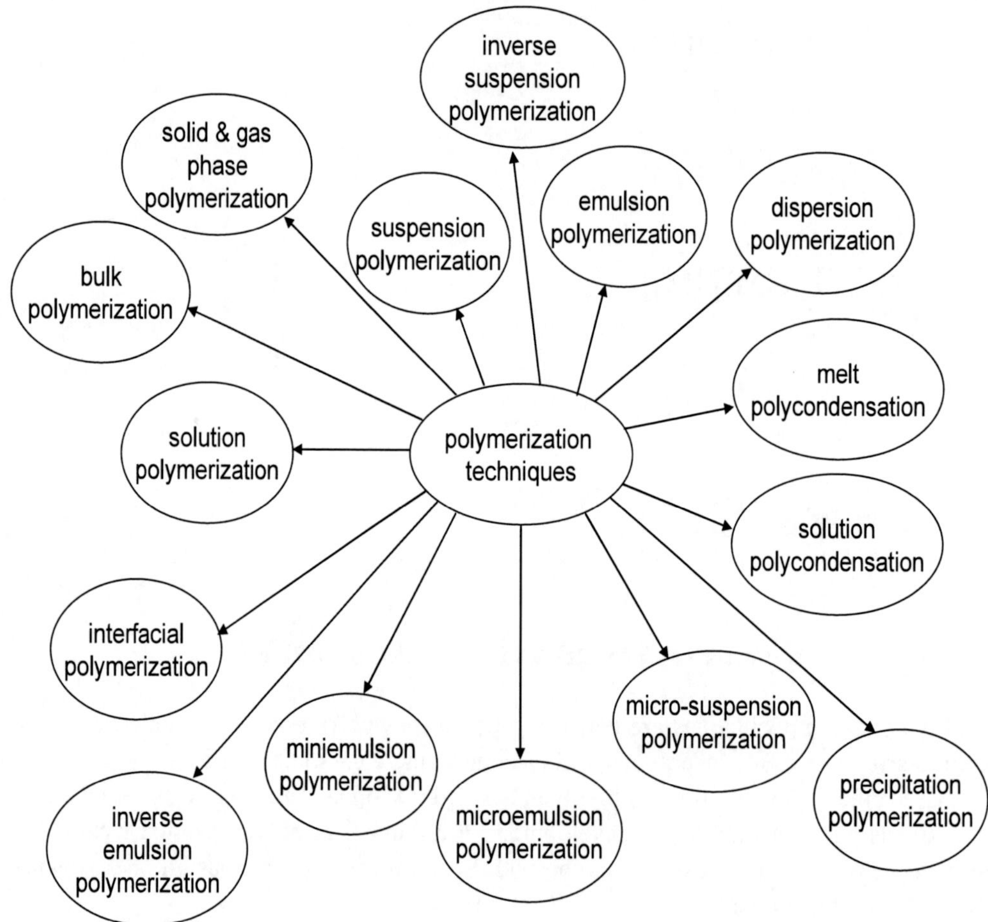

Figure 4. Various polymerization techniques to generate a wide spectrum of polymers.

E.2. Solution Polymerization

Solution polymerization is a method in which a solvent in which the monomer and polymer are soluble is used as a reaction medium. The initiator is also soluble in the solvent used. The use of solvents allows one to stir the reaction medium easily thus avoiding the viscosity problems and the heat dissipation can also be controlled easily. Thus, in many ways, the solution polymerization is a better technique as compared to bulk polymerization. However, the use of solvents is not completely problem-free. The choice of solvents must be proper; otherwise extensive chain transfer to solvent can take place resulting in only low molecular weight polymer chains. The polymer is recovered from the solvent either by evaporating the solvent or by precipitation of the polymer in a non-solvent generally methanol. Thus, it can also be troublesome occasionally if the removal of the solvent is difficult or is not complete. Commercially acrylonitrile and polyisobutylene are produced by this method.

E.3. Precipitation Polymerization

Precipitation polymerization is a form of polymerization in which the polymer is not soluble in the monomer or the reaction medium and precipitates out soon from the solution. Thus, precipitation polymerization starts as homogenous polymerization, but is soon turned into a heterogeneous polymerization. This phenomenon can be achieved both in bulk or solution polymerization. The particle nucleation keeps on occurring throughout the polymerization and the polymerization proceeds by the absorption of monomer and initiator in the polymer particles therefore leading to continuous increase in size during the polymerization. The termination is generally delayed in the reaction environment of precipitation polymerization leading to higher molecular weights in the polymer chains. Polymerization of acrylonitrile and vinyl chloride are examples of precipitation polymerization. Usually aqueous polymerizations are achieved by precipitation polymerization, but sometimes organic solvents can also be used.

E.4. Dispersion Polymerization

Dispersion polymerization also is similar to precipitation polymerization that the polymer formed is not soluble in monomer or organic solvent. It also initially starts as homogenous polymerization, but is soon converted into heterogeneous polymerization. However, the main locus of polymerization in this case is polymer particles. The particle sizes formed lie between the range of 0.1 and 10 μm, i.e., in between the ranges covered by emulsion and suspension polymerization. The polymerization also includes the addition of particle stabilizer which is generally an uncharged polymer. After the formation of polymer particles, these are stabilized by the particle stabilizer and the polymerization proceeds in these polymer particles by the absorption of monomer into the polymer particles. Styrene and methyl methacrylate are generally polymerized by this method in a variety of solvents like alcohols, hydrocarbons etc. The kinetics of bigger particle generation is similar to suspension polymerization and for the smaller particles, emulsion polymerization kinetics is applicable.

E.5. Suspension Polymerization

Suspension or bead polymerization is a method in which monomer droplets are directly polymerized to generate high molecular weight polymers. In his method, water insoluble monomer is suspended in water with the aid of suspension stabilizers. The average size of droplets ranges from 100–500 μm in diameter. The initiator used is also water insoluble or monomer soluble. The size of monomer droplets can be controlled depending on the ratio of monomer to the dispersion medium, i.e., water, the speed of agitation to generate droplets as well as by the amount of stabilizing agents. During the polymerization, the monomer droplets polymerize independently and each droplet can be visualized as a bulk polymerization happening inside the droplet. As these drops are suspended in water, therefore, there is no problem of viscosity enhancement and heat can also be dissipated without problem. The polymerization leads to narrow molecular weight distribution with high molecular weights as the termination is very low in the droplets as at a time generally one radical can exist in these droplets. In the end, the polymer particles are present as pearls or beads, therefore, this polymerization is also termed as bead polymerization and these beads can be easily recovered from water by filtration. Though the beads can be used for a number of applications after drying of water, however, the suspension stabilizers present on their surfaces can sometimes cause unwanted lowering of properties. However, the amount of these stabilizers is generally very less as compared to the surfactants in the emulsion polymerization. One thing to note here is that the suspension stabilizers do not form any micelles as the surfactant micelles generated in the emulsion polymerization. Common examples of polymers generated by suspension polymerization are polystyrene, polyvinylchloride, polyacrylonitrile and acrylic and methacrylic esters.

E.6. Inverse Suspension Polymerization

As the name suggests, the inverse suspension polymerization is the opposite of conventional suspension polymerization. In this case, a water soluble monomer is used and its droplets are generated in an organic solvent. Initiators used are also water soluble and the monomer droplets are stabilized similarly by using suspension stabilizers. The common example of such a system is the polymerization of acrylamide.

E.7. Emulsion Polymerization

Emulsion polymerization is a very versatile technique to generate small particles. With this technique, water insoluble monomers are polymerized by suspending them in water in the form of droplets. The initiators used are water soluble in contrast to suspension polymerization where water insoluble initiators are used, the most common being potassium persulphate (KPS). The monomer droplets in emulsion polymerization are stabilized by adding surfactants, the most common of them being sodium dodecyl sulphate (SDS). However, a wide range of anionic, cationic or non ionic surfactants can be used to lower the surface energy and cause stabilization of the monomer molecules. The behavior or role of the stabilizer in emulsion polymerization is different from the suspension polymerization. The stabilizer in the emulsion polymerization apart from stabilizing the monomer droplets also forms the micelles once the concentration of stabilizer exceeds critical micelle concentration. The inside environment of a micelle is totally hydrophobic and is very favorable for monomer polymerization. The structure of stabilizer is generally amphiphilic, with one part hydrophobic and the other part hydrophilic. These molecules thus arrange themselves in a way that their hydrophilic parts are in interface with water. The polymerization is initiated in micelles and these growing polymer particles are then supplied by the monomer molecules from the monomer droplets by diffusion through the aqueous phase. The termination of the radicals is quite slow as at a particular time during polymerization, there is rarely more than one radical per particle. The polymerization leads to a very high molecular weight of the polymer chains in the polymer particles. The formed latexes, as they are called, are very stable with time as compared to the suspension beads which would sediment as soon as the agitation is stopped. A variety of monomers like styrene, acrylates, methacrylates etc. can be polymerized through emulsion polymerization. A disadvantage of emulsion polymerization is the use of significant amount of stabilizer which sometimes can have detrimental effects on the surface properties of the particles. To achieve more clean latexes, emulsion polymerization is also achieved without the use of emulsifier. The nucleation and particles formation follows completely different routs and the particles sizes thus formed are generally quite larger than the particles with emulsified systems.

E.8. Inverse Emulsion Polymerization

Along the same line as inverse suspension polymerization, inverse emulsion polymerization is used for the polymerization of water-soluble monomers like acrylamide. In this case, an organic solvent is used as a dispersion medium, and a water-soluble monomer in solution in water is emulsified in the form of monomer droplets. Special emulsifiers like high molecular weight block copolymers are used for such purposes. The monomer diffuses from the droplets into the polymer particles during the course of polymerization. Therefore, it is also required for the monomer to have small oil solubility otherwise the diffusion of the monomer to the polymer particles would not take place.

E.9. Miniemulsion Polymerization

Polymerization of extremely low water soluble monomers is very difficult with conventional emulsion polymerization. The reason for this phenomenon is the same as described above. The low solubility of the monomer would not allow its diffusion to the polymer particles. Generally, cyclodextrins are added to improve the water solubility of the monomers. However, another technique of miniemulsion polymerization has been developed, in which the monomer droplets are generated by using high shear in the presence of an ionic surfactant and a co-surfactant or hydrophobe like hexadecane. The droplets are generally in the size range of 50–500 nm. When the polymerization is initiated, the monomer droplets directly polymerize to form polymer particles. Thus, in this process the number of final polymer particles is expected to be the same as the initial number of droplets; however, many times the final number of polymer particles can be quite low. The generation of miniemulsion requires a lot of shearing force and the miniemulsion before polymerization is not stable over long periods of time. Apart from that, the use of a volatile hydrophobe can also be a problem.

E.10. Microemulsion Polymerization

Similar to miniemulsion polymerization, microemulsion polymerization is also used for the polymerization of specifically water insoluble monomers. This system also uses a combination of surfactant and co-surfactant. However, the droplet size in this case is much smaller than the miniemulsion polymerization case. Droplet sizes of generally 5–20 nm are generated in microemulsion polymerization, indicating that much more energy would be required to achieve such a system. The microemulsions are, however, stable for a longer period of time.

E.11. Melt Polycondensation

Melt polycondensation is used when at least one of the monomers is in solid state. However, the reaction generally requires high temperatures (higher than the melting point of the monomers) and thus it has to be assured that the monomers do not decompose at these high reaction temperatures. The reaction has problems similar to bulk polymerization, in that the reaction system becomes very viscous during polymerization. Transfer of the polymers from the reactor also requires high temperatures, otherwise the polymer may solidify inside the reactor.

E.12. Solution Polycondensation

Solution polycondensation is similar to solution polymerization, which is used to circumvent the problems faced by bulk polymerization, in this case melt polycondensation. The use of an appropriate solvent leads to the proper dissipation of heat and keeps the viscosity of the reaction mixture low even when a high molecular weight polymer is formed. It also allows the polymerization reaction to be carried out at lower temperatures, thus avoiding the thermal stability concerns of the monomers at high temperatures. As the reaction mixtures are properly stirred, throughout the polymerization process, the thermal degradation of the formed polymer is also eliminated, which otherwise is a major concern in melt polycondensation, as at high viscosity, hotspots are generated which can have a very high temperature owing to polymers being poor conductors of heat and thus can cause serious damage to the molecular weight. Polymer discoloration is the common symptom of such a phenomenon.

REFERENCES

[1] K. Matyjaszewski and T. P. Davis (2002). Editors, *Handbook of Radical Polymerization*, John Wiley & Sons, Inc., New Jersey.

[2] G. Odian (2004). *Principles of Polymerization, Fourth Edition*, John Wiley & Sons, Inc., New Jersey.

[3] V. R. Gowariker, N. V. Viswanathan and J. Sreedhar (1986). *Polymer Science*, John Wiley & Sons, Wiley Eastern Limited, New Delhi.

Chapter 3

EMULSION POLYMERIZATION

A. INTRODUCTION

Emulsion polymerization is a heterogeneous polymerization technique in which the polymer is synthesized from water insoluble monomers in the form of particles suspended in water. Though reverse processes are also developed in which the water soluble monomers are polymerized in organic solvents, these processes are not as common as conventional emulsion polymerization. In the emulsion polymerization, the monomers, e.g., styrene, methyl methacrylate, etc., are suspended in water in which a surfactant has been added. The surfactant can be a cationic, ionic or nonionic surfactant. The emulsion polymerization leads to very high molecular weight polymer chains and, owing to the better process control, can lead to a variety of polymer morphologies and polymer particle sizes. The bulk polymerization, though, leads to cleaner polymers because of the absence of any additional foreign species such as surfactants or stabilizers; however, these processes are very difficult to control. At higher extents of polymerization, the reaction mixture becomes very viscous and almost impossible to stir. This leads to improper distribution of heat, autoaccelaration and runaway reactions. These processes lead to polymers with broad dispersity. On the other hand, emulsion polymerization is a remarkable technique in this regard, as it does not suffer from these limitations of viscosity and heat control. The polymerization in water as a dispersion medium keeps the viscosity of the reaction medium low and allows one to stir the contents easily even at high monomer conversions; the heat generated during the polymerization reaction can easily be transmitted or distributed uniformly throughout the reaction vessel and the temperature can be easily controlled by cooling (or heating) the jacket of the reaction vessel. The emulsion polymerization can be visualized as bulk polymerization of the suspended particles, as each particle in itself is a bulk polymerization reactor. However, a disadvantage of emulsion polymerization systems is the presence of a large amount of surfactant on the surface of the polymer particles. The serum also consists of an unreacted monomer or dissolved polymer or dissolved surfactant. However, it is easy to replace the serum, but cleaning the surface off the surfactant is not easy. Although for many applications it is not a critical problem, many applications, however, such as surface interactions or surface modifications of particles, require the particles to have well-defined surfaces. To circumvent these issues, surfactant-free polymerization is also used. This polymerization leads to a completely different particle nucleation mechanism, and the particles are stabilized by the negative charges on the polymer particle surfaces from the initiator. However, this technique leads to larger particle sizes, as the smaller particles are not stable and collapse on each other to form big and stable particles.

Another advantage of emulsion polymerization is the independence of the reaction rate from the generated molecular weight of the polymer chains [1]. In other polymerization techniques, the molecular weight is inversely dependant on the polymerization rate. The molecular weight of the polymer, in these polymerization techniques, can only be increased if the polymerization rate is slowed down. There are different ways to achieve this, either by reducing the amount of initiator or by decreasing the reaction temperature. The reduced polymerization rate, therefore, means longer polymerization times to achieve higher extents of conversion. However, emulsion polymerization is kinetically different from other polymerization techniques, and in this mode of polymerization the molecular weight can be increased without slowing the polymerization rate. The amount of the surfactant added during polymerization is the key, as it leads to a different number of particles and, depending on the number of the particles, the polymerization rate is correspondingly affected. Thus, emulsion polymerization can lead to fast generation of high molecular weights in the polymer chains.

B. Components and Polymerization Mechanism

The surfactants are amphiphilic molecules, i.e., one hydrophobic component and one hydrophilic component. These surfactants tend to lower the surface tension in the oil-water phases and help to achieve stable suspension of monomer droplets in water. The examples of anionic surfactants, in which the hydrophilic part is an anion, are alkali salts of fatty acid, whereas the examples of cationic surfactants, in which the hydrophilic part is a cation, are alkyl ammonium salts [2]. Non-ionic surfactants like polyols are also used in cases where the ionic surfactants may react with the monomer or other components. The surfactant at low concentrations may dissolve fully in water (though this depends on individual surfactants), but at higher concentration, the excess molecules start to form molecular aggregates. These aggregates are termed *micelles*, and the surfactant concentration at which the micelles start forming is termed critical micelle concentration (CMC). The dissolved surfactant molecules are not of interest in emulsion polymerization; it is the micelles that are of tremendous importance. Thus, the amount of surfactant initially added in the polymerization reaction mostly exceeds the critical micelle concentration. Every surfactant has a different critical micelle concentration value and one should be careful while using different kinds of surfactants. The micelles generally have a size of 10 nm and generally 100–200 surfactant molecules form a micelle [1]. Figure 1a shows a typical micelle in which the hydrophilic component of the surfactant molecules is facing the water, whereas the long alkyl chains in the surfactant molecules, which are hydrophobic in nature, radiate away from the water phase. Figure 1b is the representation of the micelle formed by commonly-used sodium dodecyl sulphate surfactant. This kind of assembly of surfactant molecules leads to a space inside the micelles that is completely hydrophobic in nature and is an attractive space for the hydrophobic monomer to enter when the monomer is added to the system. As it is generally known that the surface tension of the solution decreases with the addition of surfactant at

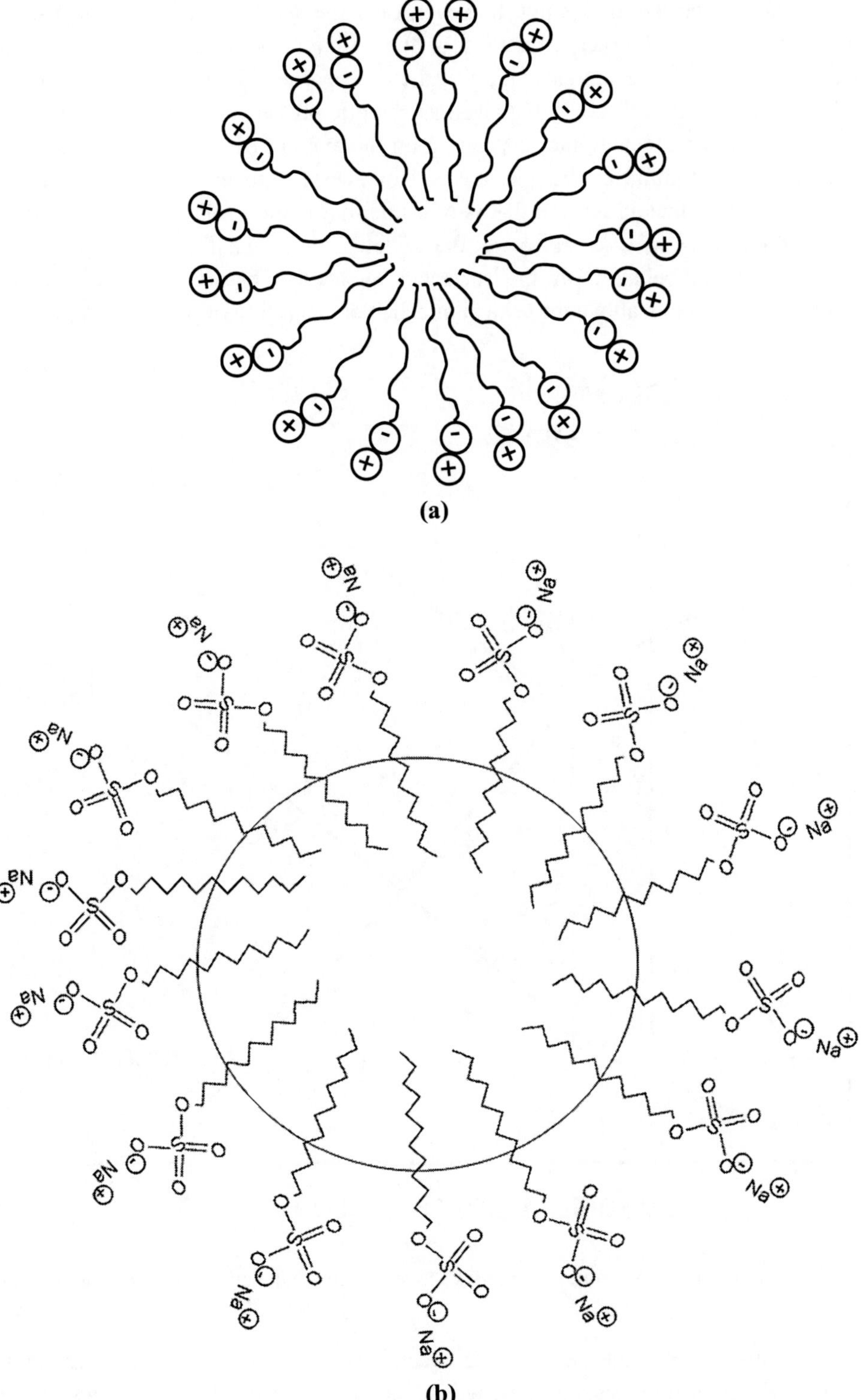

Figure 1. (a) Representation of a micelle from anionic surfactant and (b) same micelle with sodium dodecyl sulphate.

critical micelle concentration it is not, however, only the surface tension that is affected by the surfactant. A host of other properties of the solution are affected at critical micelle concentration. Figure 2 is a representation of these properties [3]. One can observe that the properties like conductivity, turbidity as well as osmotic pressure of the solution are affected upon the addition of surfactant and, depending on the amount of surfactant, these properties can be adjusted at optimum requirement levels. However, for the emulsion polymerization, it is the reduction of surfactant tension that is of prime importance, and other properties can be adjusted in reliance with it, as shown by the dotted lines in the figure. Figure 3 is also a representation of the behavior of the surfactant molecules before and after the micelle generation, i.e., below and above the critical micelle concentration of the surfactant.

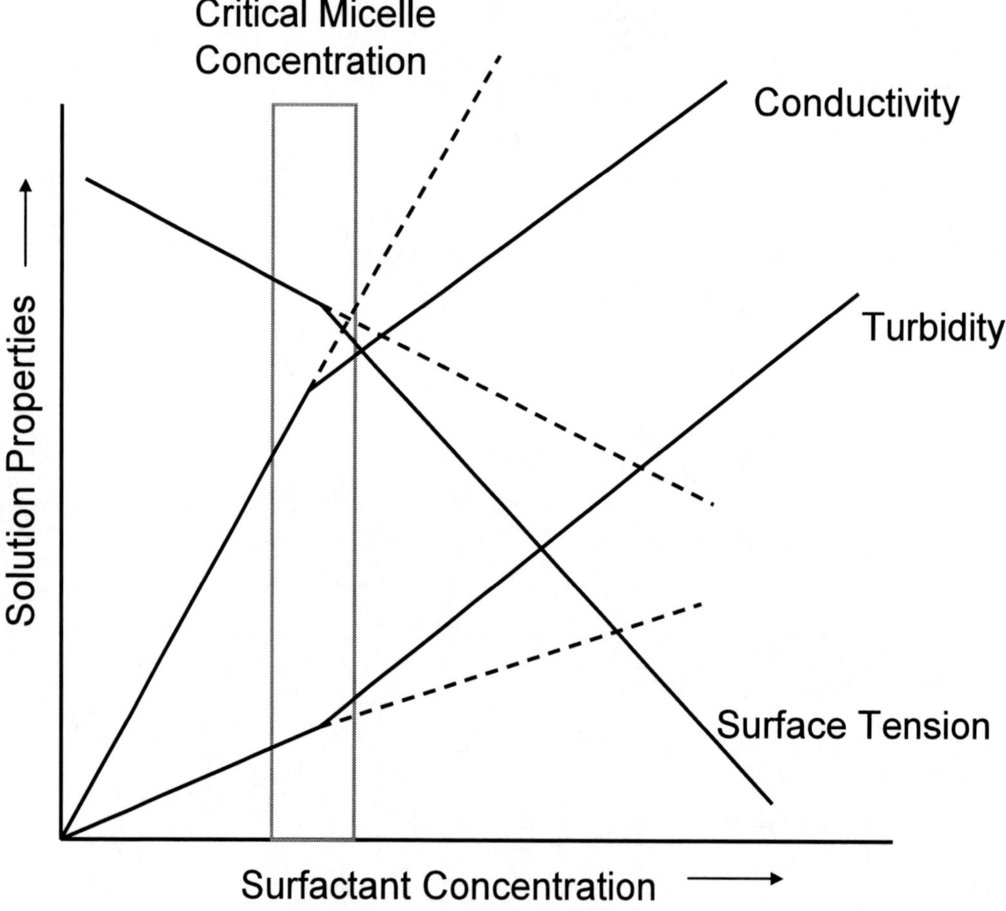

Figure 2. Behavior of different solution properties above and below the critical micelle concentration of the surfactant.

One should also note that the spherical micelles shown in Figures 2 and 3 are only one type of a multitude of micelles varieties, which can be formed by the surfactants depending on the shape parameter, P. This parameter is defined as given in equation 1:

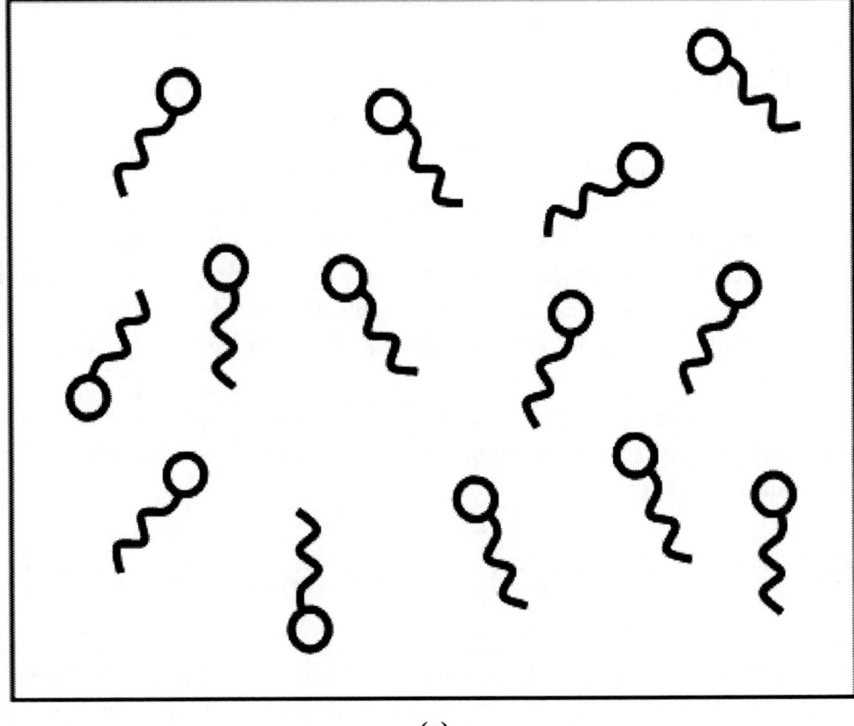

(a)

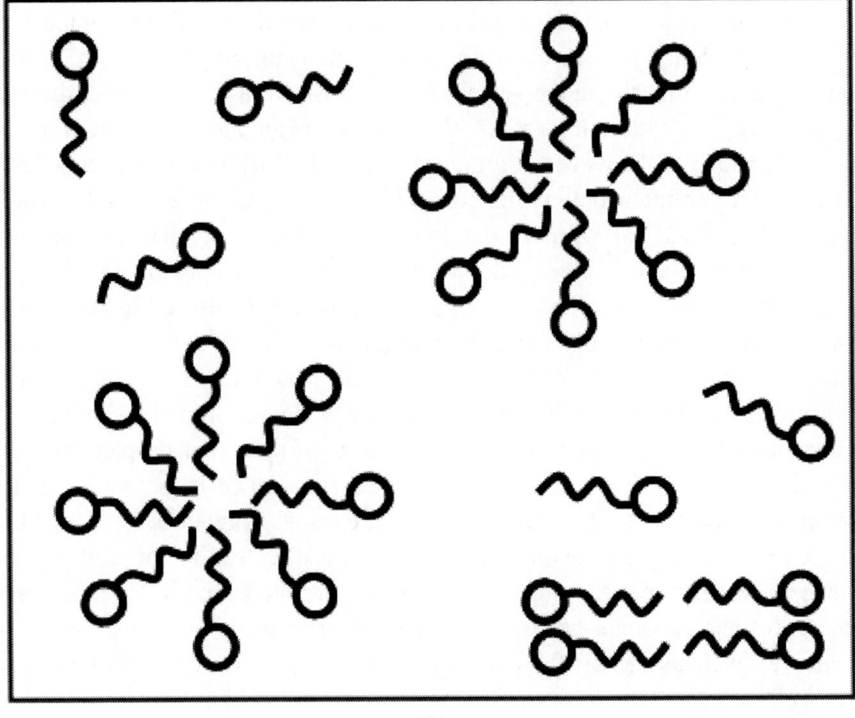

(b)

Figure 3. Association of the surfactant molecules (a) below and (b) above the critical micelle concentration.

$$P = \frac{v}{l * a_0} \tag{1}$$

where v is the volume occupied by the hydrophobic tail, l is the maximum effective length of the tail and a_0 if the surface area occupied by the head group. Shape factor of less than 0.33 leads to the formation of the spherical micelles as seen mostly in emulsion polymerization systems. Shape factor values between 0.33 to 1 lead to different shapes of micelles, like cylindrical micelles, worm like micelles, inverse micelles and bilayers, etc., as shown in Figure 4. Figure 5 also details the various different surfactants generally used in emulsion polymerization.

The initiators used in the polymerization are water soluble and most commonly used initiators are potassium or sodium persulphate. The initiation of the persulphate initiator is described in equation 2 as follows:

$$K^{\oplus}\,{}^{\ominus}O_3S{-}O{-}O{-}SO_3^{\ominus}\,{}^{\oplus}K \longrightarrow S_2O_8^{2-} \longrightarrow 2\,SO_4^{\bullet -} \tag{2}$$

At temperatures above 50°C, the O-O bond in the persulphate dissociates to generate two identical radicals which also have negative charge present on them which is helpful in providing colloidal stability to the polymer particles. Also, interesting is to note the interaction of the monomer to the surfactant micelles in water. First of all, the monomer should be water insoluble to enter the micelles. However, the monomer should have some extent of water solubility also in order for whole emulsion polymerization to proceed as during the polymerization, the monomer needs to diffuse from the monomer droplets to the polymer particles and this diffusion process through water can only take place if the monomer is partially water soluble. However, too much water solubility is also not beneficial as this would lead to more solution polymerization. When the monomer is added to the solution containing micelles, small part of monomer enters these micelles as the environment of these micelles is hydrophobic and is a very suitable place for the hydrophobic monomer to reside. Owing to the small water solubility, some part of monomer is also dissolved in water. The micelles have been experimentally shown to contain some monomer as their size increases as soon as monomer is added [1]. However, by far the major amount of monomer remains away from these micelles and is present in the emulsion as monomer droplets which are stabilized by the adsorption of the surfactant molecules. The size of monomer droplets may fall in the range of tens of micrometers. The number of micelles is much larger than the number of monomer droplets and the surface area of the micelles is also higher by orders of magnitude owing to the low number of monomer droplets. Thus, it is true to say that the monomer, initiator and surfactant are the three components of emulsion polymerization and the adjustments of their amounts relative to each other can bring about totally different morphologies as well as sizes of polymer particles. Figure 6 is a representation of this concept.

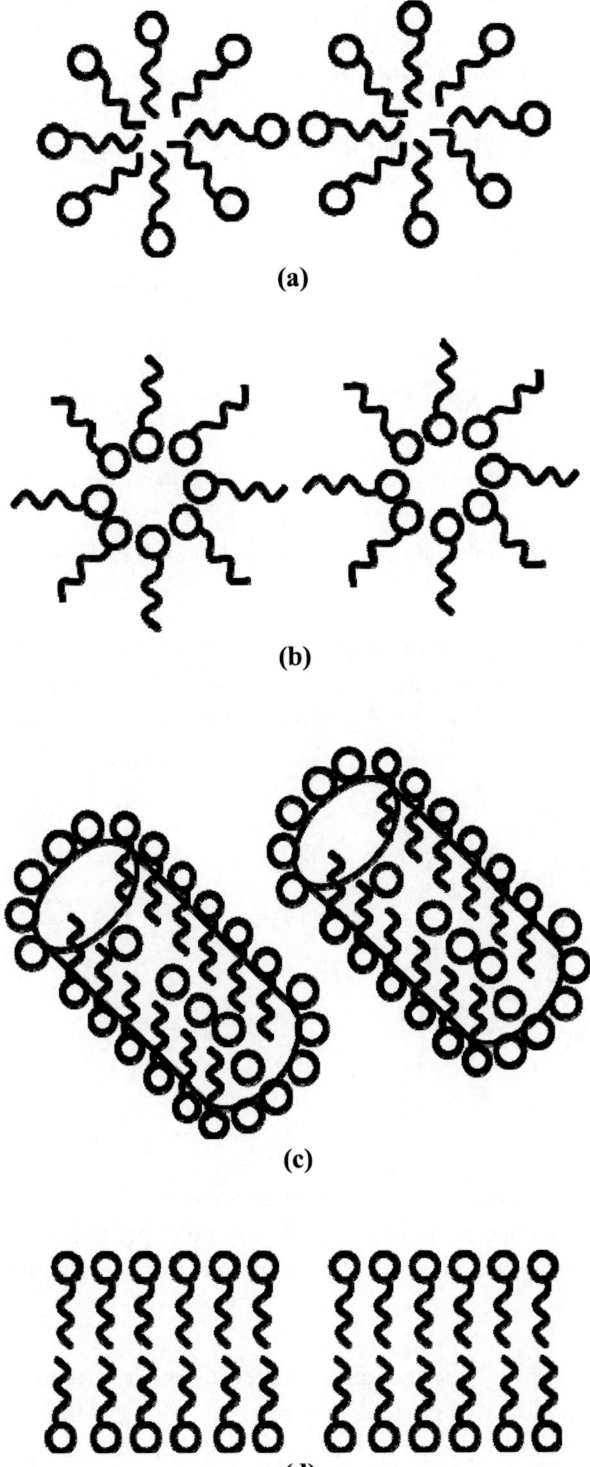

Figure 4. Various possibilities of surfactant association. (a) micelle, (b) inverse micelle, (c) cylindrical and (d) planar bilayer association.

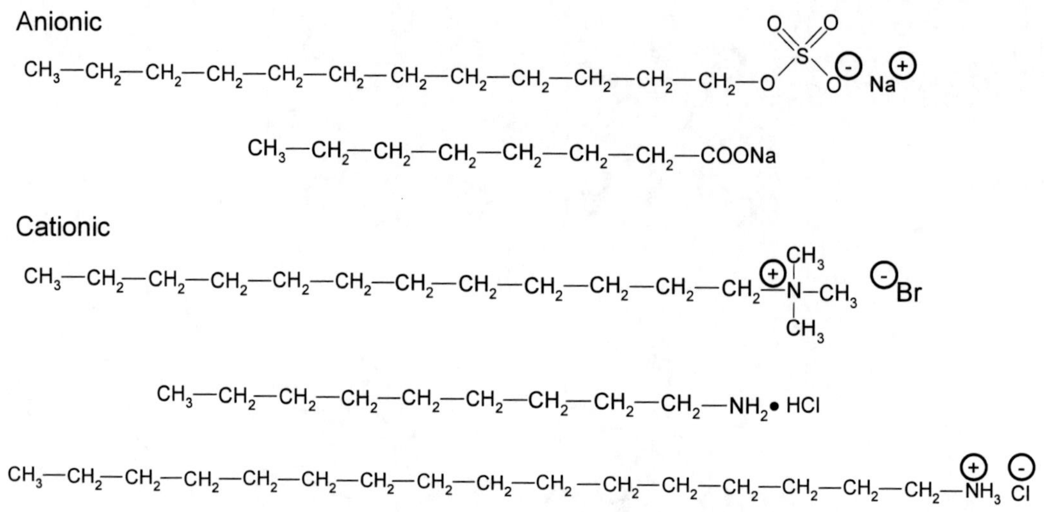

Anionic

$CH_3—CH_2—CH_2—CH_2—CH_2—CH_2—CH_2—CH_2—CH_2—CH_2—CH_2—CH_2—O—S(O)(O)—O^{\ominus}\ Na^{\oplus}$

$CH_3—CH_2—CH_2—CH_2—CH_2—CH_2—CH_2—COONa$

Cationic

$CH_3—CH_2—CH_2—CH_2—CH_2—CH_2—CH_2—CH_2—CH_2—CH_2—CH_2—CH_2—\overset{\oplus}{N}(CH_3)_3\ {}^{\ominus}Br$

$CH_3—CH_2—CH_2—CH_2—CH_2—CH_2—CH_2—CH_2—CH_2—NH_2 \bullet HCl$

$CH_3—CH_2—CH_2—CH_2—CH_2—CH_2—CH_2—CH_2—CH_2—CH_2—CH_2—CH_2—CH_2—CH_2—CH_2—CH_2—\overset{\oplus}{N}H_3\ \overset{\ominus}{Cl}$

Figure 5. Structures of various commonly-used anionic and cationic surfactants.

For polymerization of styrene in the presence of sodium dodecyl sulphate surfactant, the adsorption of surfactant on the surfaces can be described by the following Langmuir adsorption isotherm. Figure 7 is also the graphical depiction of this adsorption process.

$$C_{SC} = \frac{C_{SCmon} * C_{BCeq}}{C_{BCeq} + C} \tag{3}$$

$$\frac{1}{C_{SC}^2} = \frac{C}{C_{SCmon}} C_{BCeq} + \frac{1}{C_{SCmon}} \tag{4}$$

where,

C_{SC} is the surface concentration, moles per unit area
C_{SCmon} is the surface concentration at monolayer, moles per unit area
C_{BCeq} is the bulk concentration at equilibrium, moles per unit volume
and C is a constant.

The initiation of the polymerization is achieved by the decomposition of the water soluble initiator molecules at high temperatures (e.g. 60-70°C for potassium persulphate initiator). The initiator radicals once formed in the solution have the possibility to enter the micelles or the monomer droplets. However, as the experimental evidences point out that it is very rare that the radicals enter the monomer droplets. Thus, the locus of polymerization is always the micelles. The initiator radicals do not enter the monomer droplets as these are made up from the water soluble initiator and are not soluble in organic monomer.

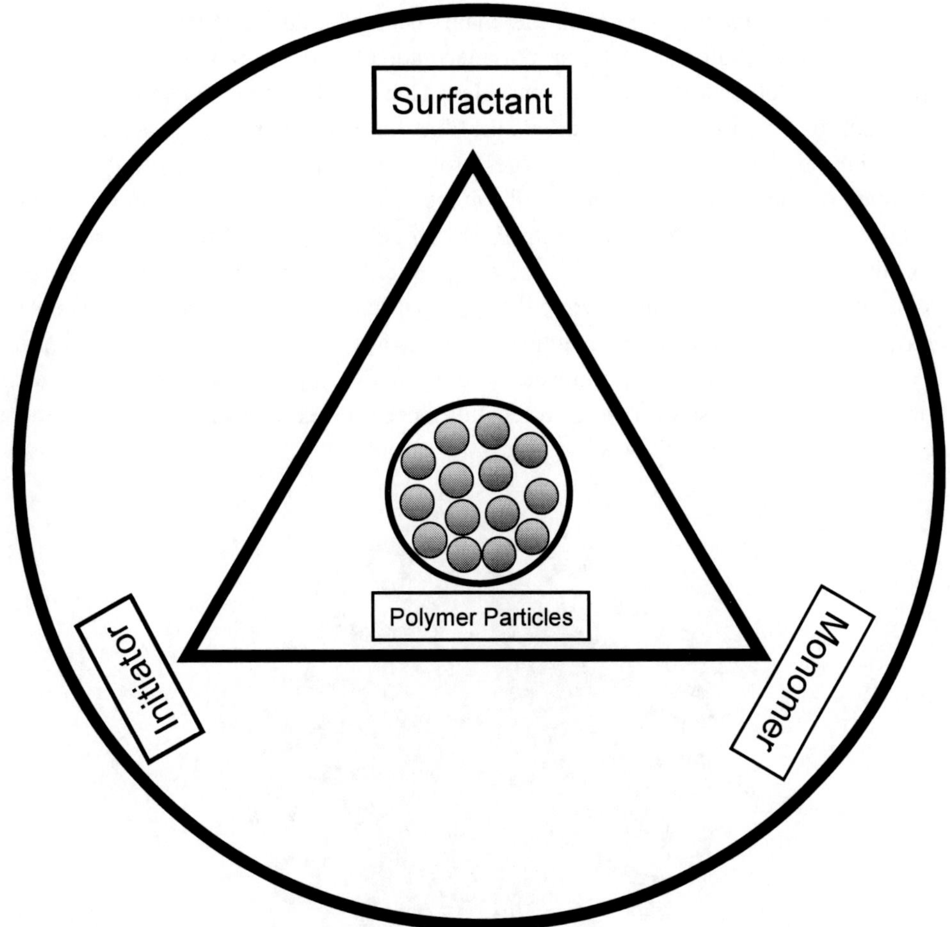

Figure 6. The three components of the emulsion system, the optimization of which leads to the better control of particles size and morphology.

On the other hand, due to the unique architecture of the micelles, the conditions at the interface are very suitable to attract the radicals and hence start polymerizing once the radical has entered the micelle. The micellar nucleation is predominant as compared to droplet polymerization also owing to the fact that number of micelles is much larger than the droplets. The micelles in which the radicals start polymerizing the monomer are no longer termed as micelles, to separate them from the non-active micelles. These active micelles are termed as polymer particles. Figure 8.1 shows the various steps involved in the micellar nucleation of the polymer particles. This kind of nucleation is also termed as heterogeneous nucleation for the generation of polymer particles. Apart from that, there is also a possibility of homogenous particle nucleation. This is generally the case when the surfactant concentration is below critical micelle concentration or when surfactant free polymerization is carried out. Apart from this, the solubility of monomer in water can also lead to significant extents of homogenous particle nucleation. However, as long as the amount of surfactant is well above the critical micelle concentration, the micellar particle nucleation is the predominant mode of particle nucleation. The homogenous nucleation occurs by the initiation of polymerization of the monomer molecules dissolved in water. After addition of a certain number of monomer units in the chain, the chains become water insoluble and come out of solution. These low molecular weight polymer radicals are very unstable and keep on collapsing on each other to form stable particles. These particles generated by precipitation get their stability by adsorption of surfactants from the solution or micelles. Also a partial stability is provided by the negative charges from the initiator charges. Figure 8.2 is the representation of the homogenous mode of polymer particle nucleation. Homogenous mode of nucleation was suggested after the Smith-Ewart model of micellar nucleation was unable to describe the polymerization processes below the critical micelle concentration of the surfactant. It was stated by this theory that the particle nucleation should stop once the concentration of the surfactant in solution falls below the critical micelle concentration. However, for many systems it was not the case and as the polymer particles could also be nucleated even in total absence of the surfactant, therefore necessitating the development of other explanations of particle nucleation process.

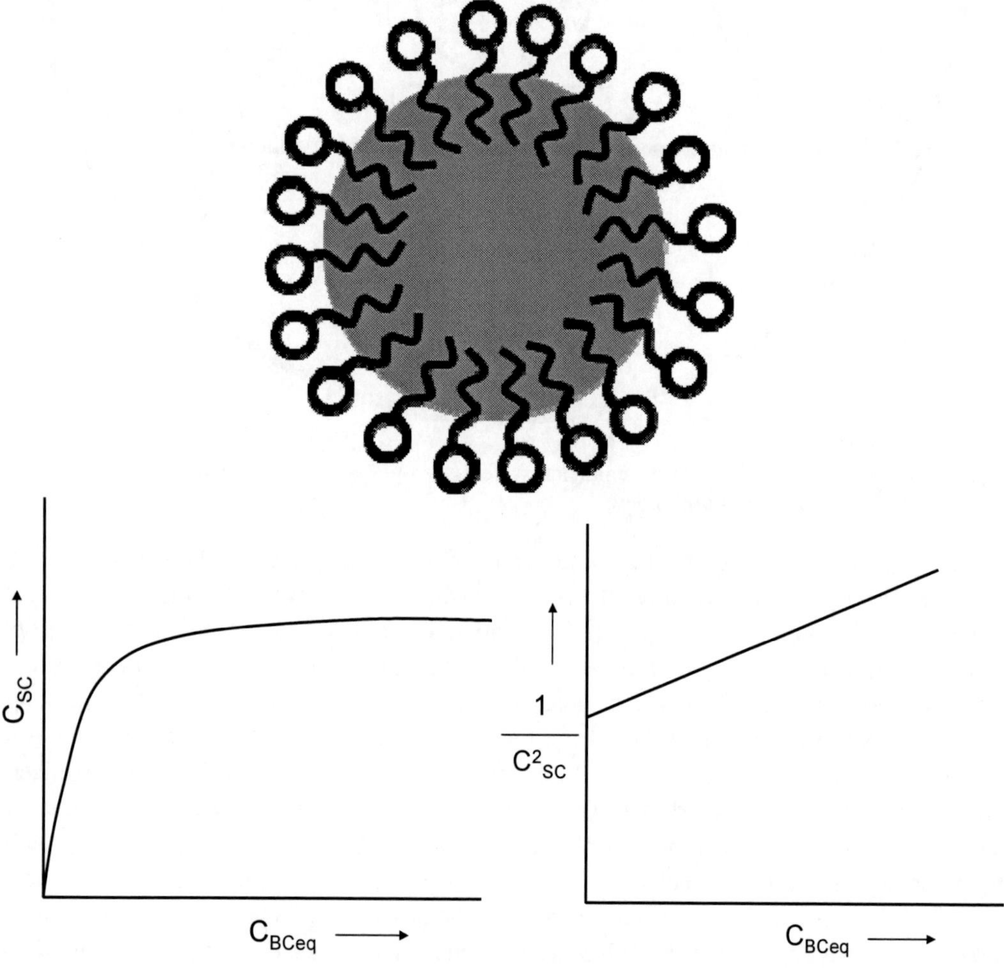

Figure 7. Adsorption process for the adsorption of surfactant on the surface.

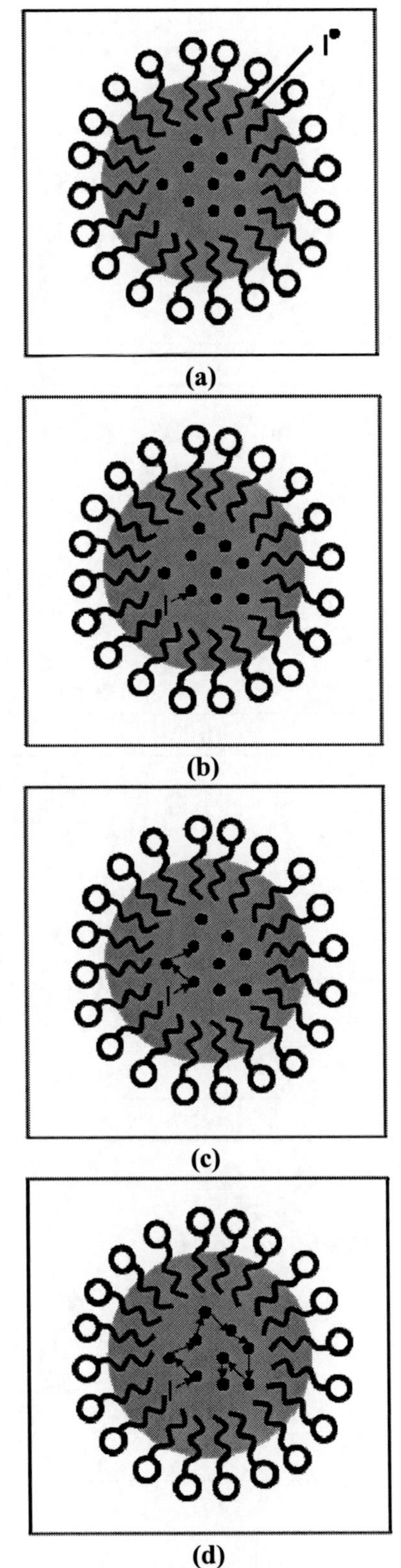

Figure 8.1. Representation of micellar nucleation mechanism.

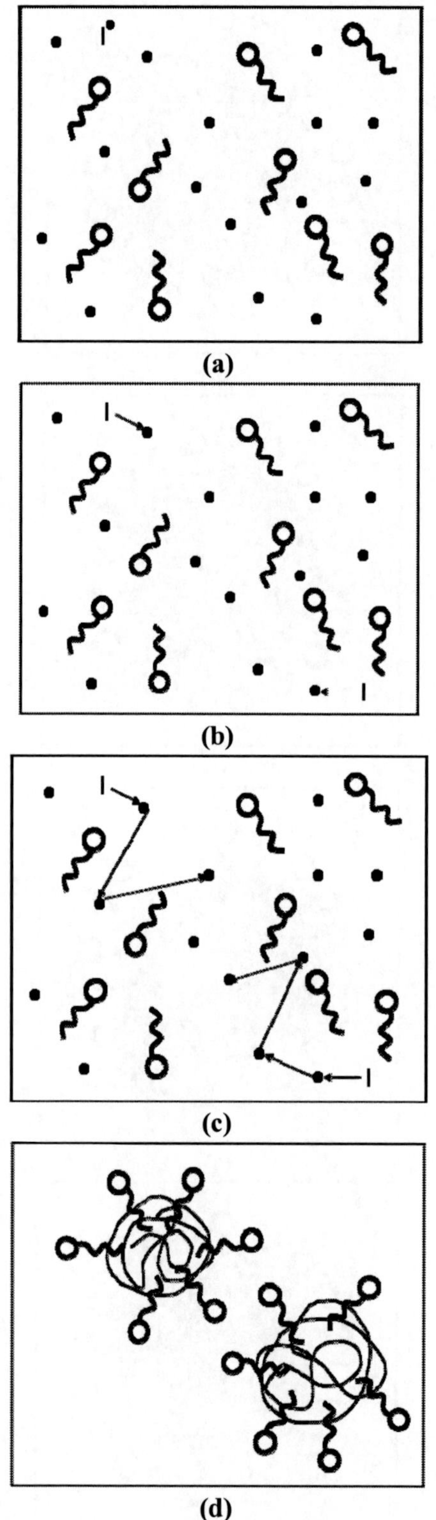

Figure 8.2. Representation of homogenous nucleation mechanism.

Based on the nature of polymerization progress, the emulsion polymerization process is generally divided into three intervals. The first interval, as shown in Figure 9a, is the particle formation phase. The initiator upon decomposition generates radicals. These radicals enter the micelles and start polymerizing the monomer molecules in these micelles thus leading to the generation of polymer particles. Thus in this regime, the number of particles keep on increasing as more and more radicals generated by the initiator decomposition keep on initiating more and more particles. This leads, therefore, also to the continuous enhancement in the polymerization rate as shown by the curve in the Figure 9a. Thus, this phase consists of three different entities, i.e., polymer particles, inactive micelles where there is no polymerization occurring and the monomer droplets. As the polymerization in the particles proceeds, their size keep on increasing and they deplete the monomer content present in them. This depletion is continuously replenished by the absorption of more monomer from the water phase, which has been dissolved in it. The water phase in return absorbs more monomer from the monomer droplets. This cycle of monomer diffusion through absorption keeps the polymerization rate running. One has to note that the final number of particles generated by the initiation of monomer in micelles is generally very low as compared to the original number of micelles in the system. Generally only 0.1% amount of the micelles form particles [1]. As the particles become larger in size, they require more and more surfactant to stabilize and the surfactant dissolved n water phase is thus continuously adsorbed on the surface of the particles. This leads to the reduction of the surfactant in solution much below the critical micelle concentration. This leads to the destabilization of the remaining micelles and they in this process completely disappear thus providing their surfactant to stabilize the growing polymer particles. By the end of the interval I, the micelles either have been converted into polymer particles or have been destabilized to lose the surfactant. As the interval one approaches its end, most of the surfactant is now in use to stabilize the polymer particles. Interval I is generally the shortest of he polymerization phases of the whole process and accounts for roughly 15% of the monomer conversion [1]. Figure 10 also is the depiction of the various intervals of emulsion polymerization.

Once no excess surfactant is available to impart stability, new particles do not nucleate. Therefore in interval II (Figure 9b), only the already formed polymer particles undergo polymerization of the monomer in these swollen particles. Therefore, the number of particles in this interval remains constant and as a result the rate of polymerization is also more or less constant. The general cycle of monomer conversion in the particles, absorption of further monomer from solution and replenishment of the monomer in the solution by absorption from the monomer droplets continues. This leads to the continuous increment in the size of polymer particles, whereas the number and size of monomer droplets keep on decreasing. At a point, which also forms the transition point of interval II to interval III, the monomer droplets completely disappear and the system is left only with the growing polymer particles still very much swollen with the monomer. This leads to the starting of interval III (Figure 9c) and the amount of conversion at this point varies with the nature of monomer. In interval III, as there is no monomer left to replenish the solution, therefore after a point, the monomer present in particles as well as solution phase is polymerized and as the concentration of monomer is decreasing continuously, the rate of polymerization also starts to drop. After the whole amount of monomer has been polymerized in these particles, the rate virtually climbs down to zero indicating the completion of the reaction. Thus, the number of particles in interval III remains same as in interval II, leading to only increment in the size of the particles. Figure 9d shows these three intervals together to sum up the whole emulsion polymerization process. Figure 11 shows the SEM micrographs explaining the evolution of size as a function of time or monomer conversion. One should also note that the emulsion polymerization is a very sensitive polymerization method where by bringing about small changes in reaction components can totally change the final outcome. Figure 12 shows the examples of such cases where small or larger particles have been synthesized, monodisperse particles or particles with bimodality or multimodality have been generated. Figure 13 is also a depiction of the various morphologies which can be generated, e.g., planar morphology, orange-peel morphology, strawberry morphology, crater morphology, etc.

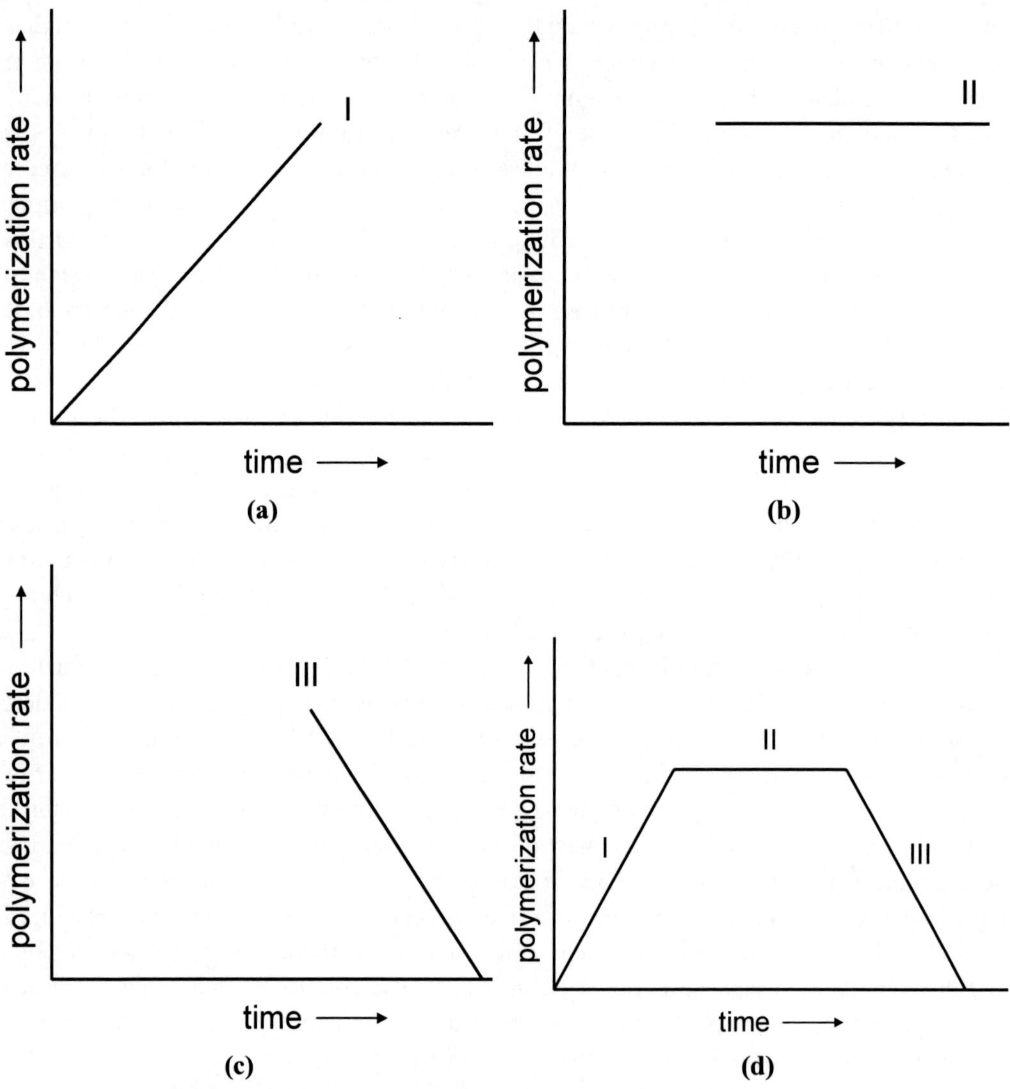

Figure 9. Various stages of emulsion polymerization process.

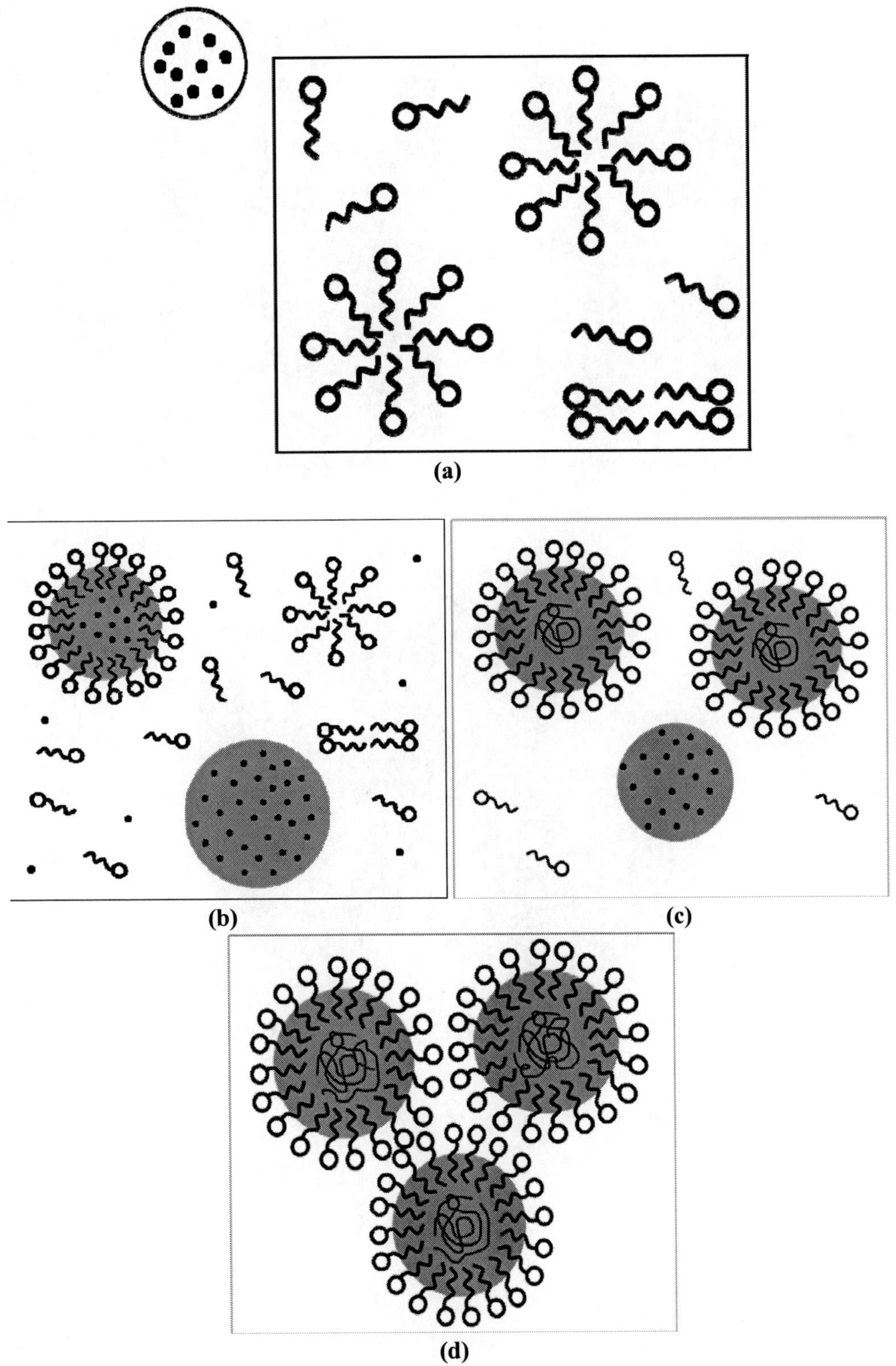

Figure 10. Representation of various intervals of emulsion polymerization.

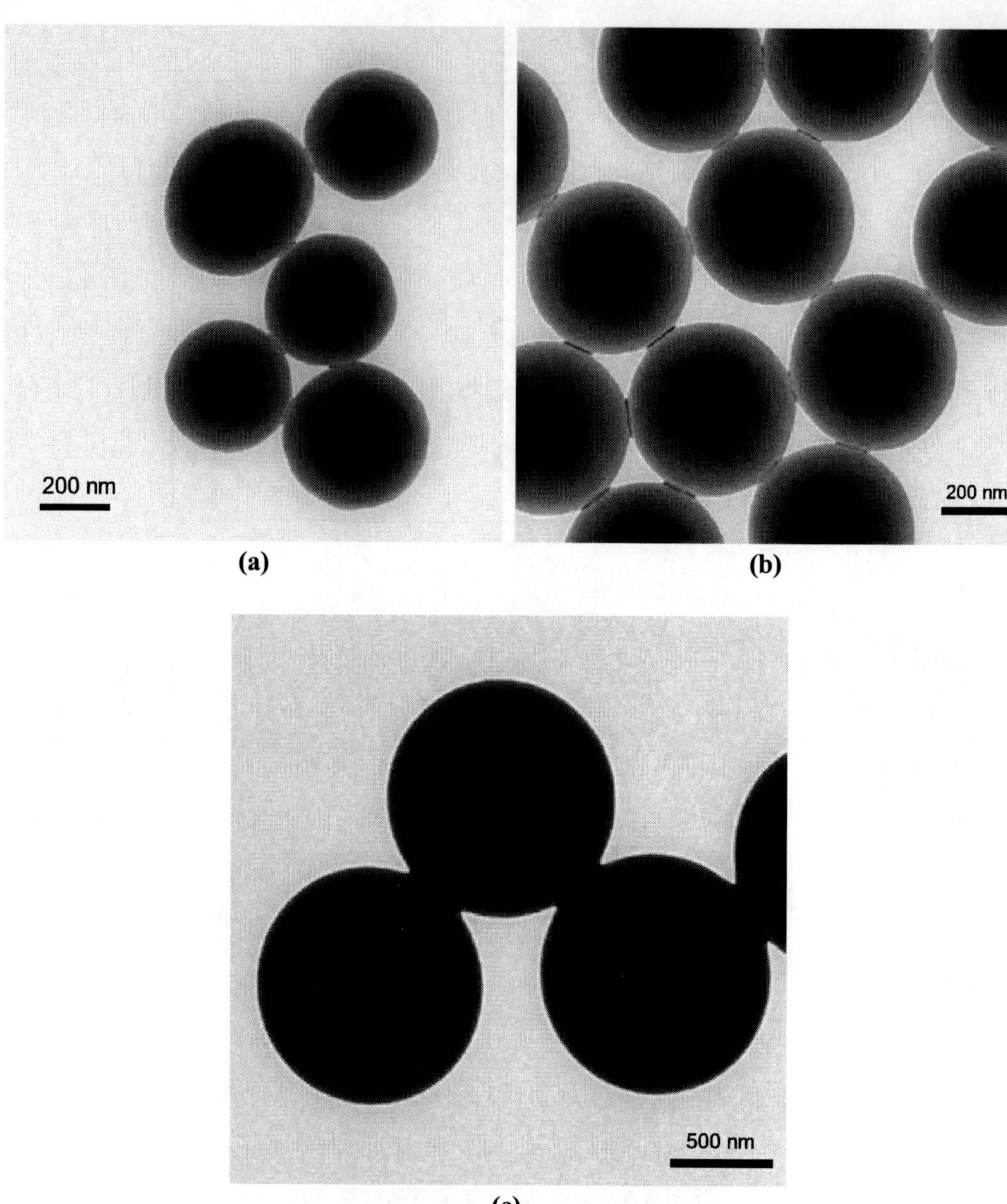

Figure 11. Evolution of particle size as a function of conversion.

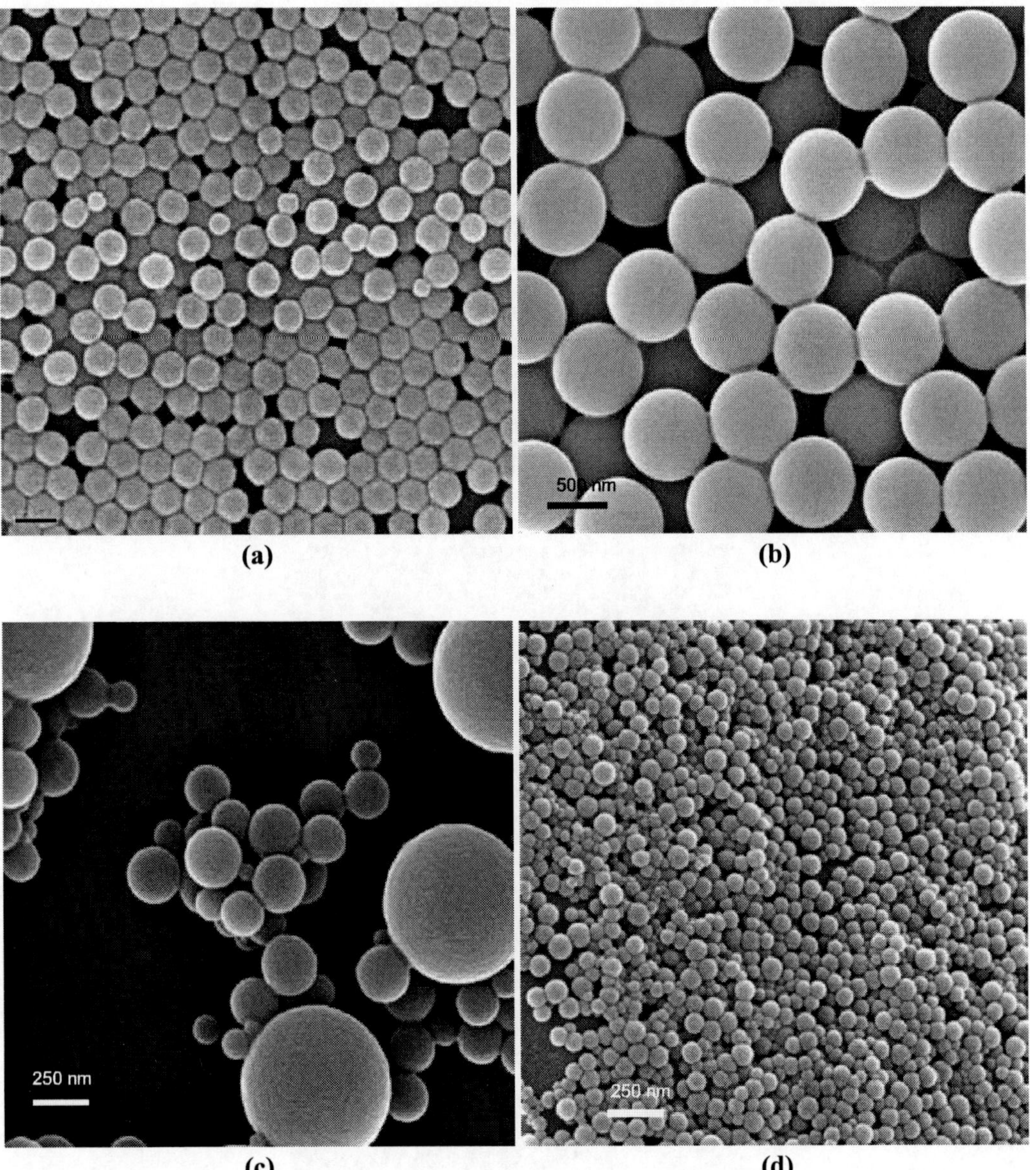

Figure 12. Different families of particle sizes generated by emulsion polymerization. Bar in (a) reads 200 nm.

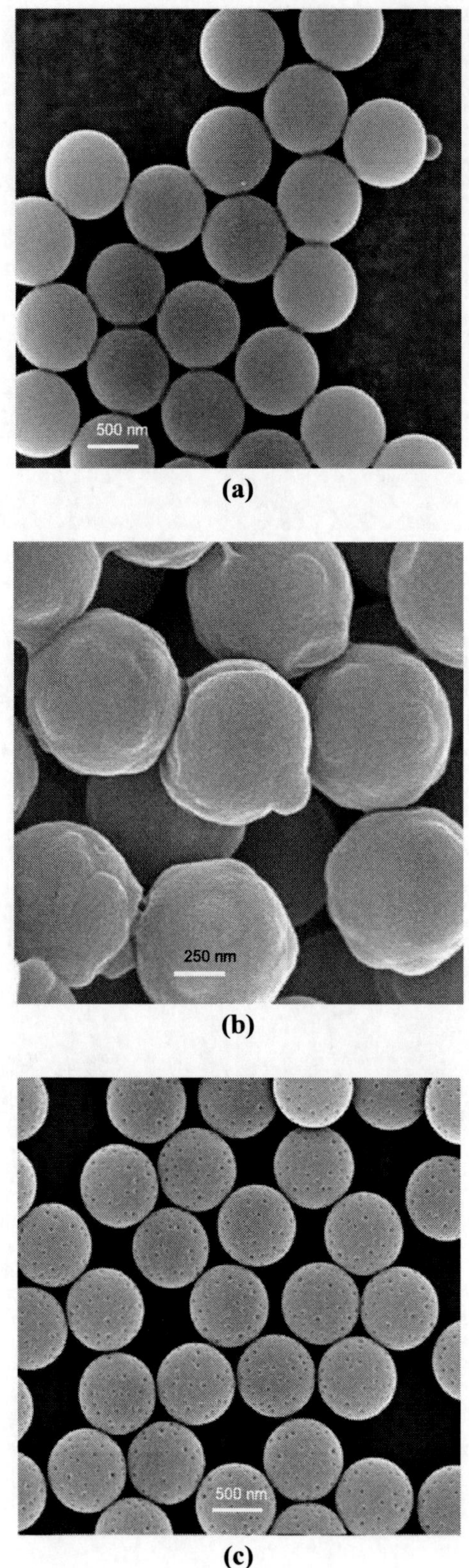

Figure 13. (Continues on next page.)

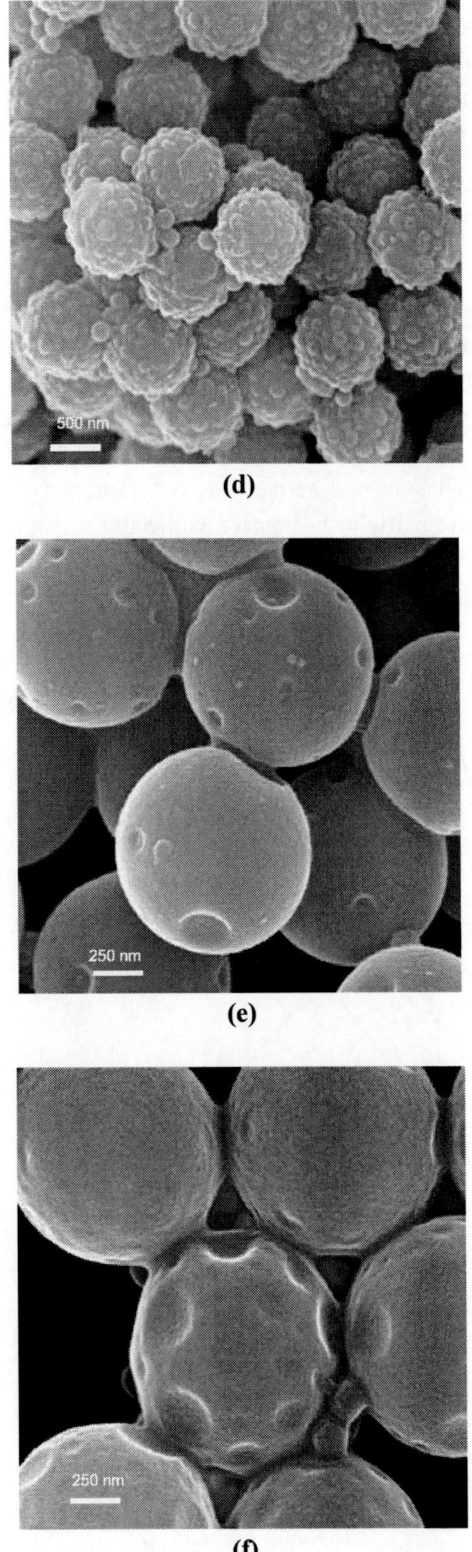

Figure 13. Various particle morphologies achieved with emulsion polymerization.

C. Kinetics of Emulsion Polymerization

One can expect the rate of polymerization inside the polymer particles to be same as the polymerization rate for conventional treatment:

$$r_p = k_p [M] \tag{5}$$

The emulsion polymerization is different from conventional polymerization that in this case, only one radical is present per particle at a time. As soon as a second radical enters the particles, it terminates with the earlier radical and the polymerization comes to an end in that particular particle to be initiated again when the third radical enters the particles. Thus the particles traverse through this cycle of activity and inactivity. Thus at a particular moment of time, all the polymer particles are not active, i.e., polymerization is not proceeding in all of these particles. Many of the particles are active while the rest are inactive and the dynamic balance of activity and deactivity keeps on changing with time.

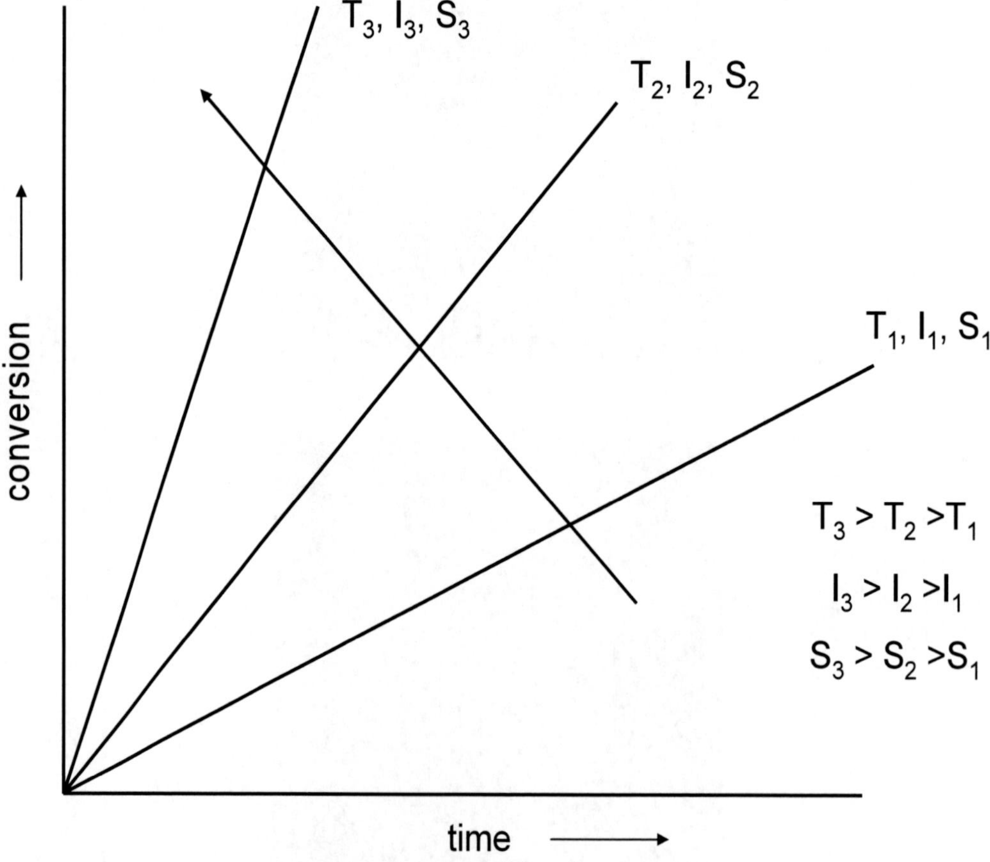

Figure 14. Effect of increasing the amount of initiator, surfactant as well as temperature on the conversion of the monomer.

As such polymerization described by equation 5 is taking place at a time in a large number of active polymer particles, therefore the overall rate of polymerization can be written as

$$R_p = k_p [M] [P^\bullet] \tag{6}$$

where $[P^\bullet]$ is the concentration of active particles. It is defined by the ratio of the product of concentration of micelles plus particles, N (or particles when considered at the end of polymerization reaction) and average number of radical per micelle and particle, n (or particles when considered after the micelles have disappeared) with the Avogadro number. Thus, equation 6 can be modified to provide the general equation for the emulsion polymerization process as

$$R_p = \frac{N \bar{n} k_p [M]}{N_A} \tag{7}$$

The value of n is critical and must be carefully considered. A number of expert studies on the possible values of this entity have been reported [1,4,5]. The usual value of n is considered to be 0.5 and it is also true in many of the cases. It has the significance that the polymer particles do not suffer from radical desorption and there are no two radicals present in the particle at a time. It is particularly true when the particles are small in size. It also indicates that the particles are active half of the time and other half of time, the particles are inactive. Therefore, a value of 0.5 is an ideal case. However, there are certain possibilities which lead to the value of n above or below the conventional value of 0.5. If the value falls below 0.5, it is an indication that the radical desorption from the particles is significant. It also means that the radicals terminate each other in aqueous phase in significant amounts. The value of n can also be higher than 0.5 indicating that some particles may have more than one radical at a particular time. This generally happens when the particles grow bigger in size and termination rate constant is low.

The rate of polymerization as derived in equation 7 is directly proportional to N, the number of particles. The number of particles generated during the emulsion polymerization reaction can be increased either by increasing the amount of surfactant in the system or by increasing the amount of initiator. Enhancing the reaction temperature would also lead to increased rate of polymerization. Figure 14 is the depiction of this phenomenon, where it has been shown that the increasing the amount of initiator, surfactant as well as reaction temperature leads to faster monomer conversion or faster polymerization rates.

The degree of polymerization in emulsion polymerization can also be written same to the conventional free radical polymerization case as

$$\bar{X}_n = \frac{r_p}{r_i} = \frac{N k_p [M]}{R_i} \tag{8}$$

It indicates that the degree of polymerization is also directly proportional to number of particles, N. Thus by increasing number of particles, the rate can be increased. As mentioned above, N can be increased by either increasing the amount of surfactant or by increasing R_i, but as R_i is present in the denominator of the equation, therefore, it would indicate that the route of increasing the R_i by increasing say initiator concentration would affect the molecular weight negatively. Best way to achieve the increased number of N is by increasing the amount of surfactant. This therefore can lead to simultaneous growth in the number of particles thus accelerating the reaction as well as increased molecular weight. This property separates emulsion polymerization from the conventional free radical polymerization.

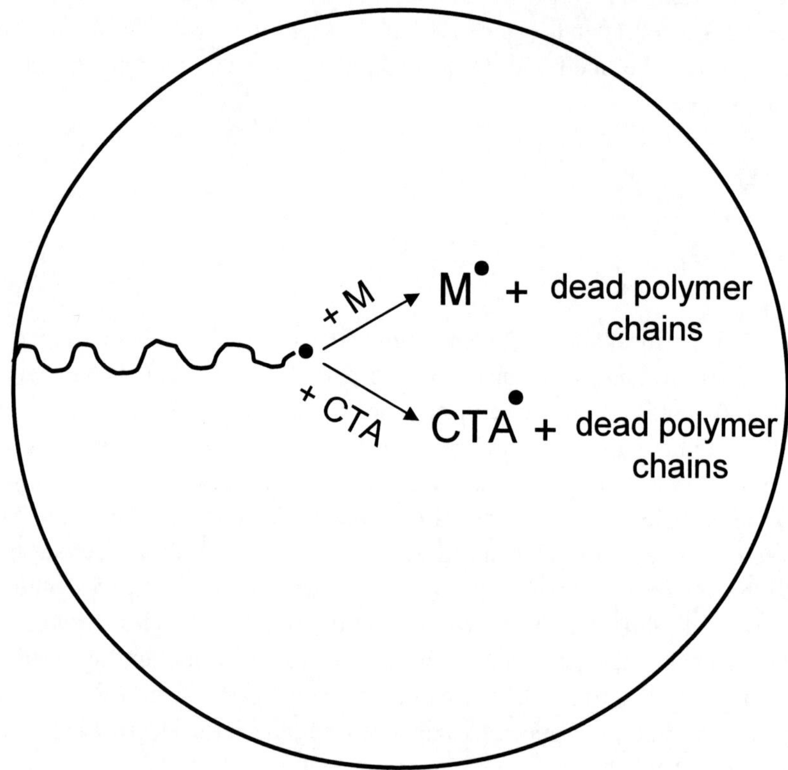

Figure 15. Chain transfer process inside the polymer particle.

The degree of polymerization when the termination also takes place by chain transfer can also be described in equation 9 as

$$\overline{X}_n = \frac{r_p}{r_i + \Sigma r_{tr}} \qquad (9)$$

where Σr_{tr} is the summation of all possible chain transfer modes. The chain transfer process achieved by externally added chain transfer agent or chain transfer to, e.g., the monomer in the absence of an externally added chain transfer agent can lead to enhancement in the rate of desorption of free radicals from the polymer particles. It thus leads to reduction in the molecular weight of the polymer chains as well as brings polydispersity in the chain length or molecular weight distribution. Figure 15 gives a description of the chain transfer process in the polymer particles. The kinetics of the chain transfer can be described by the following equations 10–13:

$$P^{\bullet} + M \longrightarrow P + M^{\bullet} \tag{10}$$

$$M + M^{\bullet} \longrightarrow MM^{\bullet} \tag{11}$$

$$P^{\bullet} + CTA \longrightarrow P + CTA^{\bullet} \tag{12}$$

$$CTA^{\bullet} + M \longrightarrow CTA-M^{\bullet} \tag{13}$$

Also, the rate constant for the transport of free radicals out of particles is given by the following equation [6]:

$$k'_d = \frac{k_{tr,M}}{k'_p}\left(\frac{12\left(\frac{\pi}{6}\right)^{2/3} D_W}{a + \left(\frac{D_W}{D_P}\right)}\right) + \frac{k_{tr,T}}{k'_p}\left(\frac{12\left(\frac{\pi}{6}\right)^{2/3} D_{Wt}}{a_T + \left(\frac{D_{Wt}}{D_{Pt}}\right)}\right) \tag{14}$$

where

a) the first component is the result of chain transfer to the monomer,
b) the second component describes the effect of chain transfer to a chain transfer agent,
c) a & a_T are partition coefficients,
d) D terms are diffusivities, and
e) k terms are rate constants

The total number of polymer particles generated during the emulsion polymerization can also be given by the following equation [1]:

$$N = k\left(\frac{R_i}{\mu}\right)^{2/5}(a_s S)^{3/5} \tag{15}$$

where

a) a_s is the interfacial surface area occupied by surfactant molecule

b) S is the total concentration of the surfactant in the system
c) μ is the rate of volume increase of the polymer particle
d) k is the constant, whose value lies between 0.37 and 0.53.

As degree of polymerization as well as polymerization rate are directly proportional to number of particles, N and as N has a 3/5 dependency on the surfactant concentration, therefore, it can be concluded that the degree of polymerization and polymerization rate are dependant on 3/5 degree of surfactant concentration.

D. Emulsion Copolymerization

Vast amounts of polymers generated by emulsion polymerization are copolymers as the properties of the individual polymers can be synergistically increased by the generation of copolymers. However, as the different monomers have different reactivities in a particular system therefore there it is always complex to predict the final composition of the copolymer chains and if it would be same as the initial monomer ratios. Apart from that, as the reactivities are different from each other, the more reactive monomer may polymerize first, thus forming core of the particles rich in this polymer followed by a outer cover of particles more rich in less reactive monomers. This leads to a gradient of concentration of different monomers in these particles. There can similarly be also differences when the water solubilities of the monomers are quite different from each other. Figure 16 is an example of comparison of the homopolymers with copolymers. The pure polystyrene particles earlier shown in Figure 11 are compared with the copolymers of styrene with water soluble monomer N-isopropylacrylamide. The generated surface morphology is totally different in these particles. One should note here that the particles were achieved without using the surfactant, i.e. particle generation was achieved by homogenous nucleation mode. The more hydrophilic monomer starts polymerizing first followed by the polymerization of more hydrophobic monomer. The hydrophobic monomer polymerize inside these particles because of hydrophobicity thus pushing the hydrophilic chains of poly(N-isopropylacrylamide) on the surface of the particles.

Monomer partioning is the term most commonly used to describe the emulsion copolymerization of two or more monomers. Owing to the different reactivity ratios of the monomers and the different ratio of monomers in the polymer particles (i.e., loci of polymerization), which is generally very different from the initial monomer ratios, the compositional drift in the copolymer composition takes place.

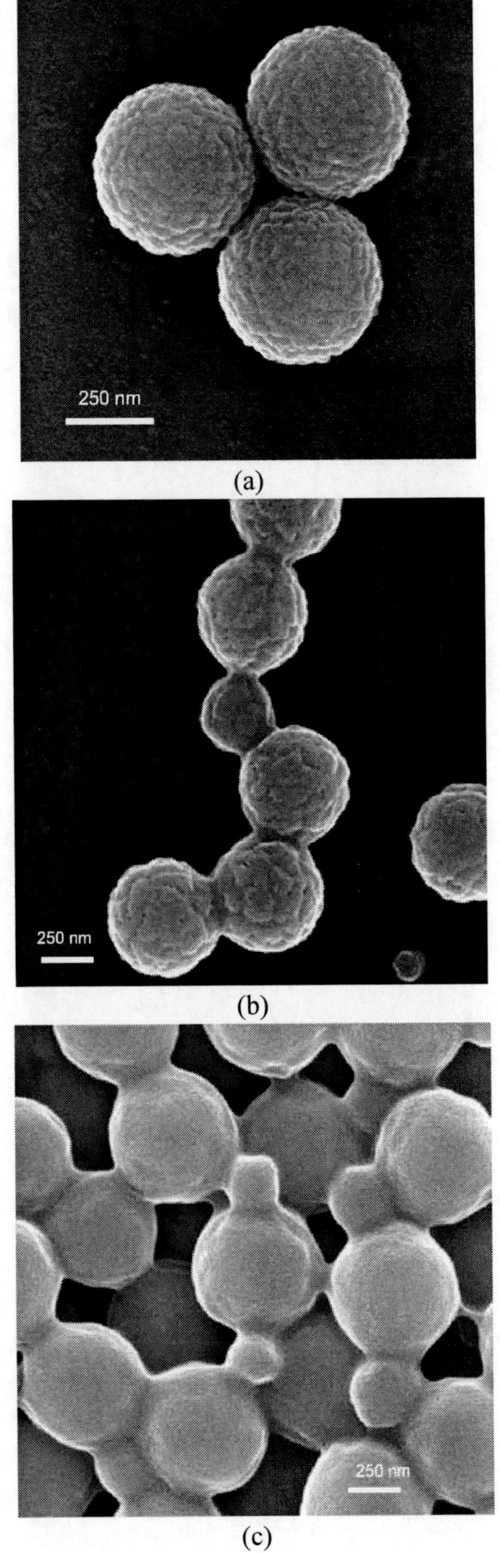

Figure 16. SEM images of styrene-co-isopropylacrylamide copolymer particles.

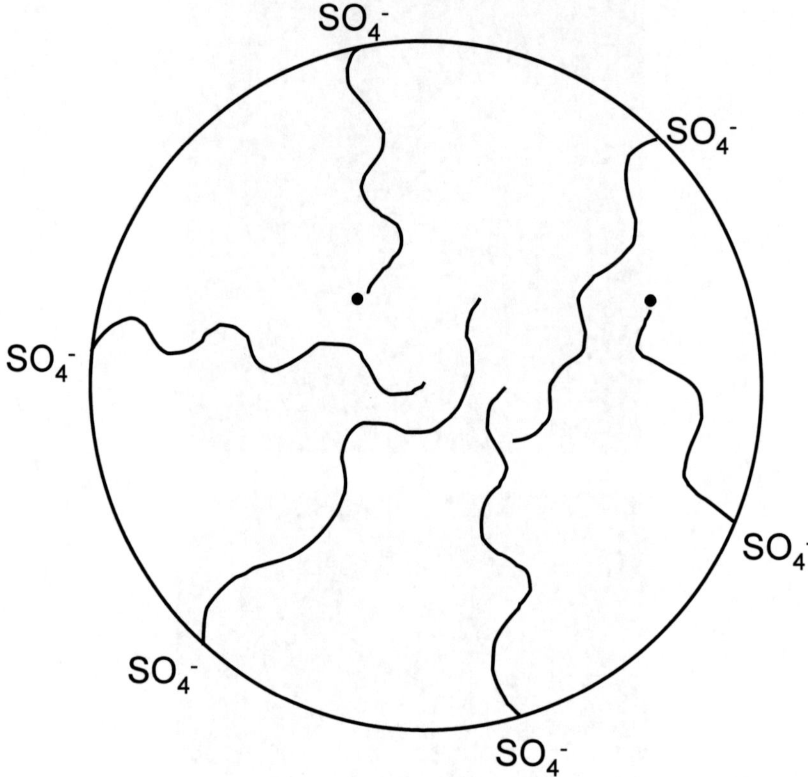

Figure 17. Representation of a polymer particle stabilized by the sulphate ions from the initiator.

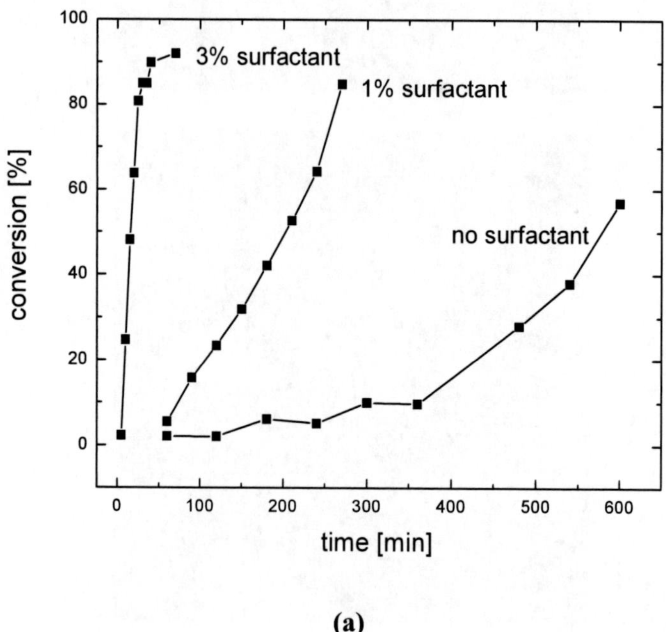

(a)

Figure 18. (Continues on next page)

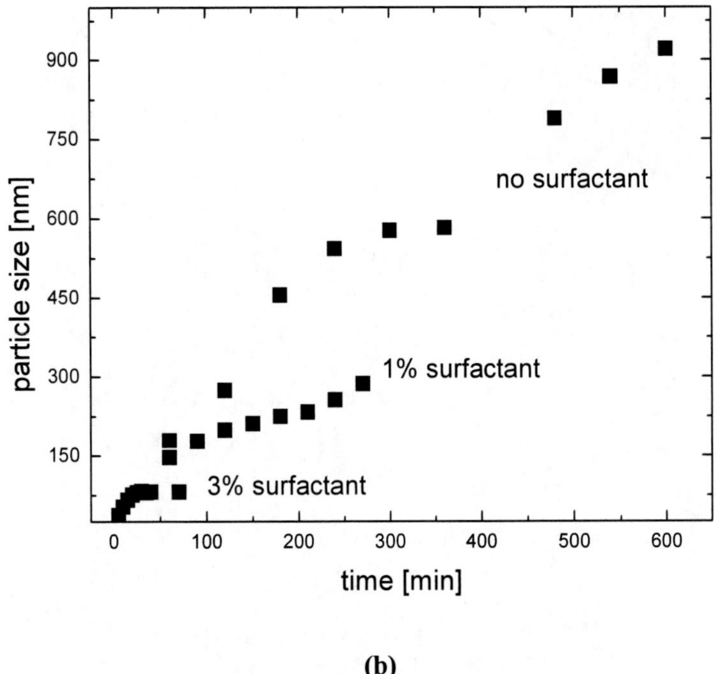

(b)

Figure 18. Conversion and particle size as a function of amount of surfactant and comparison with the case without surfactant.

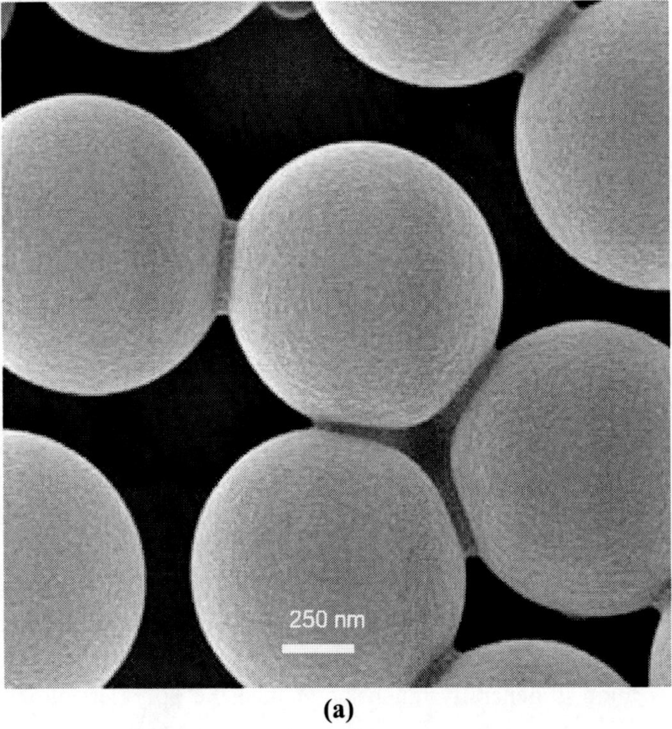

(a)

Figure 19. (Continues on next page)

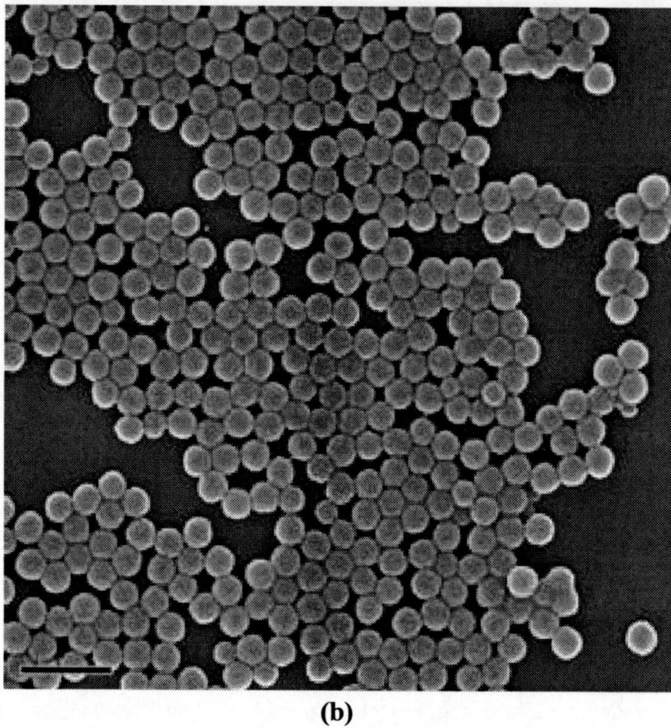

(b)

Figure 19. Comparison of the particles generated (a) without surfactant and (b) with surfactant. The bar in (b) reads 500 nm.

E. Surfactant Free Polymerization

Although majority of emulsion polymerization processes are achieved by the use of surfactants of different kinds, but there are also cases where the polymerization is achieved without the use of surfactants. This polymerization system, known as surfactant-free polymerization, is important when the particles are required for certain applications which cannot tolerate any impurities on the surface of the particles and the surface charges are accurately known. The particles in this polymerization are nucleated by homogenous particle nucleation mode and are stabilized by the negative charges on the surface from the potassium persulphate initiator commonly used in such polymerizations. Figure 17 shows such particles stabilized by the surface negative charges. As the stabilizing surfactant is absent, therefore, the number of particles which reach the stable state is generally two orders of magnitude lower than the emulsified polymerization. As the number of particles generated is less, therefore, the polymerization reaction takes long times to reach high extents of polymerization. The smaller particles generally are more unstable and keep on collapsing with each other to form larger particles. Therefore, the size of particles generated in these kinds of polymerization is generally much larger than the polymerizations with emulsifier. Figure 18 shows the generated particles sizes as well as monomer conversion with time for styrene polymerization using different extents of surfactant. Figure 19 also shows the micrographs of particles generated with and without the surfactant underlining the significantly different particle sizes when all other reaction conditions were kept constant.

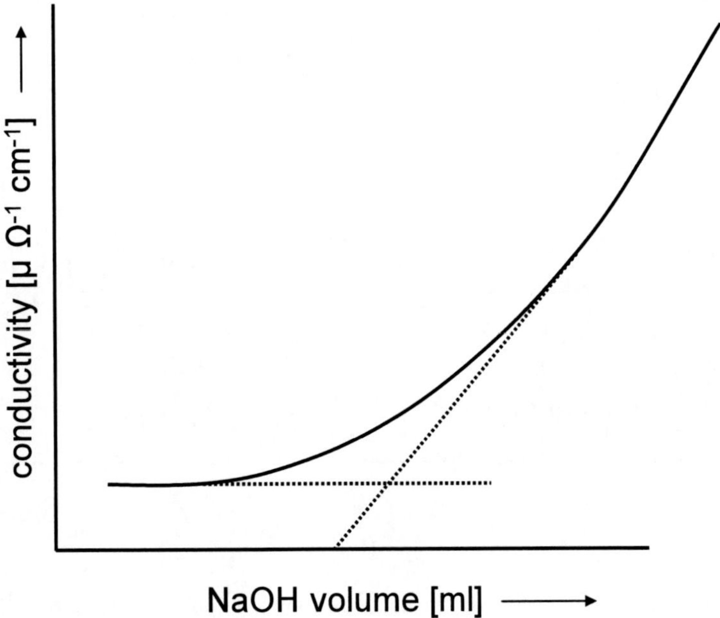

Figure 20. Titration curve to find surface charges on polystyrene particles synthesized by surfactant free polymerization.

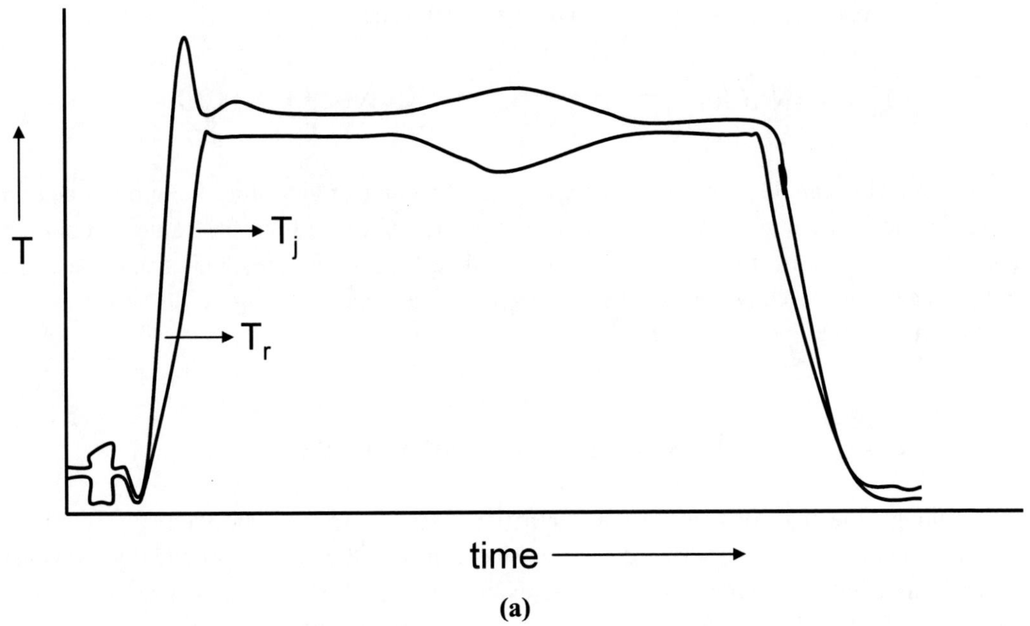

Figure 21. (Continues on next page)

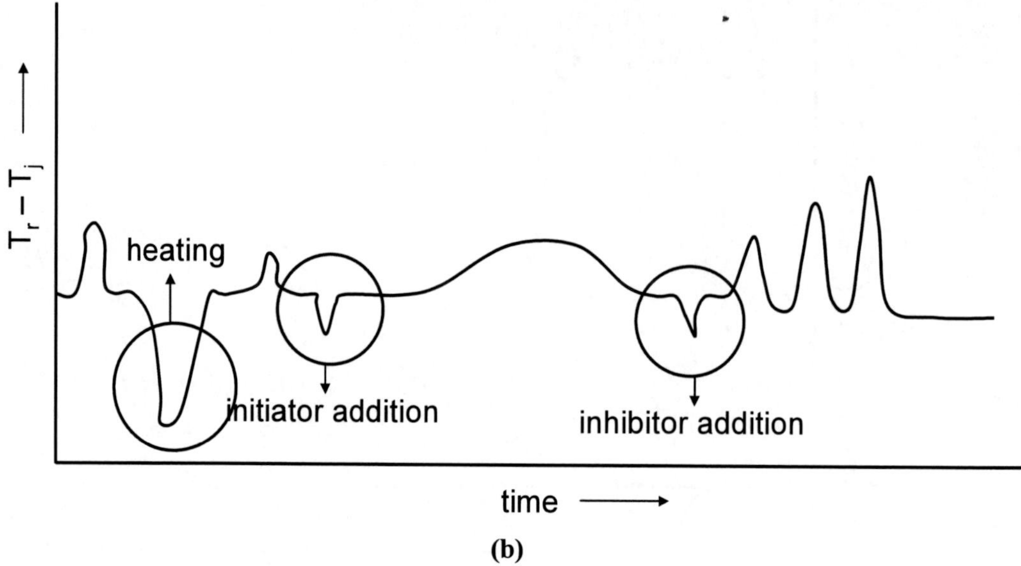

(b)

Figure 21. Reaction calorimetry curves for the accurate control of polymerization reactions. T_r and T_j are reactor and jacket temperatures, respectively.

One can also quantify the number of negative charges present on the particles by reaction of these negative charges with sodium hydroxide as explained by Hritcu et al. [7]. The SO_4^- groups on the particles surface can be reacted with NaOH as

$$—SO_4^- + NaOH \longrightarrow —SO_3Na + H_2O \qquad (16)$$

The reaction can be followed by the change in the conductivity of the solution as amount of hydroxide is increased. The tangent to the upward curve leads to the value of optimum amount of sodium hydroxide required to completely saponify the SO_4^- groups and this amount can then be converted to the amount of surface charges. Figure 20 shows such a titration curve for polystyrene particles.

F. Reaction Calorimetry

Although the reactions in the small reaction vessels can be controlled by the simple thermometers and the evolution of the conversion can be followed gravimetrically, however, larger volume polymerization contain higher extents of risk of run away reactions or reactions going out of control. Therefore in such cases, it is very important to monitor the reaction progress through reaction calorimetry which provides the online information on a number of reaction parameters like jacket temperature, reactor temperature, rotational speed of the stirrer etc. and these factors can be easily controlled according to the requirement. This data can also lead to important calculation of heat of reaction, etc., as slight changes in the reaction due either due to polymerization or due to addition of any externally added materials can be very accurately monitored and quantified as shown in Figure 21.

REFERENCES

[1] G. Odian (2004). *Principles of Polymerization, Fourth Edition*, John Wiley & Sons, Inc., New Jersey.

[2] V. R. Gowariker, N. V. Viswanathan and J. Sreedhar (1986). *Polymer Science*, John Wiley & Sons, Wiley Eastern Limited, New Delhi.

[3] P. C. Hiemenz and R. Rajagopalan (1997). *Principles of Colloid and Surface Chemistry*, Marcel Dekker, Inc., New York.

[4] M. Nomura (1982). In *Emulsion Polymerization*, Academic Press, New York.

[5] R. G. Gilbert (1995). *Emulsion Polymerization*, Academic Press, New York.

[6] K. Matyjaszewski and T. P. Davis (2002). Editors, *Handbook of Radical Polymerization*, John Wiley & Sons, Inc., New Jersey.

[7] D. Hritcu, W. Müller and D. E. Brooks (1999). *Macromolecules*, 32, 565-573.

Chapter 4

ADVANCES IN EMULSION HOMOPOLYMERIZATION BY LIVING POLYMERIZATION METHODS

A. INTRODUCTION

A limitation of conventional radical polymerization is that very high molecular weights cannot be easily achieved. Apart from that, the narrow distribution in the polymer weight and chain length is also difficult to achieve. The termination reactions take place almost out of control, although transfer reactions can be somewhat understood and controlled. But overall, these processes lead to the above-mentioned dispersities. Also, as the radical formed during the polymerization stays alive for a maximum of a second, it is therefore not possible to achieve the well-defined polymer morphologies, like block polymers or dendrimers, etc., during this time period in conventional polymerization. Thus, the simultaneous control of the polymerization rate and molecular weight is difficult to achieve in these polymerization methods. Though through emulsion polymerization the immediate termination of the radicals is delayed, as in this case, the radicals enter the micelles and form the polymer particles, and during their lifetime no other radical is present in the polymer particles, thus delaying their termination. On the other hand, the living techniques can be used to overcome these troubles, as these techniques have the main focus of eliminating the irreversible termination of the polymer chains. In these techniques, the polymer chains do get reversibly terminated for a short period of time before becoming active again, and this cycle continues and practically no termination to form dead polymer chains takes place. The reversibly terminated chains are in fact termed *dormant chains*, and various species, depending on the type of living polymerization used, can be used to cap the radicals at the end which is later uncapped to start polymerization again. These techniques thus lead to very well-defined polymer morphologies like block copolymers, polymers with special shapes like stars or dendrimers. The molecular weight distributions are also very narrow. Figure 1 explains these advantages of the living polymerization techniques over conventional polymerization. Figure 2 also shows the main characteristics of these living polymerization techniques in which the $\ln[M]_0/[M]$ linearly increases as a function of time, and the plot between M_w and monomer conversion yields a straight line [1].

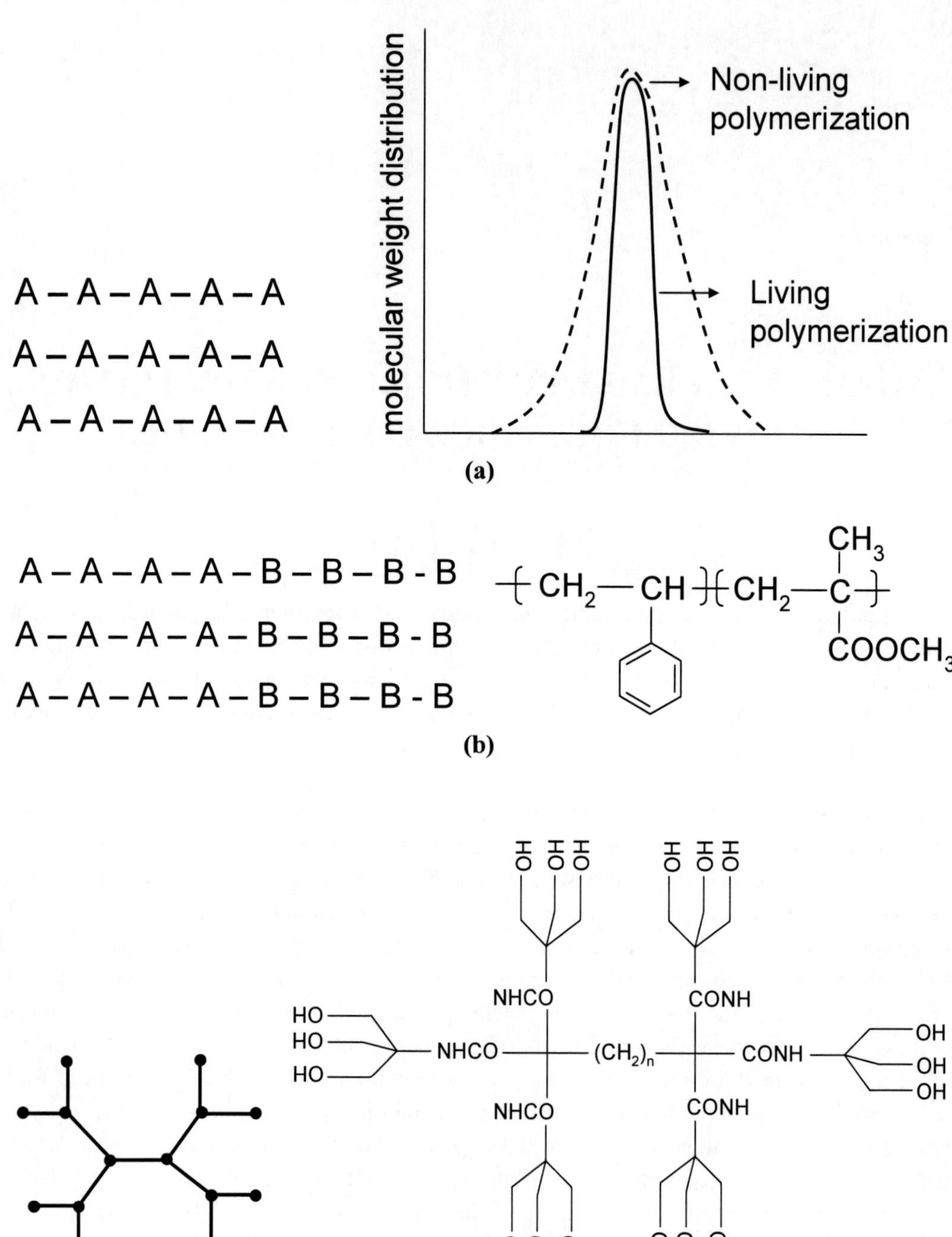

Figure 1. Advantages of living polymerization; (a) controlled molecular weight, (b) block copolymers and (c) generation of complex architecture.

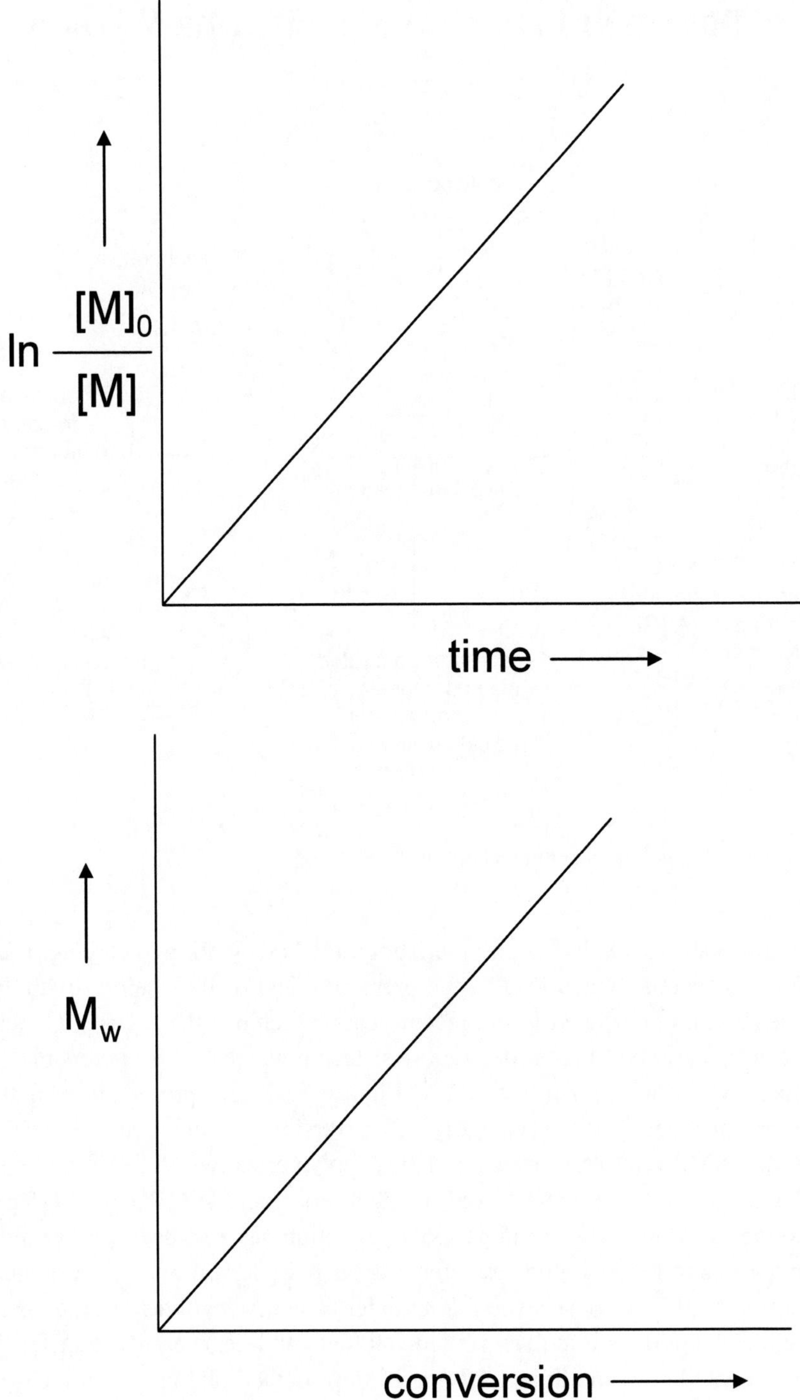

Figure 2. Characteristics of living polymerization.

B. Different Living Polymerization Methods

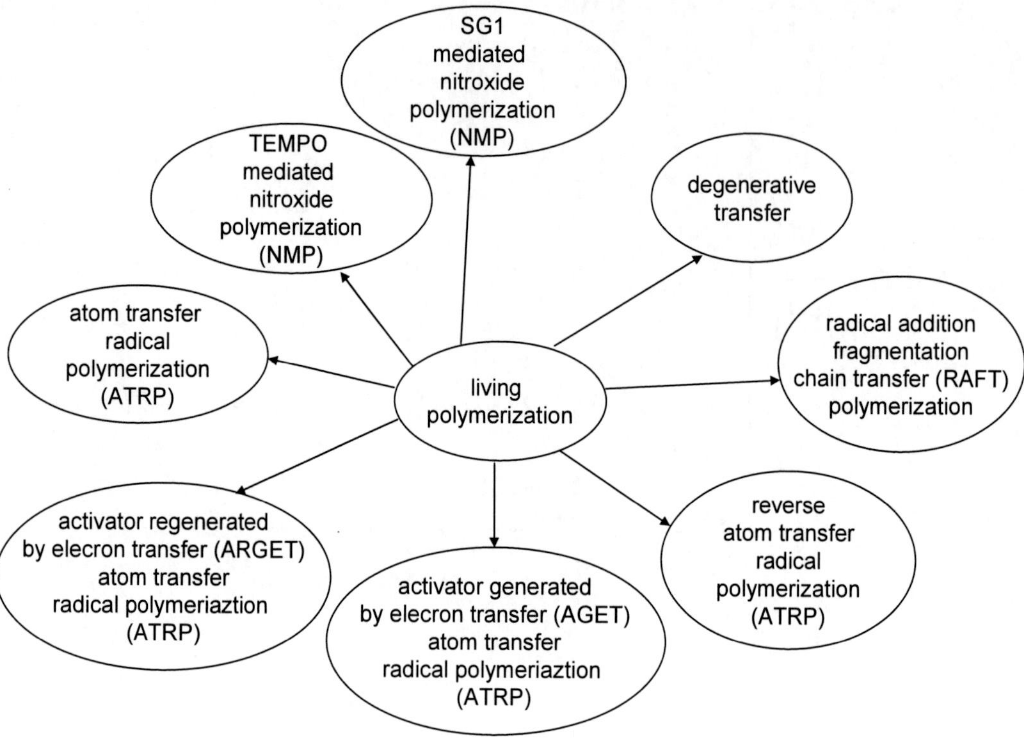

Figure 3. Various living polymerization techniques.

As mentioned above, the living polymerization methods work by delaying or minimizing the termination reactions, which lead to the premature end of the chain growth. There have been many techniques developed in the recent years to achieve these goals. Generally, these techniques can be classified into two categories: techniques based on reversible termination and techniques based on reversible transfer. Figure 3 is the representation of these living polymerization methods. In the category of reversible termination, nitroxide-mediated polymerization (NMP) and atom transfer radical polymerization (ATRP) methods are used most often. Various nitroxides can be used for such processes; TEMPO and SG1 are the most common examples. Atom transfer radical polymerization has also been further modified into other techniques, like reverse atom transfer radical polymerization, activator generated by electron transfer ATRP, etc. The reversible transfer category involves techniques like radical addition fragmentation transfer polymerization (RAFT) and degenerative transfer. Figure 4 is a general representation of the living polymerization method, along with a comparison with the conventional free radical polymerization. Figure 5 also explains how these reversible termination and reversible chain transfer processes take place [1].

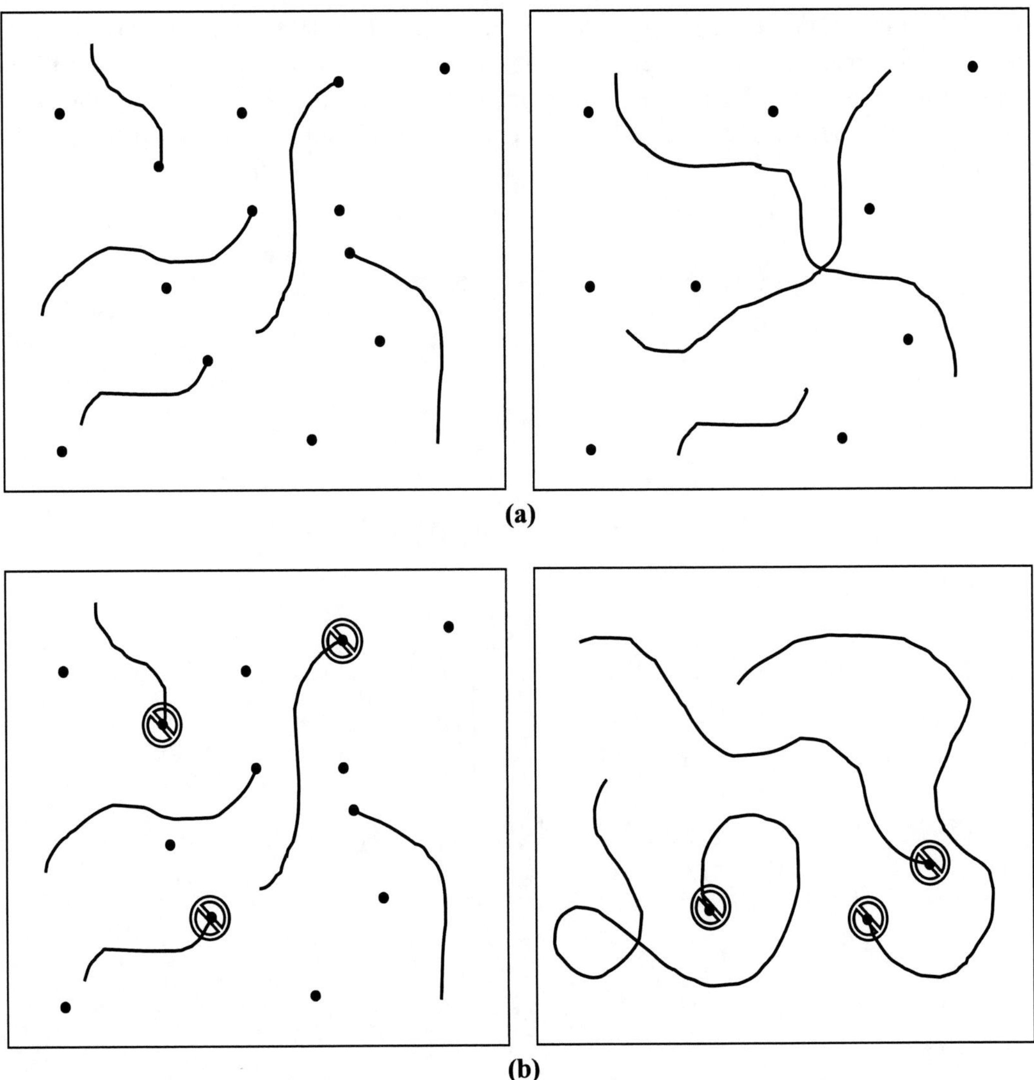

Figure 4. Comparison of (a) conventional polymerization and (b) living polymerization.

During the polymerization, the system soon arrives at equilibrium between the active and dormant species. The dormant species are the species which have been capped with the living agent (X in Figure 5a) and these dormant species are stable and are not susceptible to very fast reaction with the monomer molecules. Generally, the concentration of the stable as well as active radicals is same at the beginning of the polymerization reaction, however, the reaction keeps on shifting to the more stable species, thus, the concentration of these stable and dormant species keeps on increasing as compared to the concentration of active radicals. At later stages of polymerization, the concentration of stable species can be roughly six times the active radical concentration. This leads to significant reduction in termination and the average life time of the radical is tremendously enhanced, thus, allowing one to perform more complex synthesis to generate special morphologies in the polymer chains. The stable radical is also termed as persistent radical and its effect on the polymerization is known as persistent radical effect. Figure 5b is another representation of the living polymerization method of reversible termination, i.e., radical addition fragmentation chain transfer. In this case, a chain transfer agent containing dithioester end groups is used to transfer reversibly the radicals to different polymer chains. One active chain is made dormant in the process whereas the other dormant chain becomes active. This process continues until the monomer is fully polymerized.

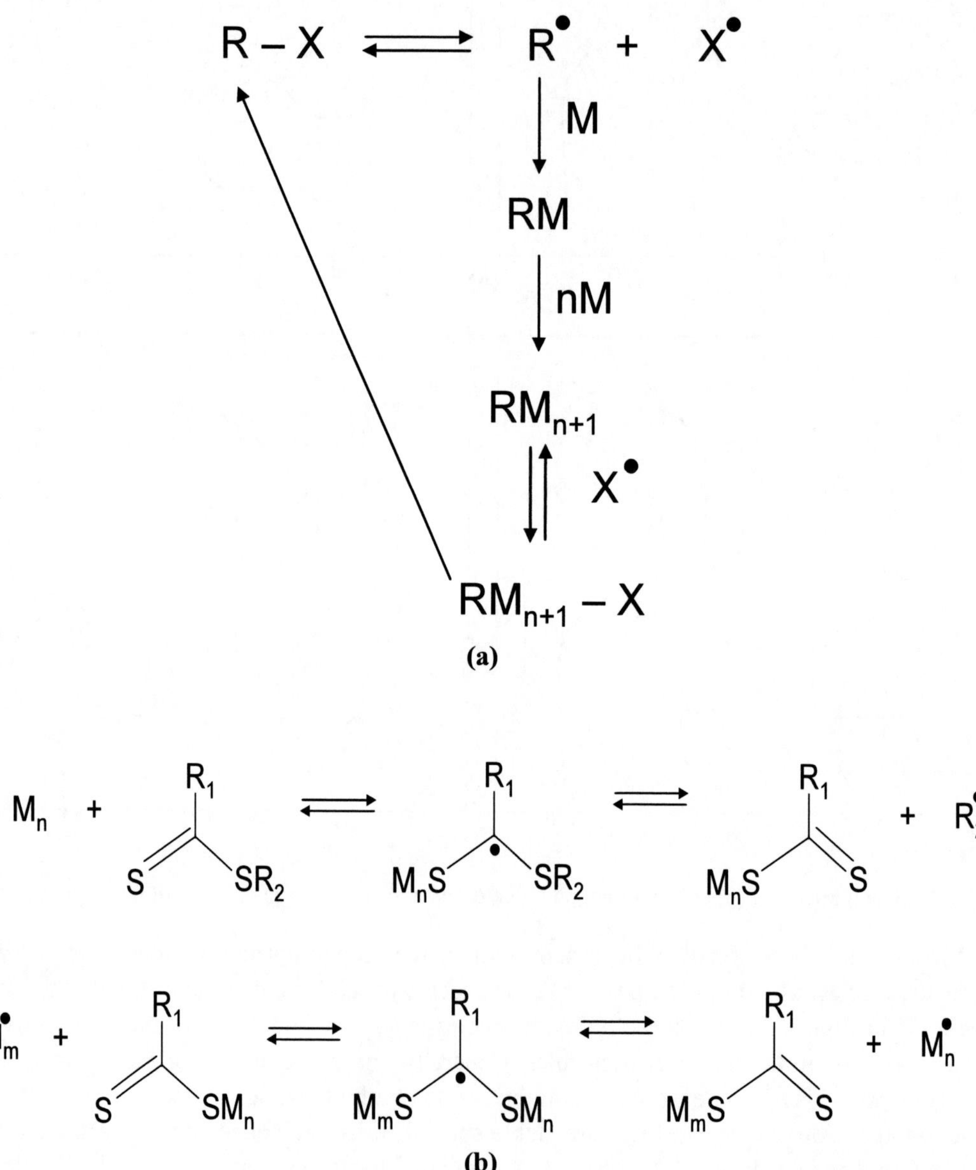

Figure 5. (a) Mechanism of reversible termination and (b) reversible transfer processes [1].

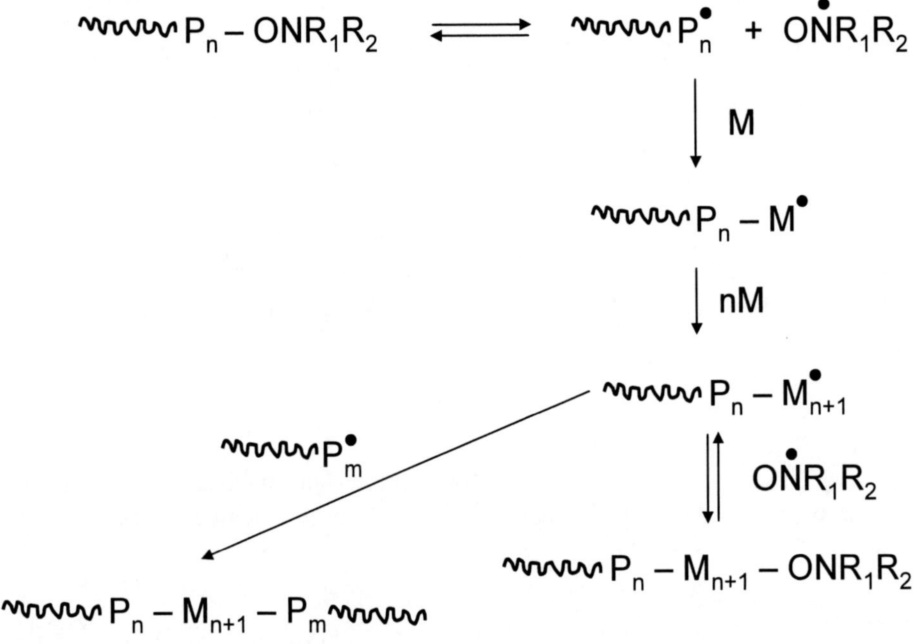

Figure 6. Detailed mechanism of nitroxide mediated polymerization process.

B.1. Nitroxide-Mediated Living Polymerization (NMP)

This polymerization type also known as stable free radical polymerization (SFRP) is a very versatile technique to achieve living polymers, especially for styrenics. In the beginning, the technique was not much successful for acrylates, however, numerous nitroxide types have

$$—CH_2—CH(C_6H_5)\{CH_2—CH(C_6H_5)\}CH_2—\dot{C}H(C_6H_5) + {}^{\bullet}O—N(C(CH_3)_2)_2(CH_2)_3 \rightleftharpoons$$

$$—CH_2—CH(C_6H_5)—CH_2—CH(C_6H_5)—CH_2—CH(C_6H_5)—O—N(C(CH_3)_2)_2(CH_2)_3 \qquad (1)$$

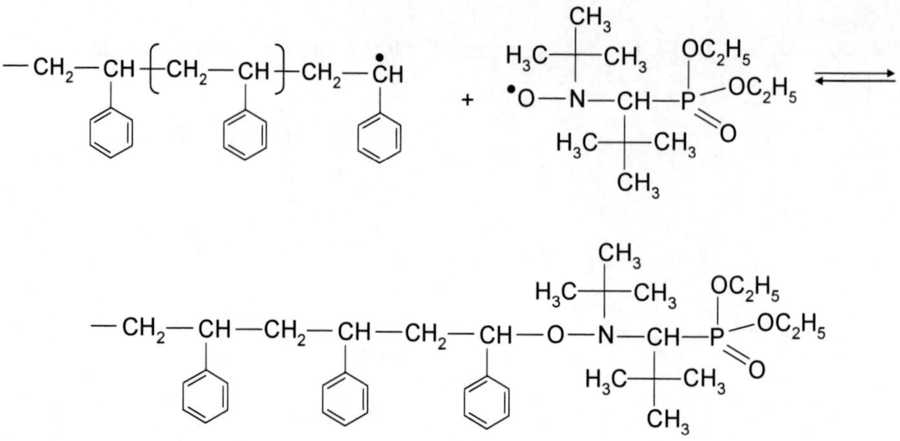

(2)

been developed which can also work very well for acrylates and methacrylates. The only limitation of these systems is that their use requires very high reaction temperatures in the

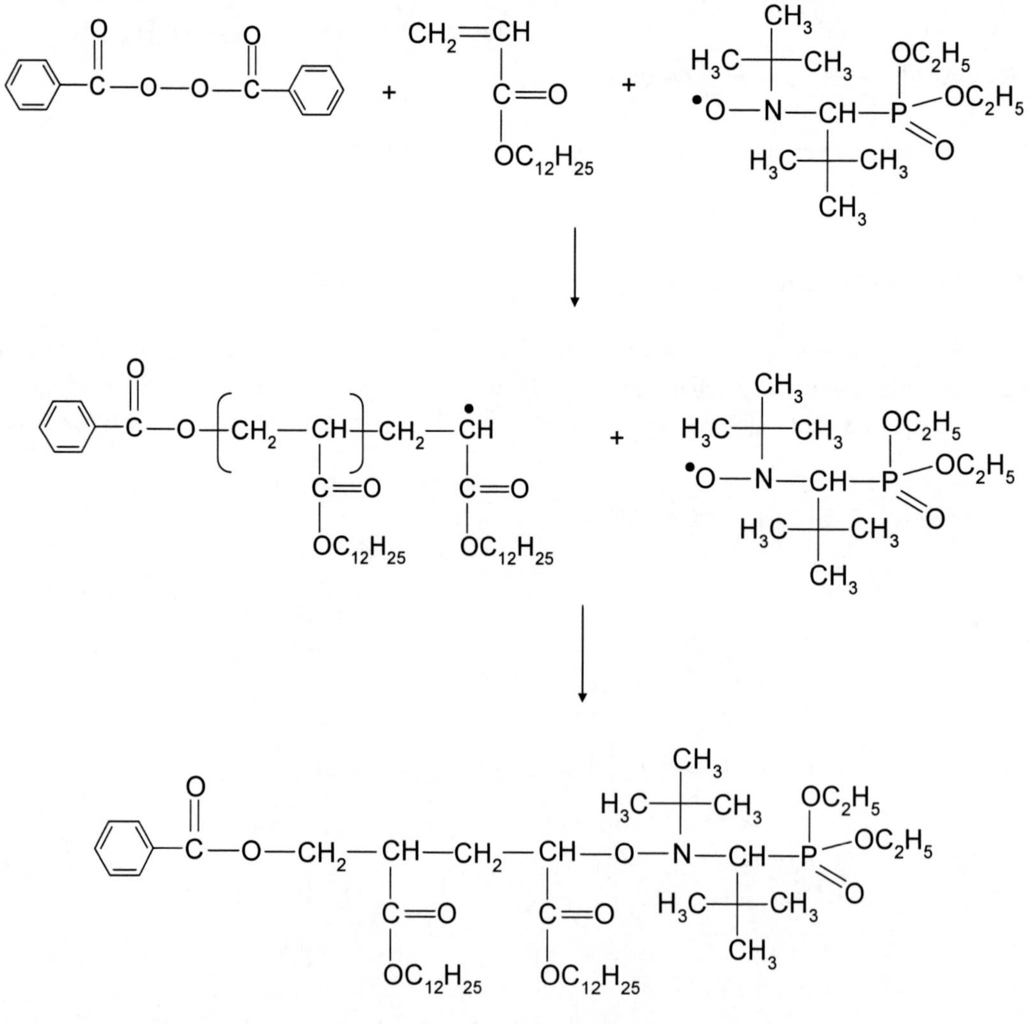

Figure 7. Polymerization of lauryl methacrylate with SG1 nitroxide.

tune of 120°C, which in some cases may not be feasible and this also forces one to proceed with the polymerization in bulk conditions. The higher temperatures are required to achieve proper balance of dormant and active polymer chains. Figure 6 is a representation of the working of the nitroxide in achieving balance between active and dormant polymer chains. P_n-ONR_1R_2 represents the polymer chain end capped with a nitroxide, which is then separated into a stable nitroxide radical and a polymer chain with a radical at end. The radical chain adds a few monomer units before it is being captured by the nitroxide radical again to form a dormant polymer chain and this process keeps on repeating itself in circles. The termination is not fully eliminated, though significantly reduced. However, by the coupling of the radicals the termination reactions can still take place, as shown in Figure 6.

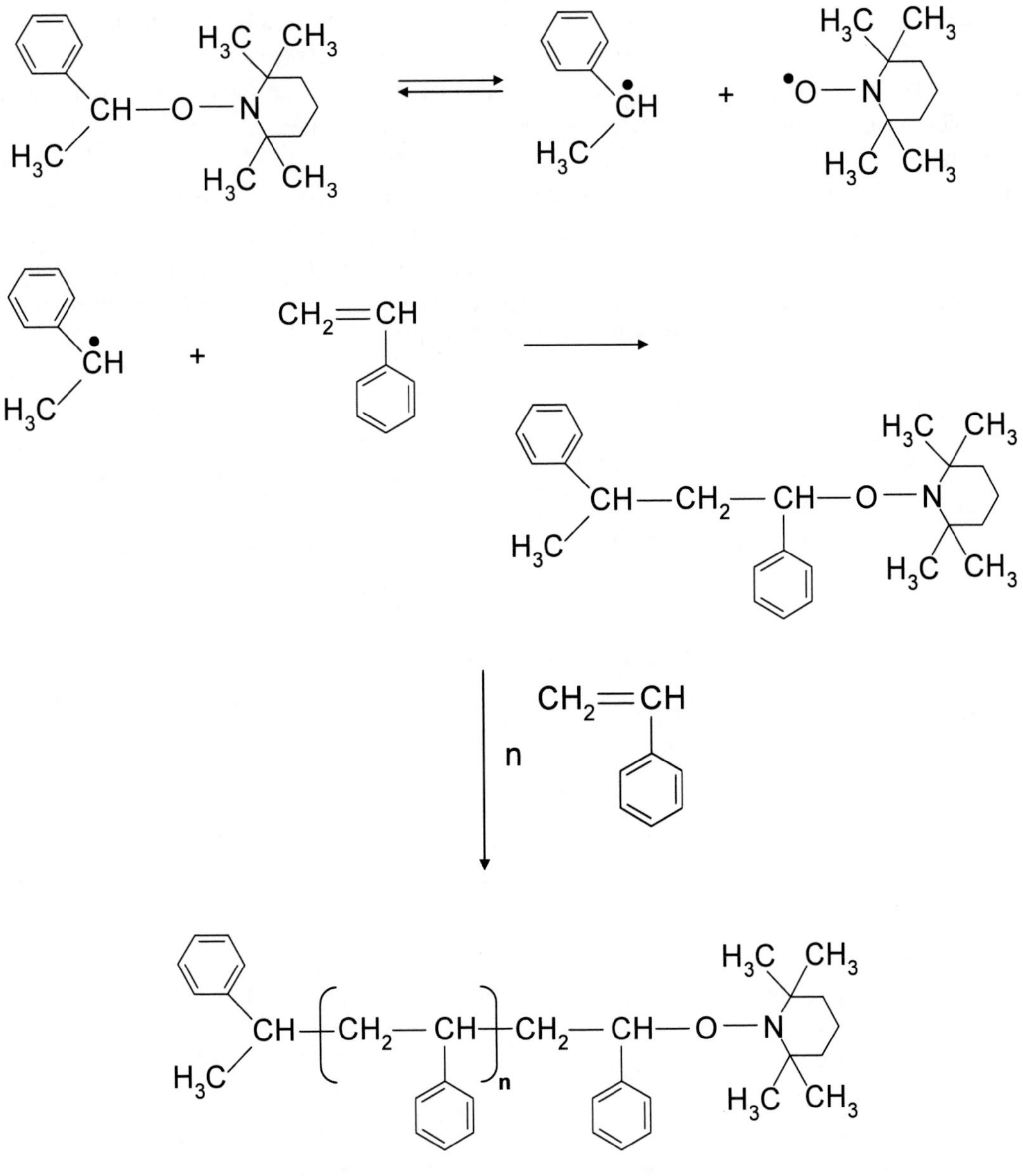

Figure 8. Polymerization of styrene by using alkoxyamine.

Two different types of methods have been developed to achieve polymerizations with nitroxides. The first method involves the use of conventional free radical initiators like AIBN and benzoyl peroxide along with additionally added nitroxide radical. 2,2,6,6- tetramethyl-1-piperidinoxyl (TEMPO) and N-ter-butyl-1-diethylphosphono-2,2'-dimethylpropyl (SG1) are two commonly used nitroxides to achieve polymerization of styrene and acrylates. Equation 1 describes the functioning of TEMPO nitroxide, whereas the method of operation of SG1 nitroxide is detailed in equation 2. Figure 7 also is a representation of such system where polymerization of lauryl methacrylate is carried out with benzoyl peroxide as initiator and SG1 as the stable nitroxide radical. The second method involves the use of alkoxyamines, the decomposition of which would lead to the generation of a stable nitroxide radical and a reactive radical. As the reactive radical can directly be used for the initiation of polymerization, therefore, no other external initiator is necessary in this case. Figure 8 shows such an alkoxyamine and the process of its decomposition along with the initiation and propagation of polymerization of styrene. A number of alkoxyamines have been developed in the recent years for the polymerization of specific monomers. These may work very well for one system, but may not be successful for the other, therefore making them more system specific rather than generally successful alkoxyamines for all systems. Figure 9 shows a few examples of alkoxyamines [2]. It has also been successfully proved that the oligomeric polymer chains can also be used as macroinitiators, thus avoiding the use of low molecular weight initiators. Figure 10 lists a few of such macroinitiators end capped with nitroxides, which can be used for further polymerization of different monomers [2].

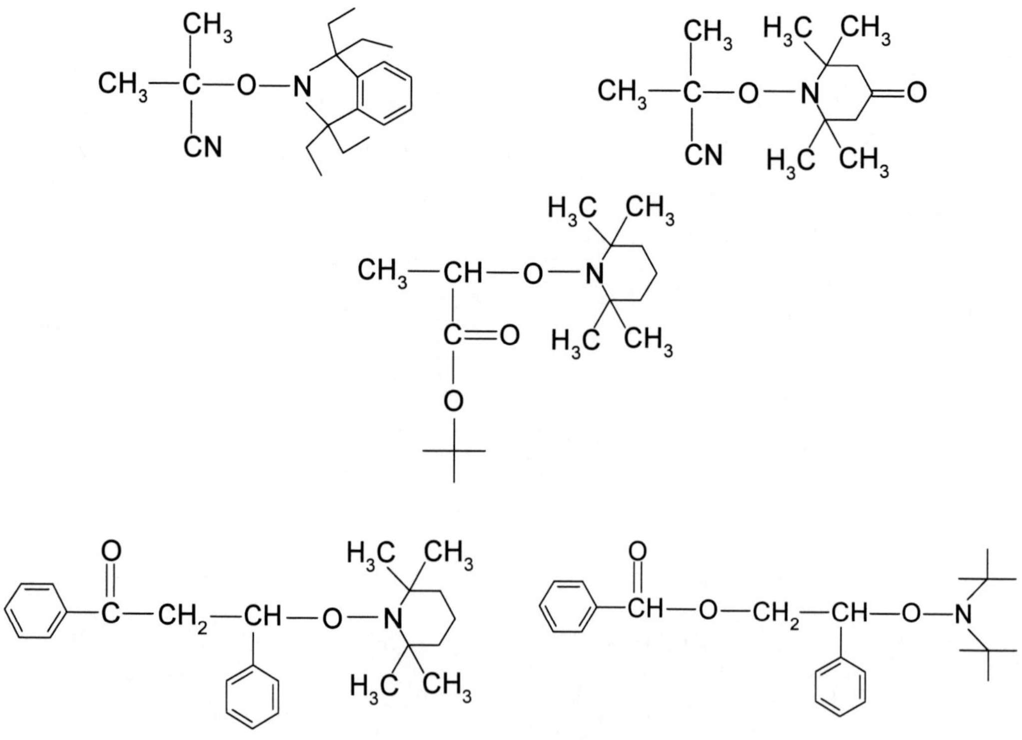

Figure 9. Commonly used alkoxyamines for nitroxide mediated polymerization.

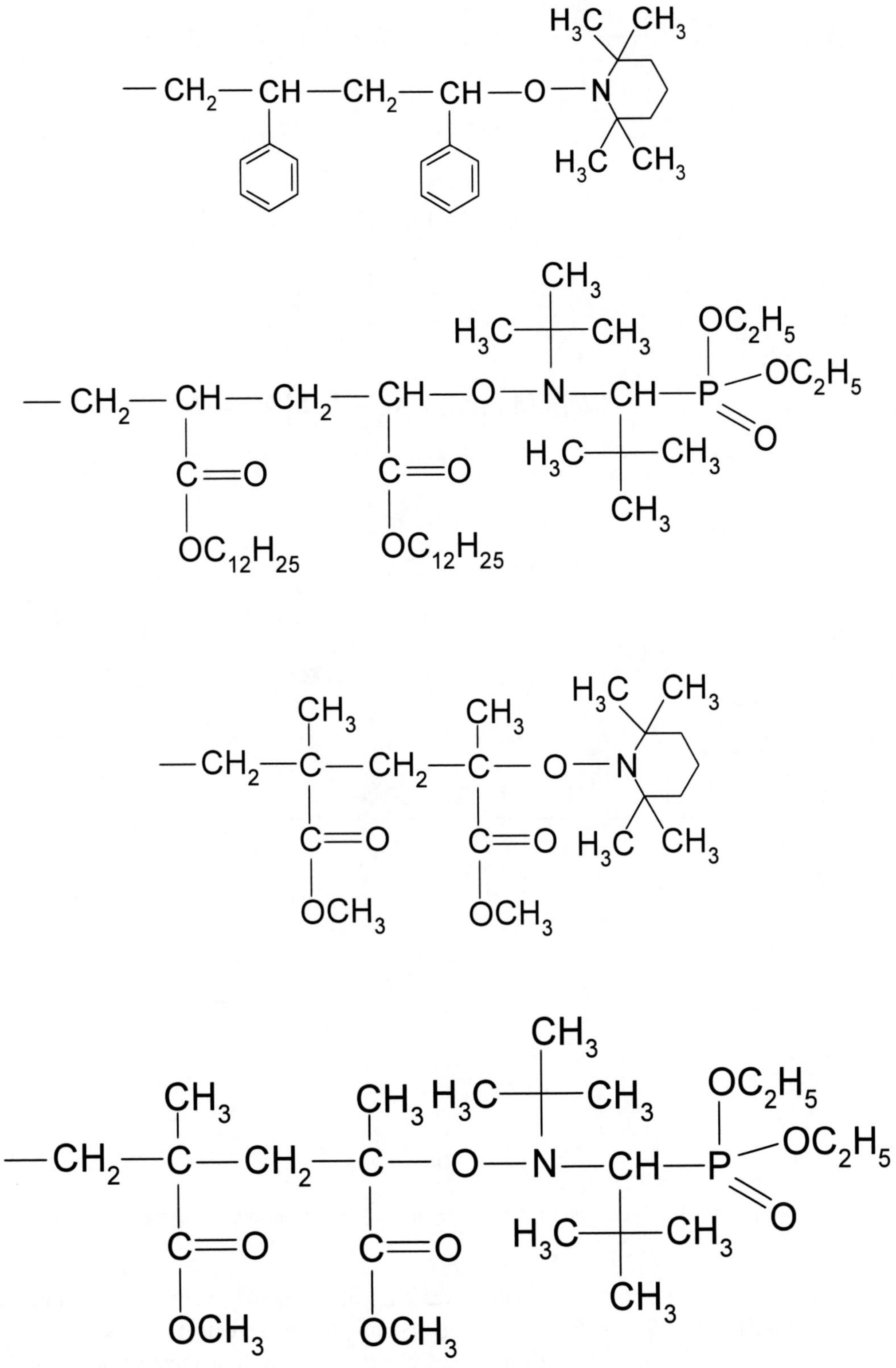

Figure 10. Macroinitiators used during the nitroxide mediated polymerization.

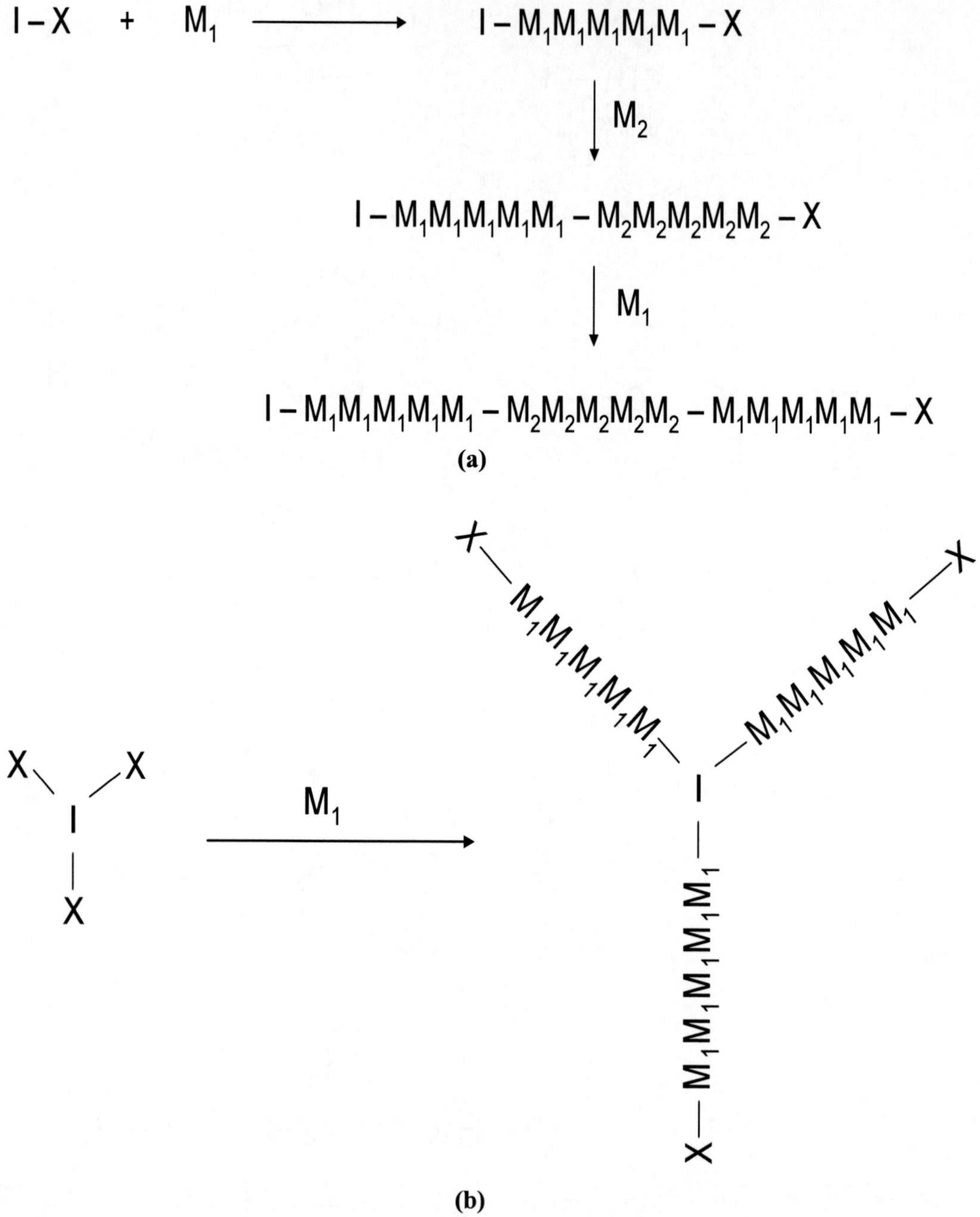

$$I-X \quad + \quad M_1 \longrightarrow I-M_1M_1M_1M_1M_1-X$$

$$\downarrow M_2$$

$$I-M_1M_1M_1M_1M_1-M_2M_2M_2M_2M_2-X$$

$$\downarrow M_1$$

$$I-M_1M_1M_1M_1M_1-M_2M_2M_2M_2M_2-M_1M_1M_1M_1M_1-X$$

(a)

(b)

Figure 11. Generation of (a) block and (b) star copolymers by the use of living polymerization methods.

As mentioned above, one of the main advantages of living polymerization systems over the conventional free radical polymerization methods is the achievement of special polymer architectures with the living polymerization e.g., generation of well defined block copolymer or branched polymers. Figure 11 depicts these possibilities for the generation of such special polymers. Figure 12 is the example of generation of polystyrene-block-poly(methyl methacrylate) copolymer by using TEMPO as nitroxide radical. Here the initiator used is benzoyl peroxide. By controlling the initially added amount of monomers, the length of blocks of each polymer can be accurately defined. Figure 13 is still another example to achieve well defined branched polymers by using multifunctional monomers. Similarly more complex morphologies like star polymers can be achieved by using different monomers and suitable nitroxides. In this particular example, no external initiator was employed as the alkoxyamine used on the system is capable of initiating the polymerization reaction by the generation of reactive radical along with the stable nitroxide radical.

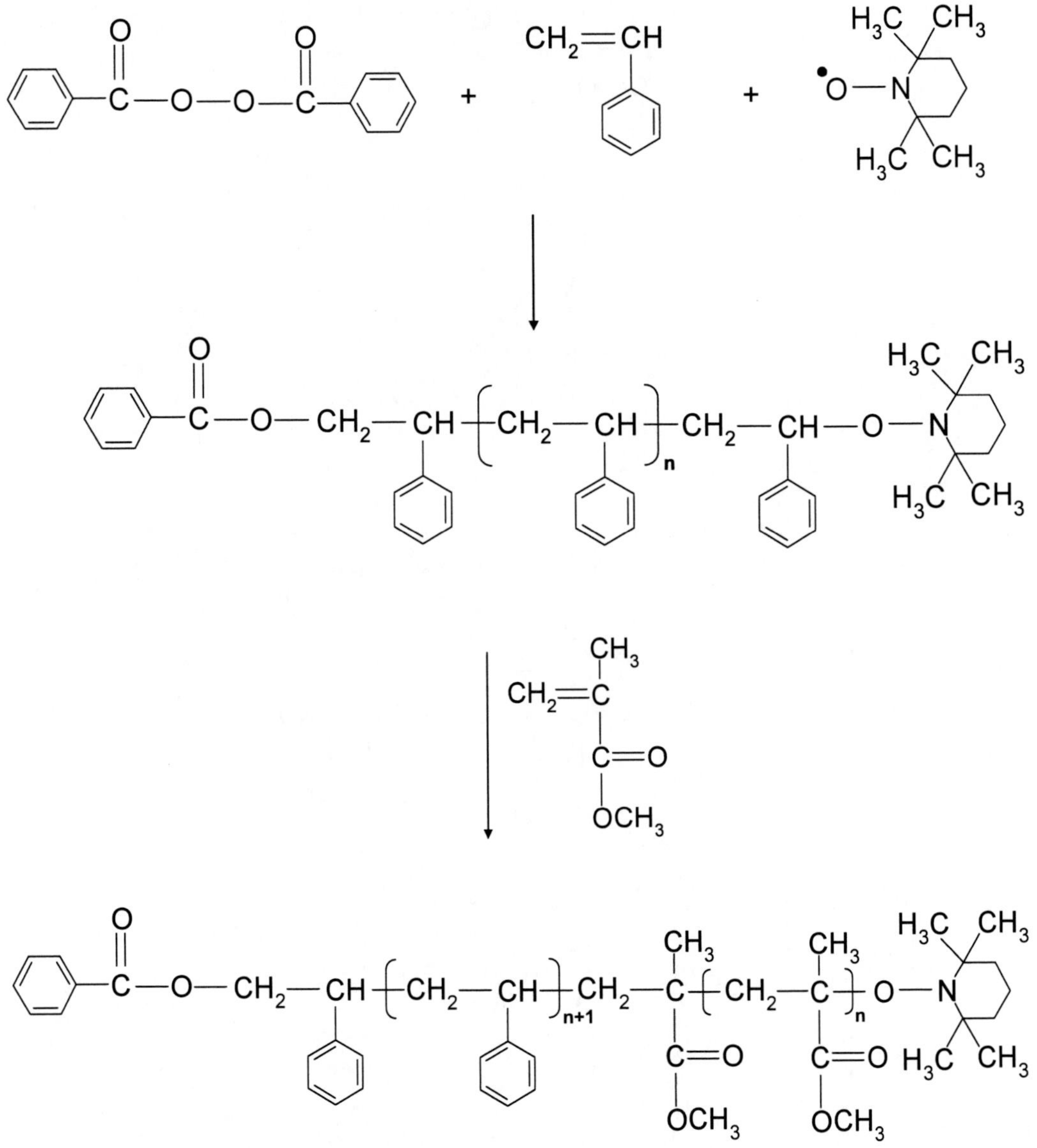

Figure 12. Use of TEMPO for the generation of block copolymer of polystyrene and poly(methyl methacrylate).

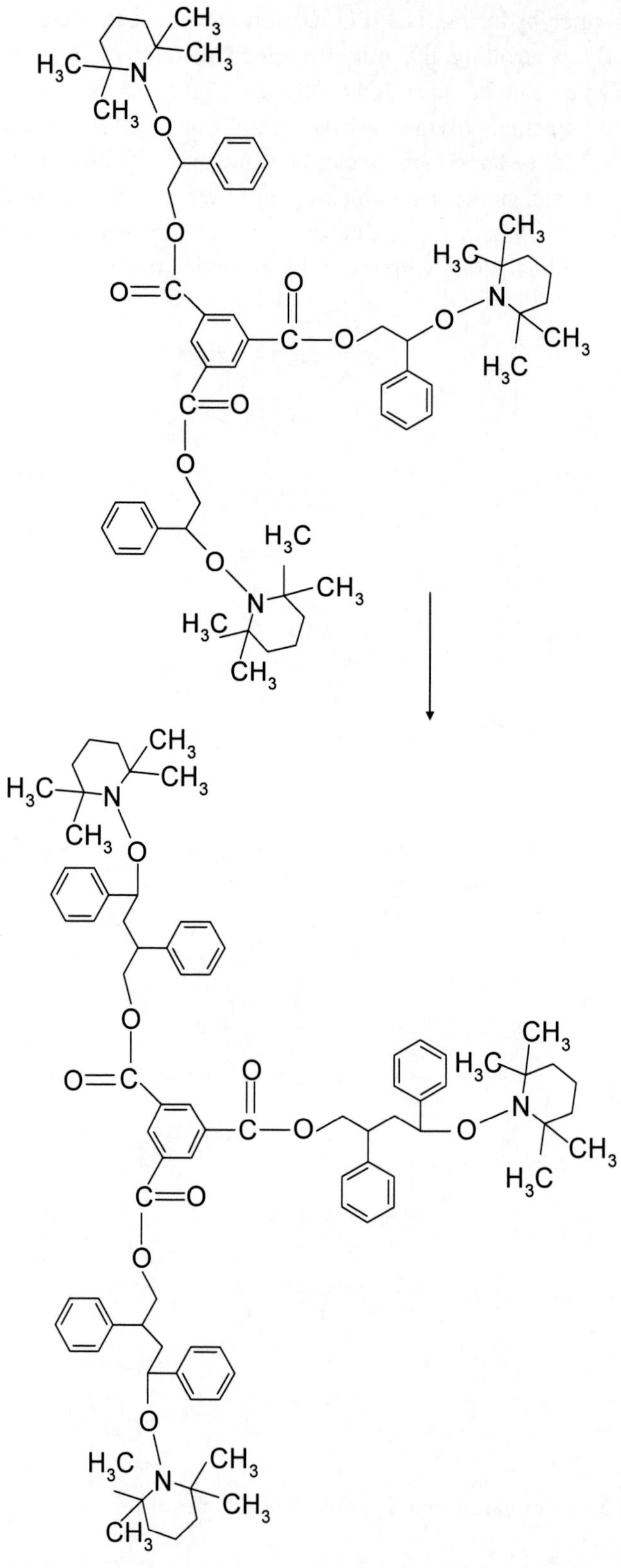

Figure 13. Generation of branched polymers by the use of multifunctional nitroxides.

B.2. Atom Transfer Radical Polymerization (ATRP)

Atom transfer radical polymerization is another versatile method of reversible termination approach to achieve living polymerization. The process involves an organic halide which keeps on terminating reversibly the polymer chains and generating reversibly active chains and the redox process is catalyzed by a transition metal compound such as cuprous chloride or bromide. ATRP has been reported to be very successful in the living polymerization of styrenics as well as acrylates or methacrylates. The polymerization can also be achieved at much lower temperatures as compared to nitroxide mediated polymerization, sometimes at room temperature. The only limitation is the presence of transition metal compounds in the end product, which though can be removed, but it is not very straightforward to achieve this cleaning. Another limitation of this method is the possible interaction of copper compounds with other components of the polymerization system, e.g. these compounds can react with the emulsifiers used in emulsion polymerization system and thus the catalyst is poisoned leading to no or a little polymerization. The polymerization in emulsion phase can though work when no surfactant is used in the system. Though the initial research focused only on the achievement of controlled polymerization in the organic solvents, however, aqueous ATRP has also been developed, though it is not very successful in controlling the molecular properties as well as compared to ATRP carried out in solvents. However, the aqueous ATRP has also been able to achieve controlled brushes or polymer chains and that too at room temperature confirming the versatility of this approach [3,4]. During these polymerization reactions, a small amount of $CuBr_2$ is also added in the beginning as it increases the deactivation of the free radicals in the system and thus decreases the polymerization rate. It leads to shifting of the reaction equilibrium more towards the dormant species. Thus the reaction rate can be controlled according to the requirement. Elemental copper is also occasionally added as it being a reducing agent helps to eliminate the oxidation and radical chain termination induced by the dissolved oxygen in the system.

Figure 14 shows the schematic of ATRP process. Cuprous salt forms a complex with ligand, L (amines of different chemical architectures) which makes it more soluble in the solvents [1]. Initiation of the reaction takes place by the dissociation of the halide atom from the initiator and leading to the generation of a free reactive radical. The bromide atom is captured by cuprous halide ligand complex and it forms $CuBr_2$ ligand complex. This compound is very stable and hence is called deactivator. The generation of this compound thus leads to reduction in the concentration of the free radical species in the system. The growing radical keep on adding monomers to the polymer chain and at some point comes in contact with $CuBr_2$ ligand complex and the growing radical is temporarily terminated by the formation of RM_{n+1}-Br compound. It is also possible to carry out the reverse ATRP process similarly (Figure 14b). In this process, a conventional free radical initiator like AIBN or benzoyl peroxide is used to initiate the polymerization reaction which is the controlled by the addition of $CuBr_2$ ligand complex. The radicals add few monomer units and during this process come in contact with this complex to form the dormant species.

A lot of effort has been devoted to the generation of different ligands as well as ATRP initiators. A few of these are shown in Figures 15 and 16 respectively. Every ligand and initiator may not be successful for a number of polymer systems, but is more specific to the studied systems. Apart from that, the macroinitiators can also be used in the same way as in

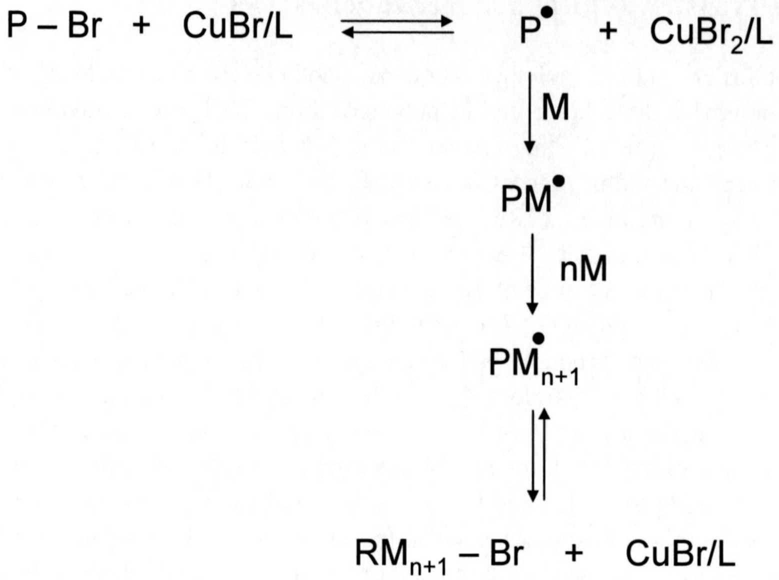

$$P-Br + CuBr/L \rightleftharpoons P^{\bullet} + CuBr_2/L$$

$$P^{\bullet} \xrightarrow{M} PM^{\bullet} \xrightarrow{nM} PM^{\bullet}_{n+1}$$

$$PM^{\bullet}_{n+1} \rightleftharpoons RM_{n+1}-Br + CuBr/L$$

(a)

$$C_6H_5-C(=O)-O-O-C(=O)-C_6H_5 + CuBr_2/L \rightleftharpoons C_6H_5-C(=O)-O-Br$$

$$\xrightarrow{M} C_6H_5-C(=O)-O-M^{\bullet} \xrightarrow{nM} C_6H_5-C(=O)-O-M^{\bullet}_{n+1}$$

$$C_6H_5-C(=O)-O-M^{\bullet}_{n+1} \underset{}{\overset{+\ CuBr_2/L}{\rightleftharpoons}} C_6H_5-C(=O)-O-M_{n+1}-Br + CuBr/L$$

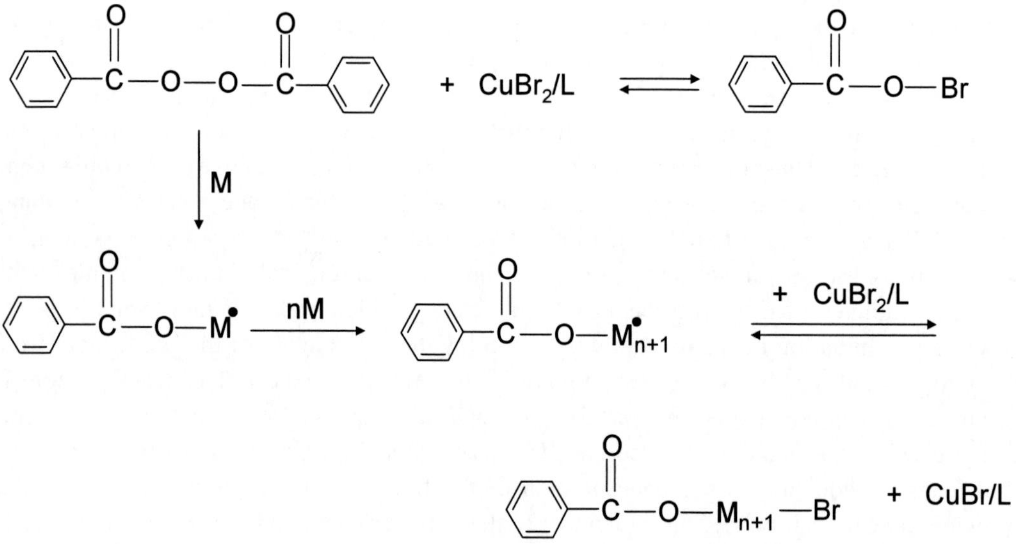

(b)

Figure 14. (a) ATRP and (b) reverse ATRP processes of living polymerization [1].

Figure 15. Various commonly used ligands for the ATRP process.

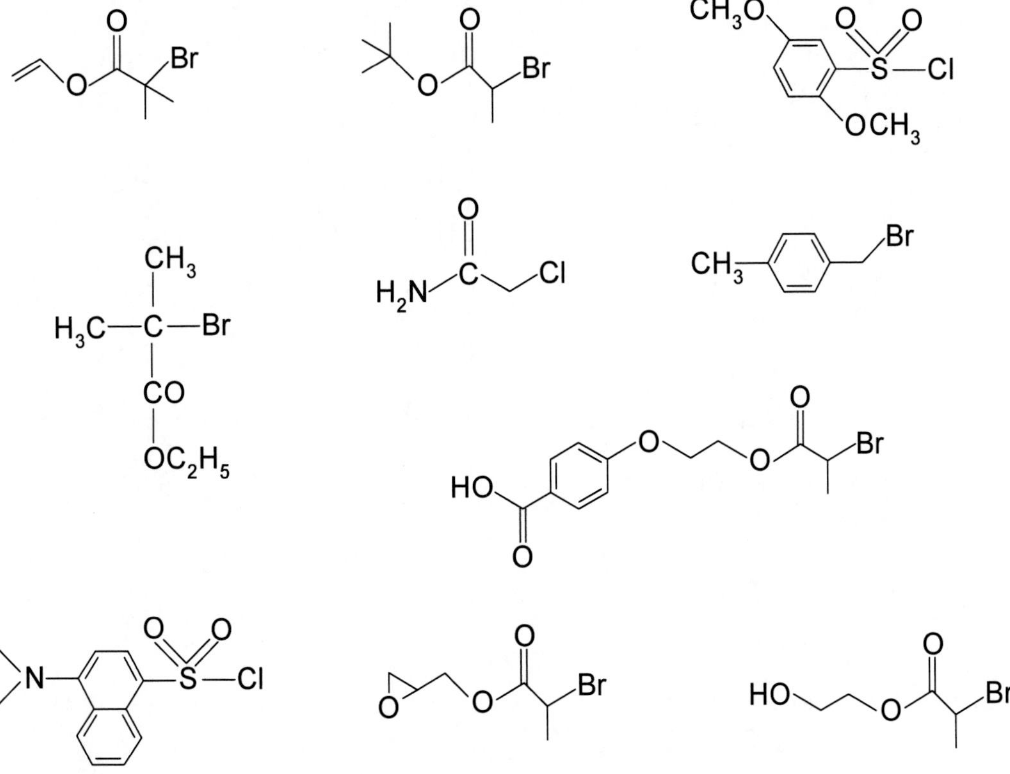

Figure 16. Various commonly used initiators for the ATRP process.

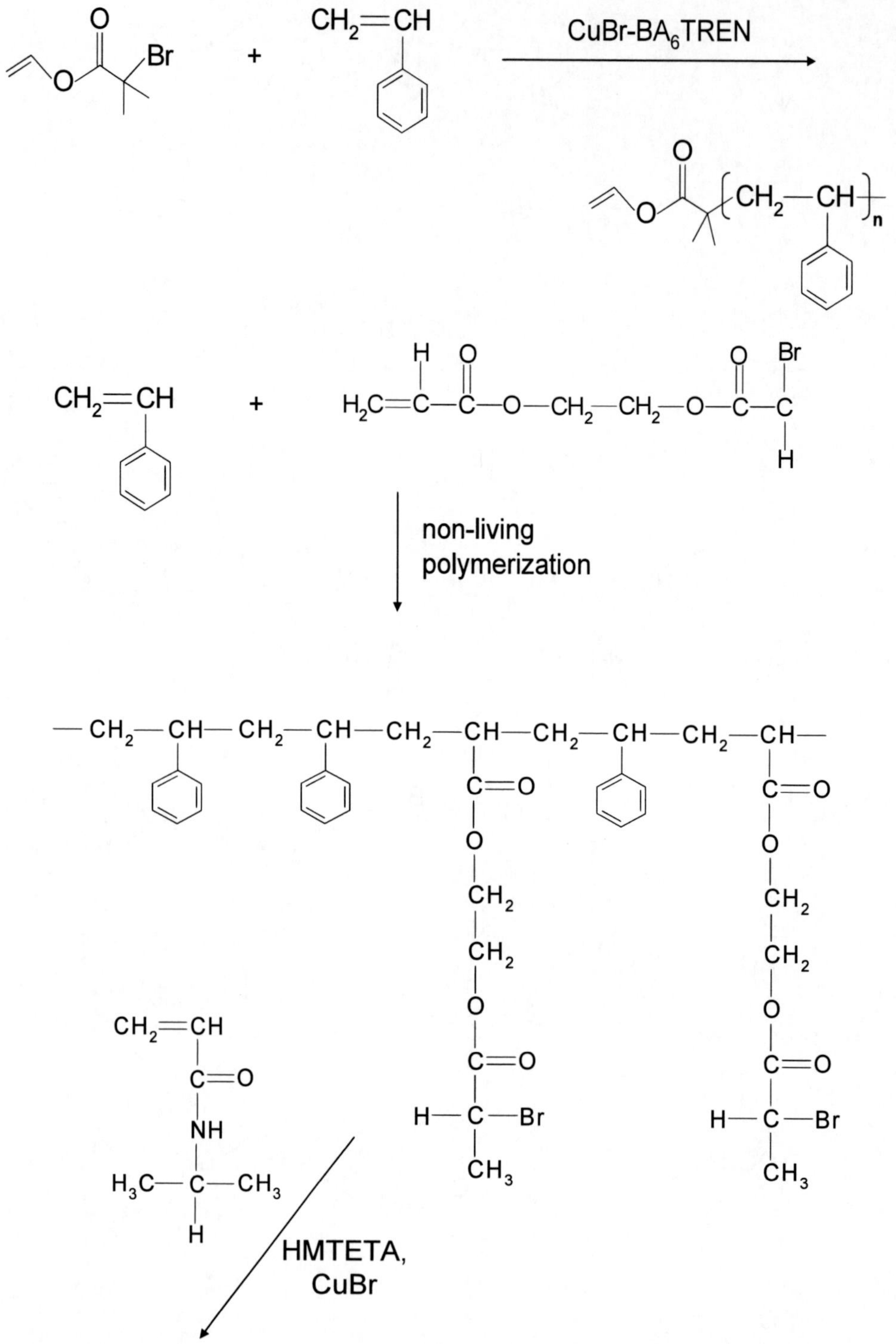

Figure 17. (Continues on next page.)

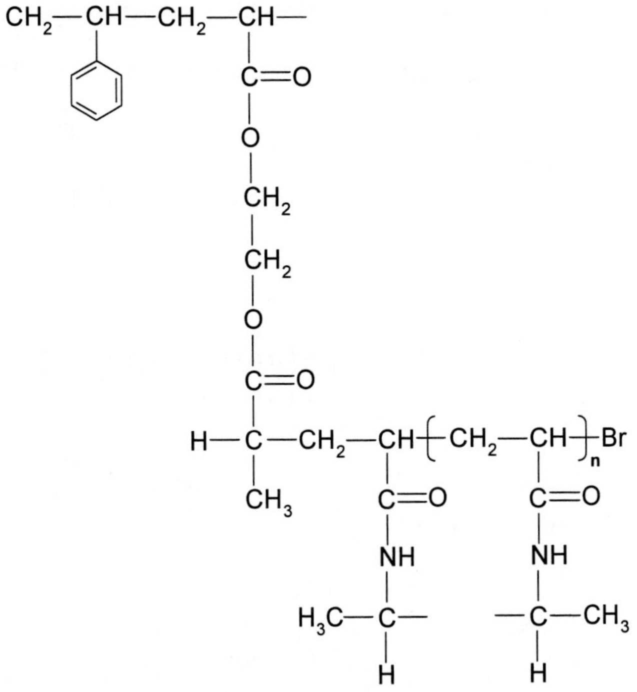

Figure 17. Various kinds of macroinitiators used in the literature for ATRP.

the case of nitroxide mediated polymerization as shown in Figure 17. A large number of possibilities to achieve different polymer architectures like the generation of block copolymers or star polymers can also be very efficiently achieved in the case of ATRP similarly as in nitroxide mediated controlled radical polymerization. Figures 18 and 19 respectively detail the ATRP processes to achieve poly(methyl methacrylate)-block-poly(butyl acrylate) copolymers as well as polystyrene-co-poly(methyl acrylate) branched copolymers. One must be careful that a number of initiators, catalysts as well as ligands have been developed for the controlled polymerizations of various monomers, mostly specifically usable for one system, therefore, choice of these initiators, catalysts as well as ligands is not straightforward. Generally the system which leads to much higher concentration of dormantspecies as compared to active species, roughly in the range of six orders of magnitude, is considered optimal for the living polymerization process. As fast initiation is required in these cases in order to achieve chains of almost same length (to decrease polydispersity in the chain length as well as polymer molecular weight), therefore, one must also use more reactive cuprous halides like copper bromide. It was mentioned above that $CuBr_2$ is also generally added in the beginning of the polymerization to reduce the polymerization rate as it shifts the reaction equilibrium towards the dormant species. Apart from this, it also functions towards the elimination of normal bimolecular termination of the radicals thus helping to decrease polydispersity in the chain length. Elemental copper as said above is also added as it can react with the dissolved oxygen in the system to reduce the termination as otherwise the dissolved oxygen traps the radicals leading to their irreversible termination. Apart from reducing termination, elemental copper is beneficial in improving the polymerization rate as excess of $CuBr_2$ reacts with it to form CuBr which can then be used to initiate the polymer chains.

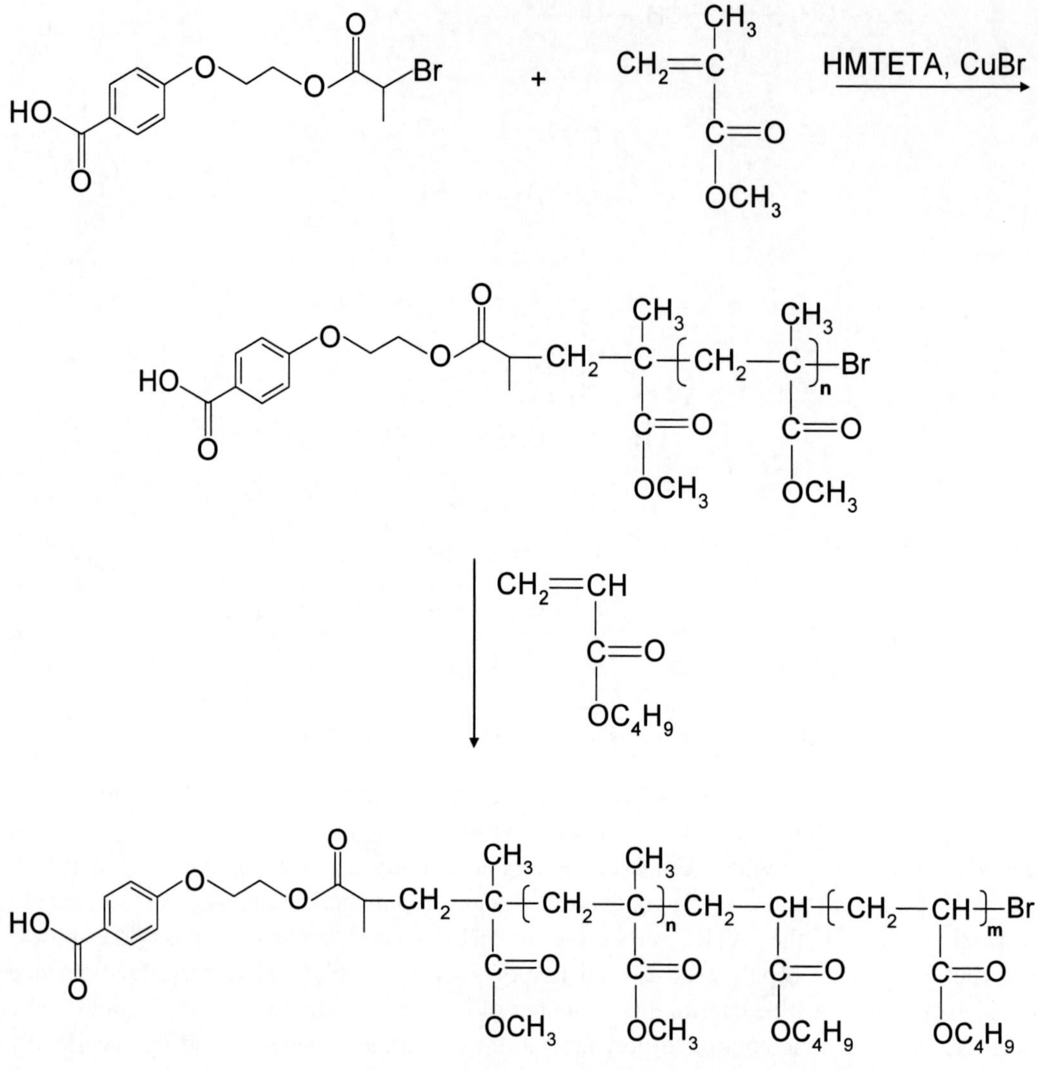

Figure 18. Use of ATRP for the generation of block copolymers of poly (methyl methacrylate) and poly (butyl acrylate).

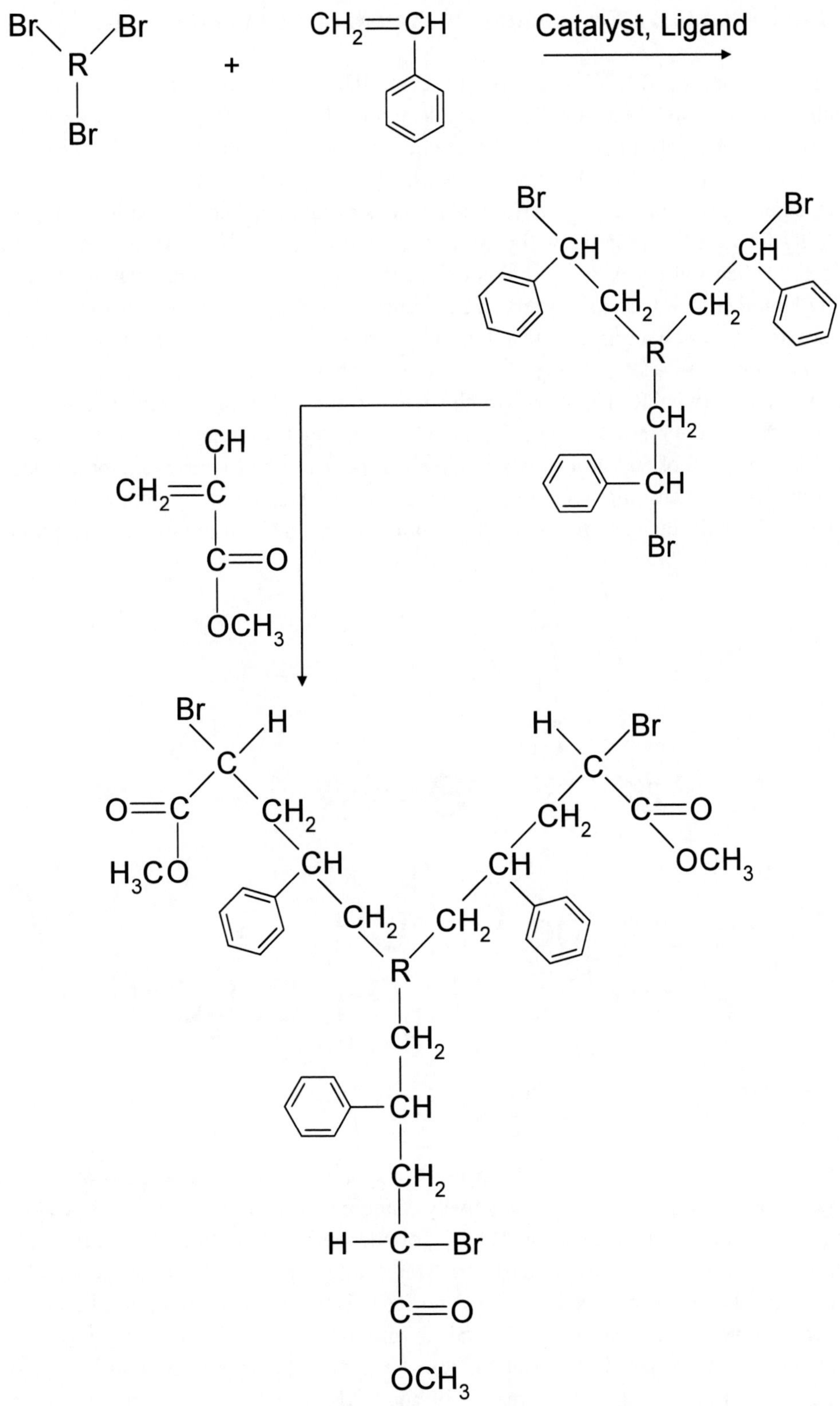

Catalyst, Ligand

Figure 19. Use of ATRP for the synthesis of branched copolymers.

B.3. Radical Addition Fragmentation Transfer (RAFT) Polymerization

As mentioned above, RAFT is a form of controlled polymerization which operates by reversible transfer mode and not by reversible termination as ATRP or nitroxide mediated polymerization [1,2]. As mentioned in Figure 5b and Figure 20, the core of this process is a RAFT agent which contains a dithioester groups. The living polymerization takes place as the transferred end group in the polymeric dithioester is as labile as the dithioester group in the starting RAFT agent. The initiator for the polymerization can be the conventional initiators like AIBN or benzoyl peroxide. RAFT is one of the most versatile techniques for controlled living polymerization of a wide variety of monomers. Another important advantage of RAFT is the possibility of polymerization reaction to be carried out at lower temperatures. However, as also mentioned for ATR polymerization, there are always concerns regarding the presence of excess or remainder RAFT agent which owing to the presence of sulphur also leads to color and odors to the product. However, there has been a lot of research effort to convert the residual groups to thiol end groups which have less odor or color concerns. Apart from that, most of the RAFT agents are not commercially available, thus requiring their synthesis in the lab. Figure 21 details the example of polymerization of styrene by using the RAFT method.

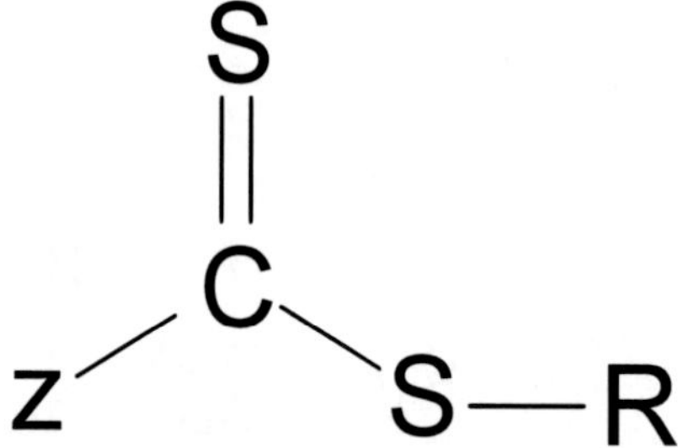

Z can be Ph, CH_3
R can be CH_2Ph, CH_2CN

Figure 20. General structure of RAFT agent.

RAFT polymerization has also been used to synthesize different polymer architectures as shown in Figures 22–24, where respectively random and block copolymers as well as star polymers have been generated using RAFT polymerization. One must be careful that the chosen RAFT agent must have the chain transfer activity appropriate to the monomer to be polymerized. The chain transfer activity of the RAFT agent is determined by the electronic and stereoelectronic properties of R and Z groups. A significant retardation in the polymerization rate occurs if these chain transfer functions are not optimum: e.g., if the growing radical has more affinity to react with the RAFT agent than the monomer itself, the polymerization rate would decrease significantly.

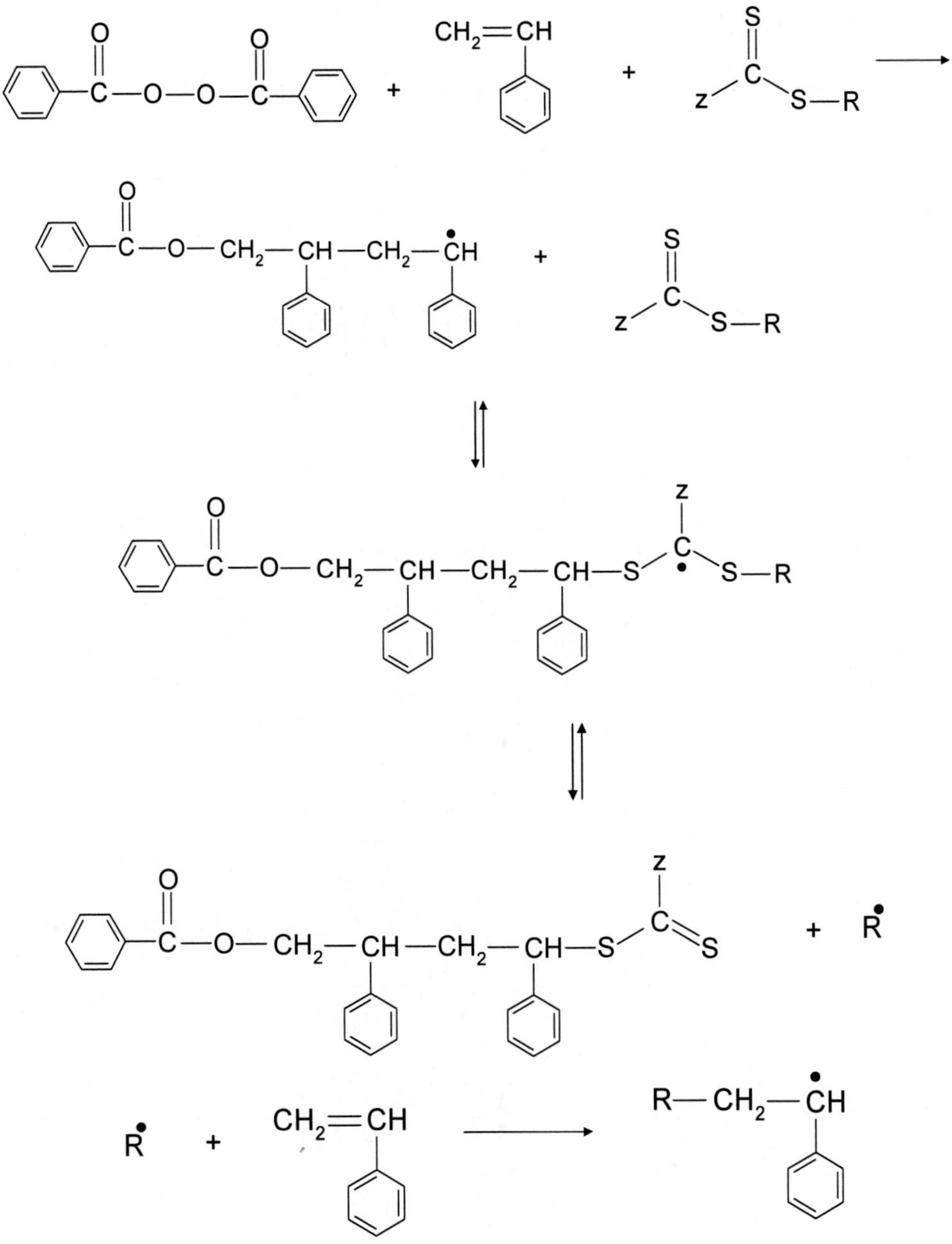

Figure 21. (Continues on next page.)

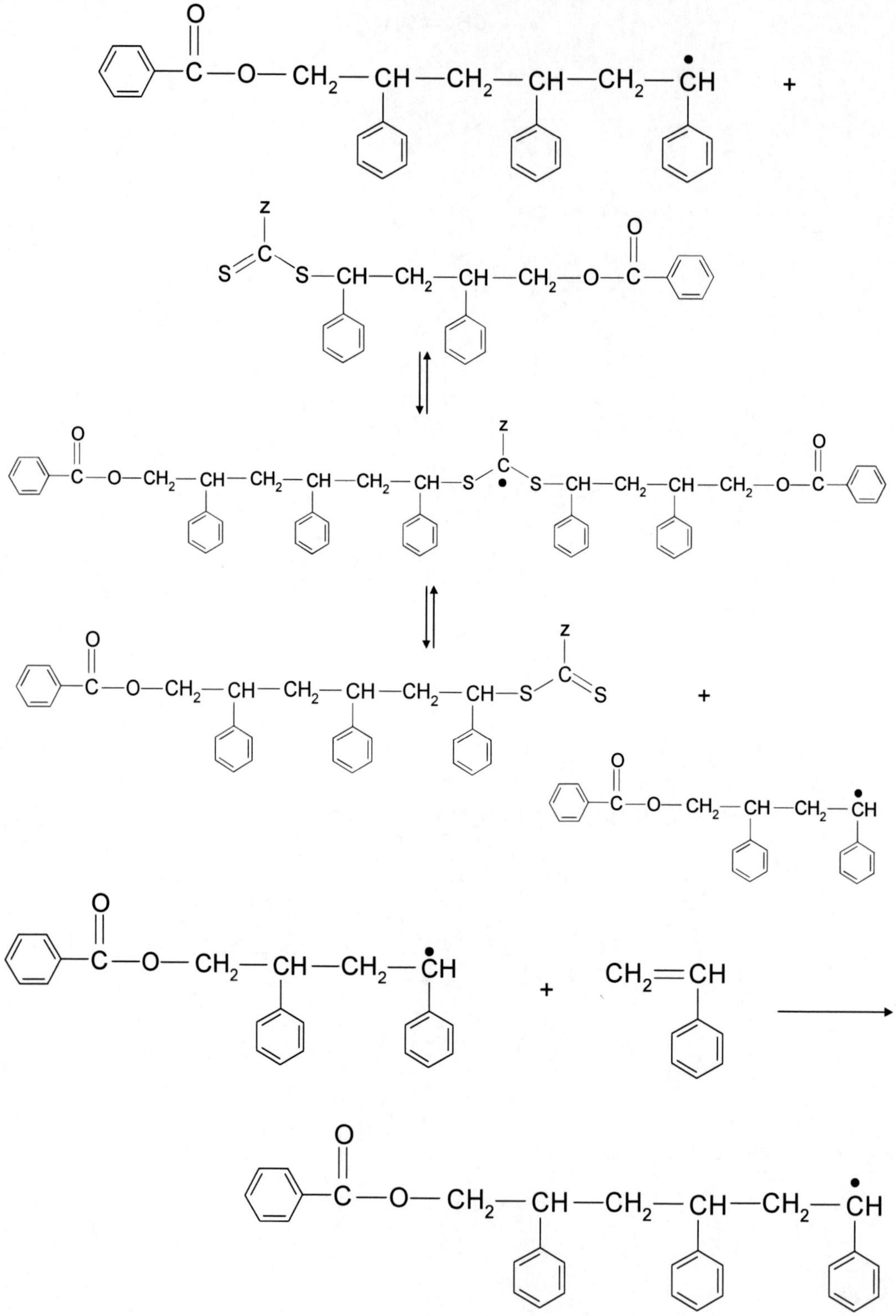

Figure 21. Polymerization of styrene by using RAFT.

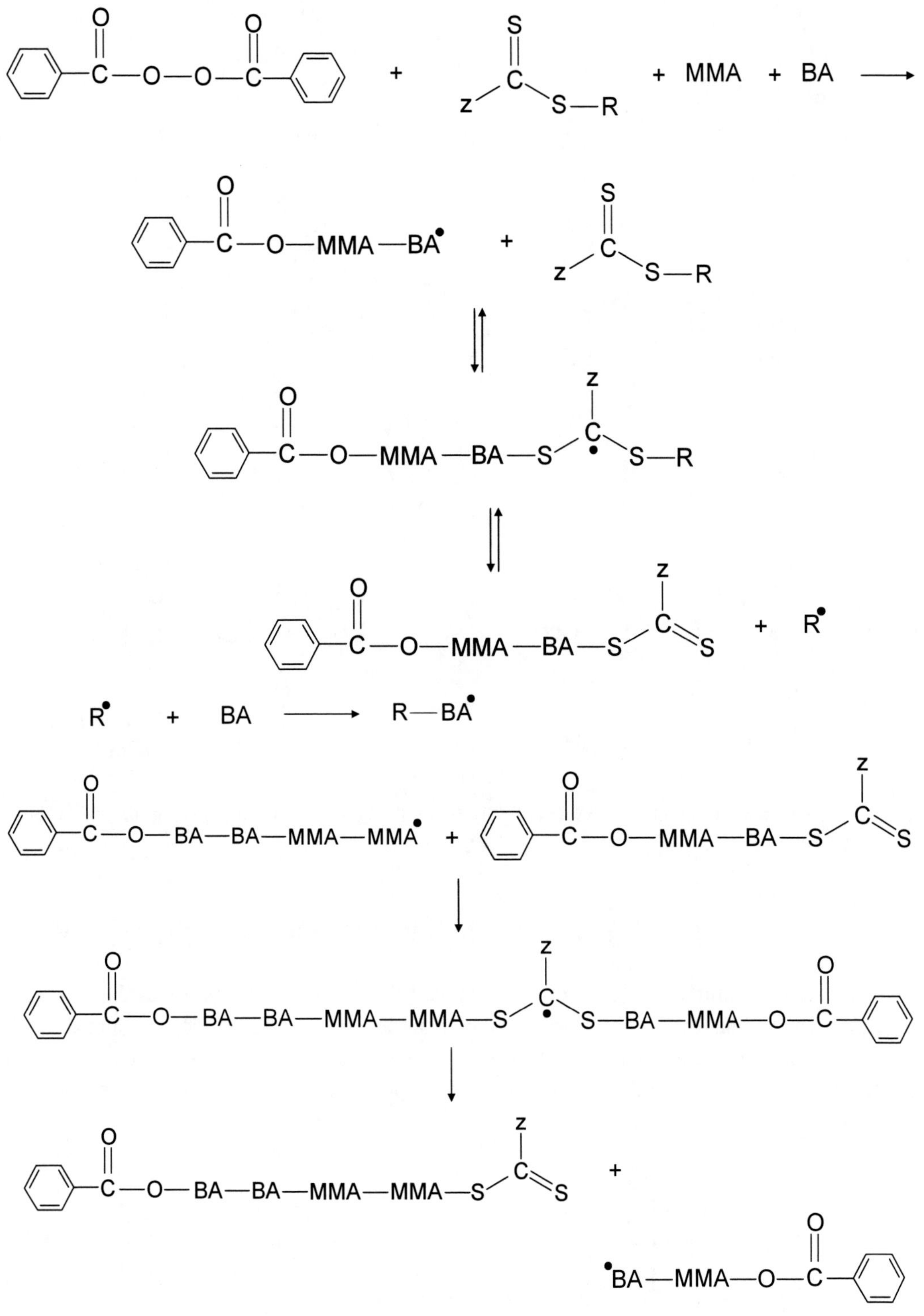

Figure 22. Generation of random copolymers by using RAFT.

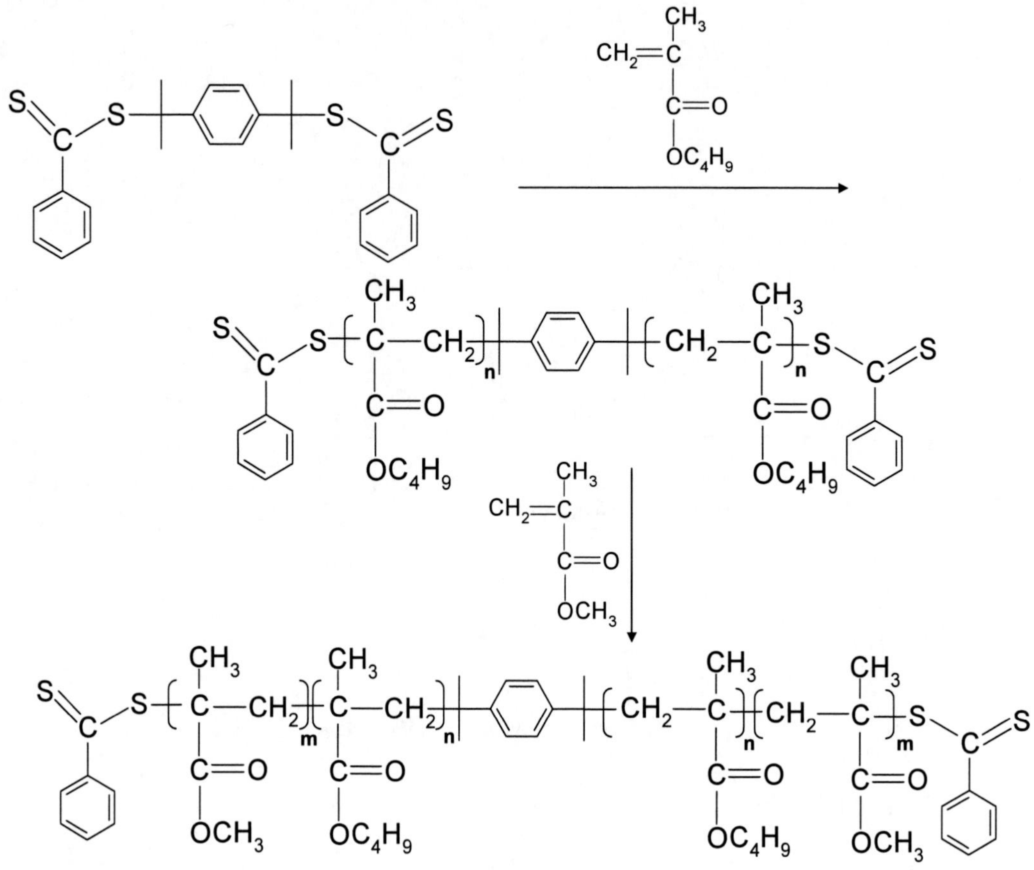

Figure 23. Generation of block copolymers of poly(methyl methacrylate) and poly (butyl methacrylate) by using RAFT.

C. KINETICS OF LIVING POLYMERIZATION

The free radical nature of SFRP and ATRP is well established, indicating that the effects of initiators, chain transfer agents, retarders, solvents, monomers, etc., would be approximately the same as those in conventional non-living polymerization. As equilibrium exits between the active and dormant chains, therefore, the rate of polymerization in this case was described as

$$R_p = \frac{k_p [M][P_n^{\bullet} - Y]}{k [Y^{\bullet}]} \tag{3}$$

where $P_n^{\bullet}$ is the active radical concentration
$Y^{\bullet}$ is the concentration of active chains
Y is the concentration of dormant chains

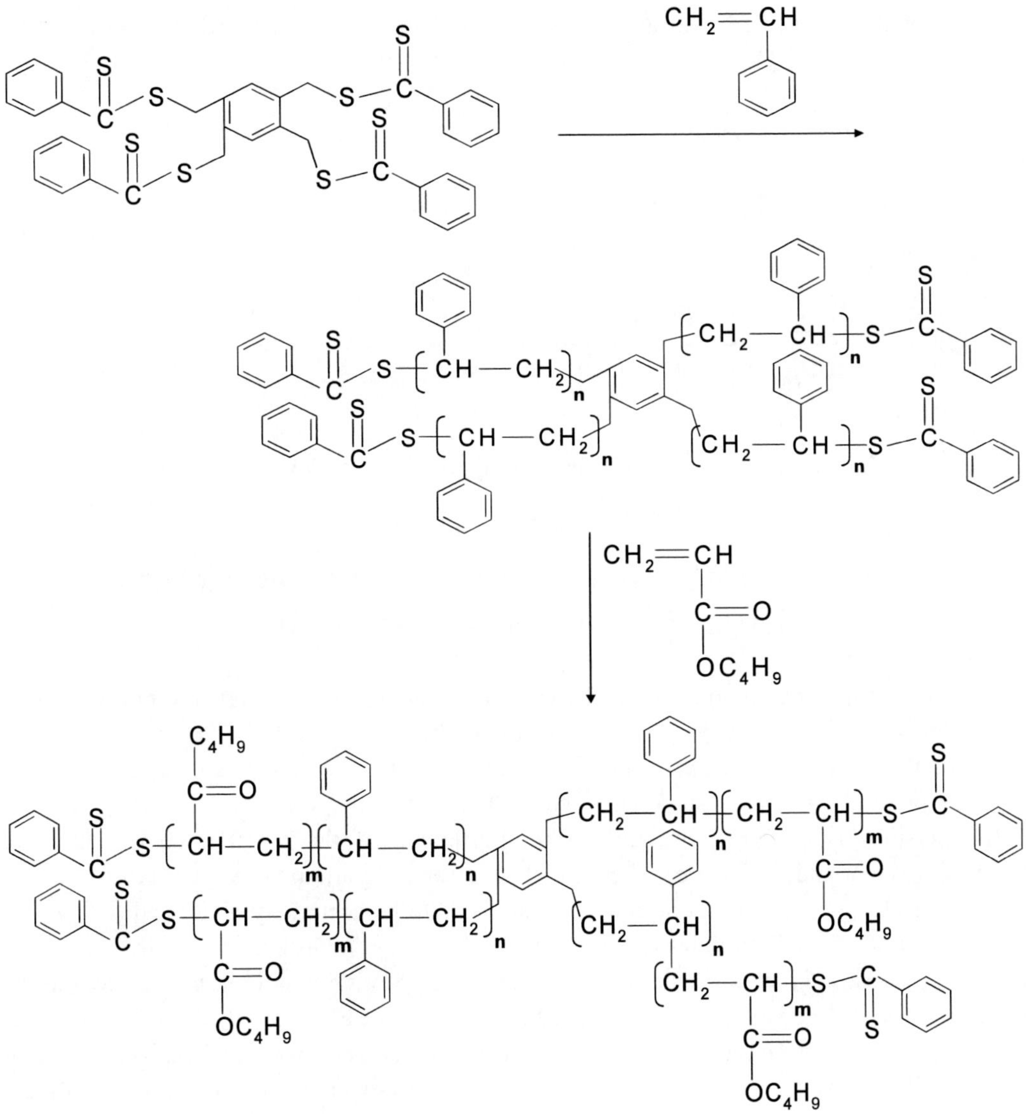

Figure 24. Generation of branched copolymers by using RAFT.

Especially in the case of ATRP, the rate of polymerization can be written as [1,2]

$$R_p = \frac{k_p [M] [I] [Cu^+]}{[Cu^{2+}]} \tag{4}$$

$$\frac{[M]_0}{[M]} = \frac{k_p K [I] [Cu^+]}{[Cu^{2+}]} t \longrightarrow \text{straight line} \tag{5}$$

where [I] is the initiator concentration. Degree of polymerization in ATRP as well as RAFT can then be detailed as [1,2]

$$\overline{X}_n = \frac{[M]_0 - [M]}{[I]_0} \tag{6}$$

$$DP = \frac{[M]_0 - [M]}{([RAFT]_0 - [RAFT]) + f([I]_0 - [I])} \tag{7}$$

where f is the initiator efficiency and [I] is the initiator concentration.

D. Living Polymerization in Heterogeneous Emulsion Homopolymerization

The use of living techniques in the heterogeneous emulsion polymerization of monomers is an interesting area, as it represents a specialty application of the living techniques keeping in view the number of additional factors to deal with when it comes to emulsion polymerization. Therefore, achieving controlled molecular weights as well as low polydispersities is not very straightforward. Partioning of the reaction components like controlling agent and initiator between the aqueous and organic phases must be considered. Colloidal stability is also a major concern that needs to be accurately addressed. It involves the coagulation of small particles to form large particles, and it has been generally understood to be the root cause in the nucleation of polymer particles in living polymerization conditions. In the non-living conventional polymerizations, the radical enters the micelles and the radicals propagate inside the polymer particles without much desorption of the radical outside the particles. However, in living polymerization, the short chain propagating radicals can easily desorb out of the particles.

The kinetics of living emulsion polymerization is also significantly different from that of the conventional emulsion polymerization method. When reversible termination methods like nitroxide mediated polymerization or atom transfer radical polymerization are carried out, as expected, the overall number of radicals or propagating radical chains is much lower inside the particles, as the NMP or ATRP agents trap the excess number of radicals to form dormant species; thus, the concentration of dormant species is quite high as compared to the reactive species. The major difference in comparison with conventional emulsion polymerization is the absence of a compartmentalization effect in the living polymerization systems. The final molecular weight achieved by these polymerization methods is less controlled in polydispersity and is thus broadly distributed. Also, the living systems do not generate molecular weights as high as those that can be generated in conventional living polymerization. One may think that the use of living polymerization systems is thus without any advantages in the case of emulsion polymerization. However, these techniques are extremely helpful in generating surface functionalized particles, especially when the achievement of a controlled surface functionalization is required. Also, the generation of the controlled block copolymers is only possible by using these methods, as the conventional non-living polymerizations are unable to control the architecture of the copolymers because the life of the radical in these polymerizations is too short. The kinetics of living polymerization achieved with reversible transfer process is also different from the conventional non-living polymerization as well as from the living methods operated with a reversible termination approach. Here the radical concentration is not affected, which is opposite to that of the case of the reversible termination processes. This, therefore, leads to higher polymerization rates. The compartmentalization effects present in conventional emulsion polymerization are also present and, thus, the final molecular weight is much higher.

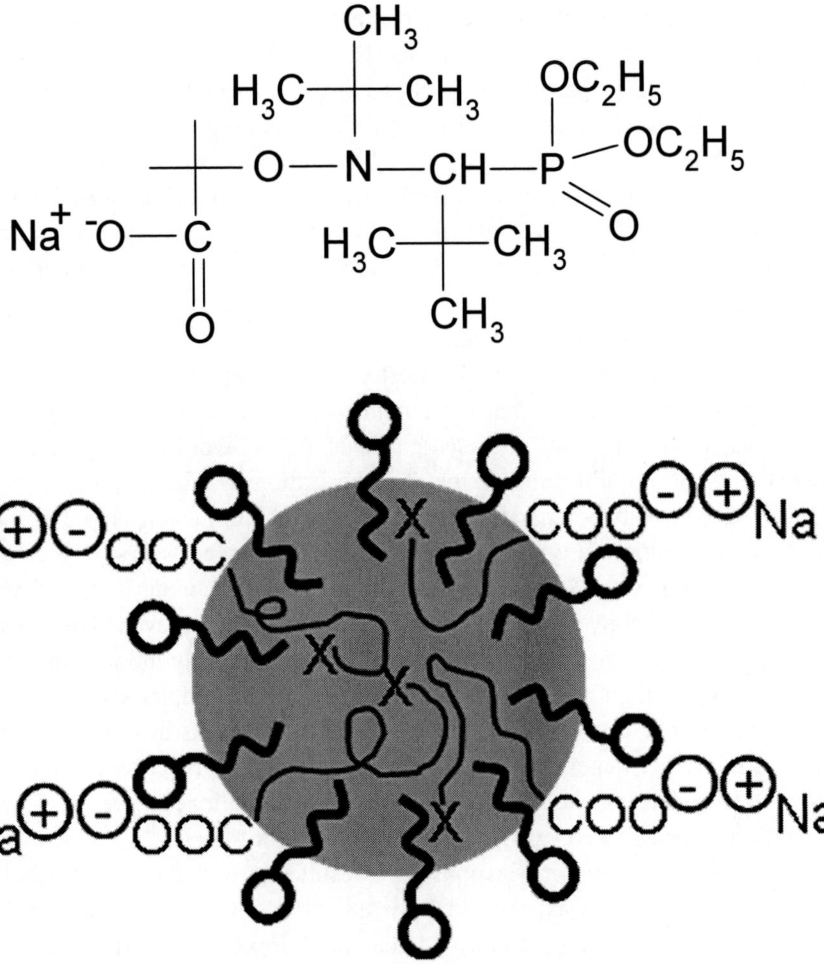

Figure 25. Nitroxide mediated living polymerization in emulsion [5].

D.1. Nitroxide Mediated Polymerization

As mentioned earlier, the initially developed nitroxides like TEMPO were suitable mainly for styrene and also required the use of higher polymerization temperatures, but a number of nitroxides have been further developed that can also be used for the polymerization of acrylates and methacrylates, as well as allow use at much lower reaction temperatures. The reactions initially carried out with styrene in emulsion led to poor colloidal stability, which resulted in a large amount of coagulum generated in the polymerization reactions. It was claimed that the particle nucleation, as well as polymerization in droplets, were among the reasons for this behavior. The seed method has also been described for emulsion polymerization with SG1 as nitroxide. In this case, a seed is generated first with low solid content, and the seed particles are then swollen in monomer and followed by subsequent polymerization of these seed particles. This helps to avoid the generation of monomer droplets and thus polymerization in droplets. Figure 25 represents this method to generate seed and subsequent latex particles with SG1 as nitroxide and styrene as monomer. In this process reported in the literature [5], SG1 nitroxide was used in the form of a water soluble alkoxyamine which was named MAMA-Na. To proceed with such a process, an aqueous emulsion of monomer and surfactant was first generated and heated at high temperature. To this emulsion MAMA-Na was added, which then initiated the polymerization reaction. The polymerization is allowed to proceed for a certain period of time in order to achieve the required seed characterizations, like particle size and conversion. To perform the polymerization further, the seed was reheated in a reactor to the required temperature, and a batch of the monomer was fed to the heated latex. This reinitiated the polymerization reaction. One has to be careful that the amount of monomer added in this step should be carefully adjusted so as to swell the seed particles and to avoid forming monomer droplets, otherwise the whole equilibrium of the polymerization would be lost. The authors also showed that instead of two separate steps to achieve the emulsion polymerization, one can achieve the same reaction in one step. In this process, the seed is first formed as earlier, but the seed latex is not cooled, and a certain amount of monomer is added after a certain characteristic size and conversion in the seed latex has been achieved. The reaction can then be performed until the high conversions. One must also carefully monitor the progress of the reaction especially at high conversions, as at very high conversions, chains start to terminate each other and the polydispersity in the chain length as well as molecular weight increases. Therefore, it is always beneficial to stop the polymerization reaction a little below the full conversion. The latexes achieved by the above-mentioned seeded latex emulsion polymerization process were very stable and did not have the problem of secondary nucleation or the droplet polymerization. As a result, the particle size distributions were in control and no coagulum was generated by the unstable polymer particles formed from monomer droplets. It was also confirmed that the alkoxyamines based on SG1 are more optimally operatable to achieve controlled polymerization of a wide range of monomers in comparison with TEMPO nitroxide. It was also shown in these studies that a diamine can also be generated based on SG1 nitroxide, and it was found to form more controlled polymers in comparison with the monoamine. One very important difference between these amines was the size of the resulting polymer particles. Diamine nitroxides were able to generate smaller particle sizes in comparison with monoamine.

D.2. Atom Transfer Radical Polymerization

Atom transfer radical polymerization in emulsion has faced problems similar to those of nitroxide mediated polymerization, as in the generation of coagulum. Figure 26 shows an example of polymerization of 2-ethylhexylmethacrylate by ATRP when non-ionic surfactants were used. Ethyl 2-bromoisobutyrate was used as ATRP initiator and copper bromide was complexed with 4,4′-dinonyl-2,2′-bipyridyl to form the catalyst system [6]. Non-ionic surfactants like Tween 85 can be used. The chemical structure of this surfactant has been described in Figure 27. The polymerization was achieved by first mixing together copper salts with 4,4′-dinonyl-2,2′-bipyridyl, to which the monomer was added. The solution was allowed to mix and surfactant was added. To this solution, water was added under vigorous stirring to form the emulsion. To this emulsion was then added the initiator to initiate the polymerization reaction. The reaction conditions needed to be very accurately controlled, and the reaction was overall very sensitive to minor changes in the reaction parameters. A similar seeded polymerization was also used in this case. One must be careful that the commonly-used surfactant in the emulsion polymerization of sodium dodecyl sulphate is not suitable for use in ATRP, as it reacts with copper (II) bromide to change it to copper (II) sulphate. This results in losing control over the deactivation process of the reaction, and thus the chains grow without control, leading to broad molecular weights and chain lengths. Therefore, in emulsion polymerization with ATRP, one must use nonionic surfactants, although they also are not completely successful in removing the large amount of coagulum formed during the polymerization. In the study mentioned above, the authors also performed ATR polymerization in the presence of cationic surfactants apart from non-ionic surfactants. Dodecyl trimethyl ammonium bromide and myristyl trimethyl ammonium bromide were used as cationic surfactants, and their effect on the latex stability, amount of coagulum and the polydispersity in the molecular weight was quantified. The first surfactant, though, allowed good control of polydispersity; however, the whole system was observed to coagulate after the initiation of polymerization. In the case of a second surfactant, the latex stability was better, but the polydispersity in the molecular weight or chain lengths was very high. It indicated that, one way or another, the cationic surfactants were not suitable for use in the emulsion polymerization by ATRP. The authors also reported that there exists a critical or threshold value of the surfactant below which the system suffers a significant amount of coagulum formation. The values of the surfactant required for the stability were quite high, e.g., the use of Tween 80 in the amount above 15.4% of the weight of monomer was suggested. The amount of surfactant required to achieve a respectable amount of colloidal stability in this case is thus quite high in comparison with the conventional emulsion polymerization, and can be a cause for concern in certain applications of the particles with such a high amount of surfactant concentration on the surface. As an example, the higher amount of surfactant present on the surface of the particles hinders the barrier performance of the films made by the fusion of the particles. The surfactant molecules, being low in molecular weight, are not a good barrier to oxygen or water, thus disturbing the capability of the particles to protect the coated surfaces. The partioning of catalyst complex with ligand is also very important for the livingness of the system. If the complex is more water soluble and thus partitions in the aqueous phase, it would lead to less control over the generation of a balance between the dormant and active species. On the other hand, if more hydrophobic complex is created so that it does partition in the organic phase, it would lead to a proper balance between the radicals and the chains reversibly terminated, thereby controlling the molecular weight of the polymer chains.

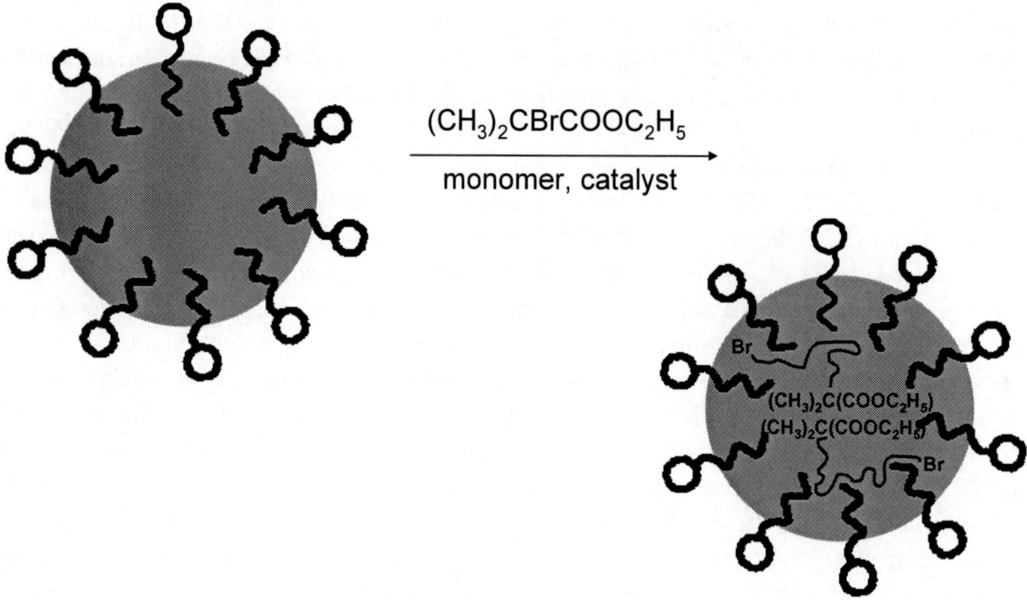

Figure 26. Use of ATRP in the emulsion system [6].

Figure 27. Chemical structure of the commonly-used non-ionic surfactant TWEEN 85.

It was also reported [7] that the addition of $CuBr_2$ was useful in controlling the polydispersity in the molecular weight of the formed chains. The particle sizes were also affected as the amount of $CuBr_2$ was increased in the system. However, after a certain amount, the latex stability was seriously affected, resulting in a vast amount of coagulum in the system. Similar studies for the emulsion polymerization of other monomers with ATRP have also been reported [8,9].

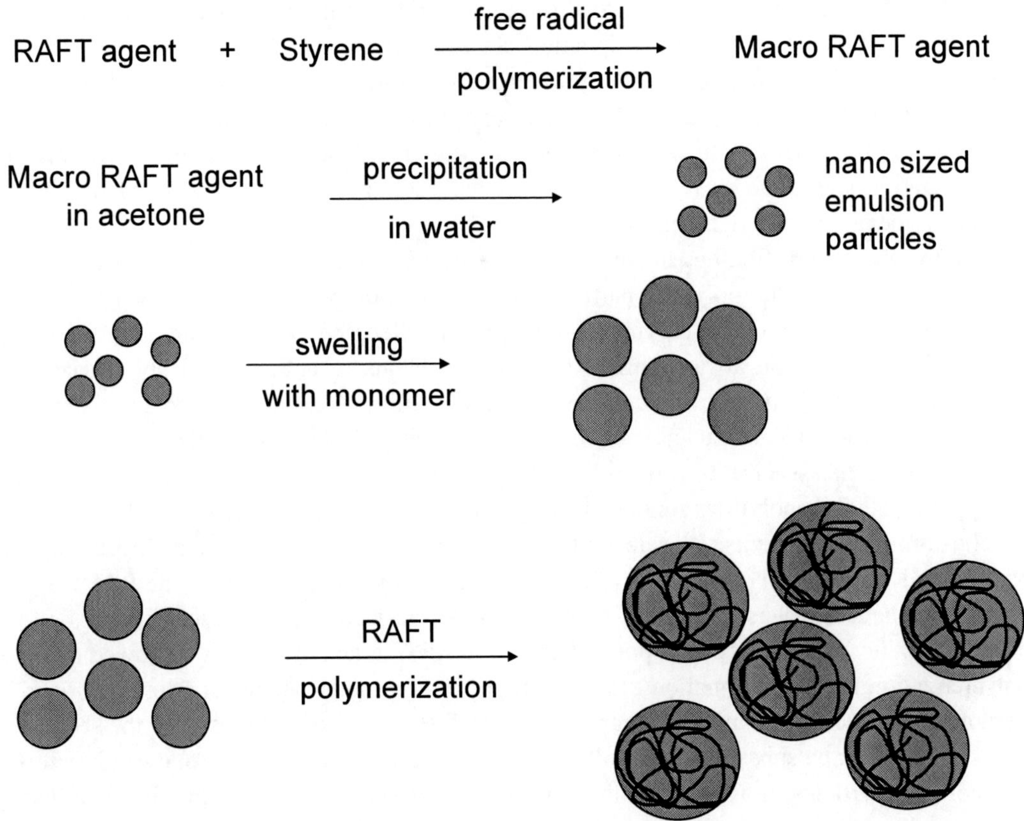

Figure 28. Nanoprecipitation process for the generation of controlled polymerization in emulsion by using RAFT system [10].

D.3. Reversible Addition Fragmentation Chain Transfer

Similar to NMP and ATRP, initial trials with RAFT also were faced with difficulties in coagulum generation. The RAFT agent was difficult to transport to the polymer particles through the aqueous phase. It was also observed that the water solubilities of the RAFT agent as well as the initiators were related in a complex way and affected the particle size, molecular weight, polymerization rate as well as polydispersity in the chain length or molecular weight. In general, the control of the polymerization reaction was not very strong and the molecular weight distributions were broad. It was also observed that, apart from the solubility of the RAFT agent in water, the solubility of the monomer also affects the diffusion of the RAFT agent from the monomer droplets to the polymer particles. Seeded polymerization was also used in such circumstances, an example of which is depicted in Figure 28 for the polymerization of styrene. The process named nanoprecipitation process was carried out by forming nano-sized particles by the precipitation of the acetone solution of the macro RAFT agent in the aqueous poly(vinyl alcohol) solution [10]. The macro RAFT agent was prepared by conventional free radical polymerization. The formed nanoparticles were subsequently swollen with monomer and were polymerized in the living manner. This nanoprecipitation method was also named seeded polymerization, because in this case the nano-sized particles formed by precipitation act as seeds to form polymer particles. Both water- and oil-soluble initiators were used. When the initiator used was oil soluble, it was premixed with the RAFT agent. On the other hand, the water soluble initiator was dissolved in PVA solution. Both with water-soluble as well as oil-soluble initiators, the rate of polymerization was quite slow, and increasing the reaction temperature was not helpful in increasing the rate of polymerization.

In any case, the difficulties in partitioning the RAFT agent in the aqueous phase, as well as desorption from the polymer particles, make the homopolymerization in the living conditions through emulsion polymerization quite challenging. Apart from that, ab initio emulsion polymerization always suffers from the problems of uncontrolled polymerization, colloidal unstability and very low conversions, leading to the formation of creamy layers on the surface, indicating unreacted monomers. It was also shown by Gilbert et al. that the diffusion problem associated with the RAFT agent can be totally eliminated by the use of amphipathic block copolymers functionalized with the desired RAFT agent, which also act as micelle-forming surfactants. This leads to the presence of the RAFT agent directly in the particle and thus the polymerization rate is not affected by the diffusion of the RAFT agent through the aqueous phase. The RAFT agent was synthesized by the polymerization first of the hydrophilic monomer end-capped with the RAFT agent. This was followed by the polymerization or block formation of the hydrophobic monomer which, as a result, formed a rigid micelle which automatically contains the RAFT agent; these micelles form the seed [11-13]. The seed can be subsequently swollen with the hydrophobic monomer and can be grown into polymer particles. It also avoids the generation of monomer droplets and thus eliminates the possibility of polymerization in monomer droplets and the generation of coagulum.

REFERENCES

[1] G. Odian (2004). *Principles of Polymerization, Fourth Edition*, John Wiley & Sons, Inc., New Jersey.

[2] K. Matyjaszewski and T. P. Davis (2002). Editors, *Handbook of Radical Polymerization*, John Wiley & Sons, Inc., New Jersey.

[3] J. N. Kizhakkedathu, A. Takacs-Cox and D. E. Brooks (2002). *Macromolecules*, 35, 4247–4257.

[4] V. Mittal, N. B. Matsko, A. Butté and M. Morbidelli (2007). *European Polymer Journal*, 43, 4868-4881.

[5] J. Nicolas, B. Charleux and S. Magnet (2006). *Journal of Polymer Science, Part A: Polymer Chemistry*, 44, 4142–4153.

[6] H. Eslami and S. Zhu (2005). *Polymer*, 46, 5484–5493.

[7] M. F. Cunningham (2008). *Progress in Polymer Science*, 33, 365-398.

[8] E. Makino, T. E. Tokunaga and E. Hogen (1998). *Polymer Preprints*, 39, 288.

[9] S. G. Gaynor, J. Qiu and K. Matyjaszewski (1998). *Macromolecules*, 31, 5951-5954.

[10] A.R. Szkurhan, T. Kasahara and M. K. Georges (2006). *Journal of Polymer Science*, Part A: Polymer Chemistry, 44, 5708–5718.

[11] R. G. Gilbert (2006). *Macromolecules*, 39, 4256–4258.

[12] C. J. Ferguson, R. J. Hughes, B. T. T. Pham, B. S. Hawkett, R. G. Gilbert, and A. K. Serelis (2002). *Macromolecules*, 35, 9243–9245.
[13] C. J. Ferguson, R. J. Hughes, D. Nguyen, B. T. T. Pham, R. G. Gilbert, and A. K. Serelis (2005). *Macromolecules,* 38, 2191–2204.

Chapter 5

LIVING AND NON-LIVING MINIEMULSION POLYMERIZATION

A. INTRODUCTION

Miniemulsion polymerization has generated tremendous research interest in recent years. In this polymerization, it is actually the monomer droplets that are the loci of polymerization. The monomer droplets, much smaller in this case, are generated by the addition of a hydrophobe, also known as costabilizer, along with the surfactant and the use of a high shear to generate the droplets [1-3]. Commonly-used costabilizers are hexadecane or cetyl alcohol. The costabilizer is required to be very hydrophobic to prevent the collapse of small droplets of monomers. The amount of surfactant is generally below the critical micelle concentration to avoid the micellar nucleation. To some extent micellar nucleation still occurs, but it is negligible compared to the extent of droplet polymerization. The water soluble initiators are generally used, and the radicals generated by the decomposition of initiator enter the monomer droplets and start the particle nucleation and subsequent polymerization. The resulting particle size in the miniemulsion polymerization is in the range of 50–500 nm, which is similar to the size of monomer droplets generated in the beginning. Most of the surfactant is present on the surface of these small monomer droplets, thus providing them with colloidal stability.

B. MINIEMULSION, MICROEMULSION AND INVERSE MINIEMULSION

Miniemulsion polymerization has several advantages over conventional emulsion polymerization, the most prominent being the elimination of the need of the monomer diffusion through the aqueous phase. Because the monomer droplets are directly polymerized, it is therefore not required to feed the monomer to polymer particles by diffusion through the aqueous phase. This can solve many of the problems faced in emulsion polymerization of highly hydrophobic monomers. Another major advantage of miniemulsion polymerization over the conventional emulsion polymerization is in terms of nucleation. Conventional emulsion polymerization is carried out by micellar nucleation, which is extremely sensitive to a large number of factors such as amount of surfactant, amount of initiator, agitation speed,

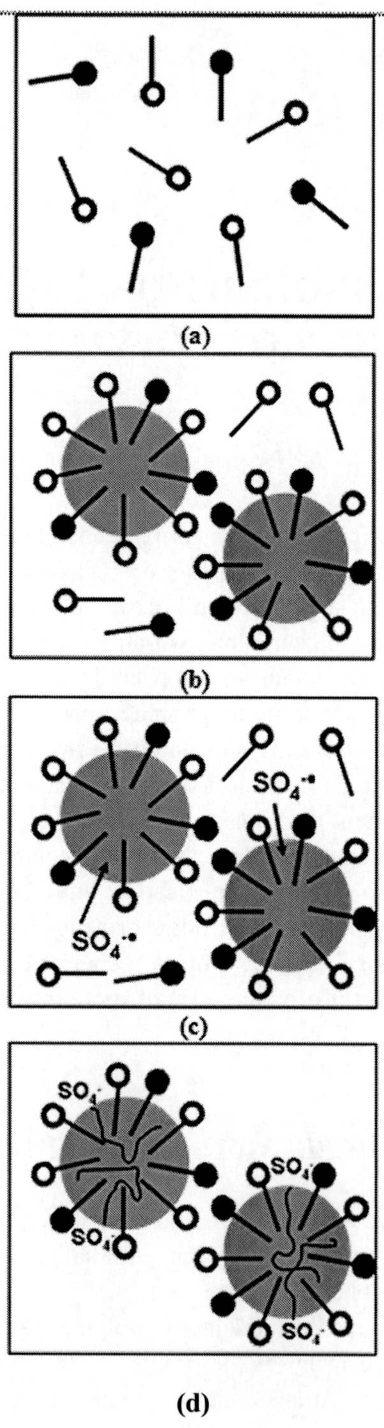

Figure 1. Process of miniemulsion polymerization. The molecules with open and filled circles represent the surfactant and costabilizer, respectively.

temperature of the polymerization reaction, mode of addition of the monomers, etc. The number of final particles in miniemulsion polymerization is dependent on the amount of surfactant and the costabilizer along with shear, but is independent of the initiation system, thus making the control of the particle numbers in miniemulsion polymerization much easier. Miniemulsion polymerization has also been quite advantageous in the living polymerization systems in comparison with the conventional emulsion polymerization. The various living polymerization methods like nitroxide mediated polymerization (NMP), atom transfer radical polymerization (ATRP) and reversible addition fragmentation chain transfer (RAFT) have been shown to be beneficial in miniemulsion polymerization to generate specialty polymers or polymers with a special architecture like block copolymers. A significant advancement has been achieved, therefore, in these living polymerization methods in miniemulsion polymerization. The colloidal stability is also much better in miniemulsion polymerization as compared to the conventional emulsion polymerization, which makes it a technique of choice.

Figure 1 explains in detail how a miniemulsion is achieved and how the particles are generated by carrying out the miniemulsion polymerization. The monomer is added along with the stabilizer as well as co-stabilizer, and the whole emulsion is then homogenized under high shear to break the bigger monomer droplets into monomer droplets in the size range of 10–500 nm. The size of the droplets depends obviously on the amount of stabilizers as well as the extent of sharing forces. Figures 2 and 3 show the comparison between the miniemulsion polymerization and the conventional emulsion polymerization. The conventional polymerization relies on the polymerization of the monomer in micelles to nucleate polymer particles and subsequent diffusion of the monomer from the aqueous phase to these growing polymer particles, whereas, as mentioned above, the miniemulsion polymerization has no such diffusion process. The particle size therefore increases during the polymerization in the case of conventional emulsion polymerization, whereas the particle size is more or less the same as the size of the monomer droplets in the case of miniemulsion polymerization. The rates of polymerization of the two polymerization techniques are also quite different, as seen in Figure 4. As the polymerization in the conventional emulsion polymerization is nucleated in the micelles, the rate continues to grow until the polymer particles are being formed. Once the surfactant has depleted (along with the inactive micelles), the polymer particles stop growing in number owing to the instability of the newly formed particles in the absence of any surfactant. The rate of polymerization also stops growing and becomes constant with time, and the size of the particles continues to grow following the import of more and more monomer from monomer droplets through the aqueous phase. Once the monomer droplets are also depleted, the rate of polymerization falls with time as the amount of monomer present in the polymer particles decreases with time. However, as there is no diffusion of the monomer taking place in miniemulsion polymerization and, as the droplets are directly polymerized, there is effectively no presence of a constant rate of polymerization period in miniemulsion polymerization. Figure 5 shows the SEM micrographs of polystyrene latex particles generated by miniemulsion polymerization at different magnifications. Apart from the many advantages of miniemulsion polymerization, a limitation of such polymerization is the presence of hydrophobe or costabilizer in the final latexes, which may be totally unacceptable for some applications.

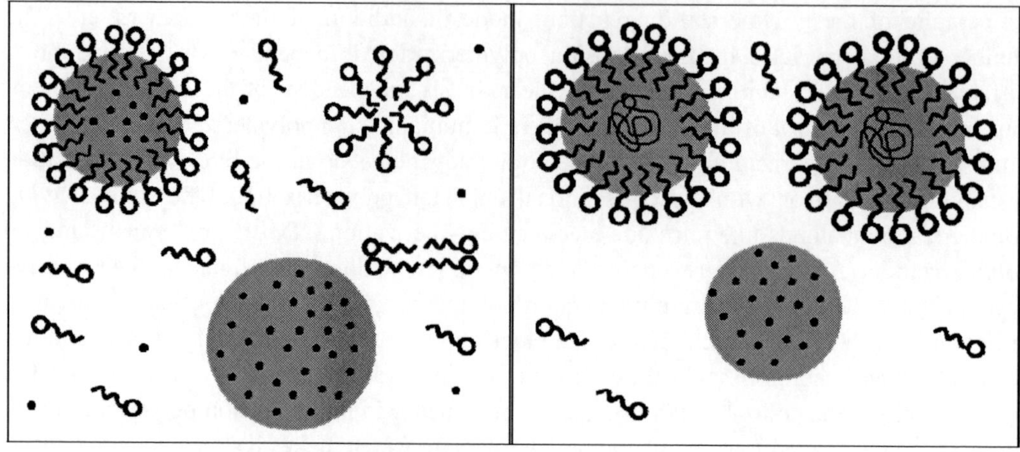

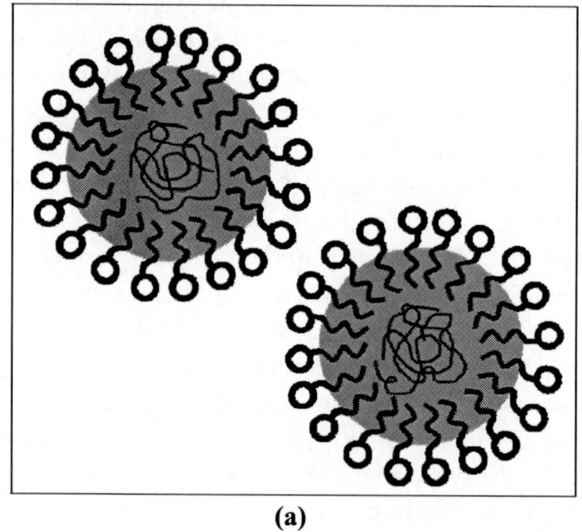

(a)

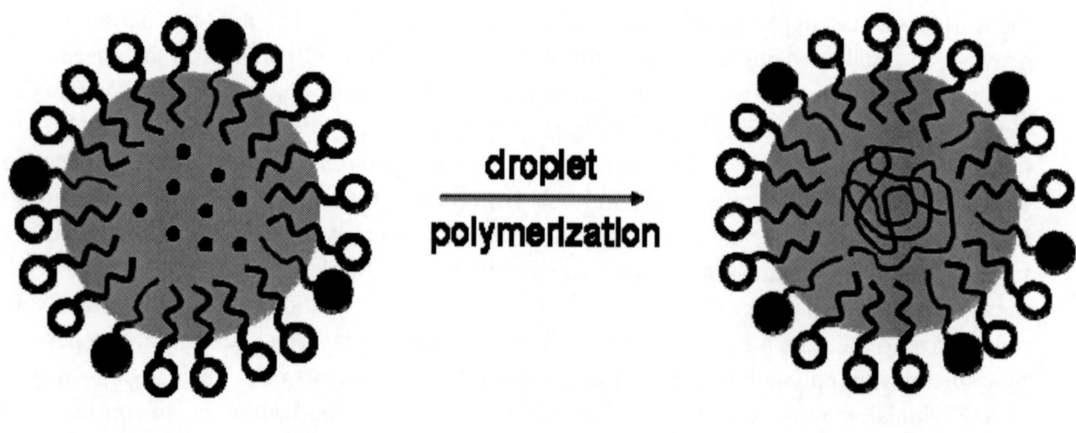

(b)

Figure 2. Comparison between (a) conventional emulsion polymerization and (b) miniemulsion polymerization.

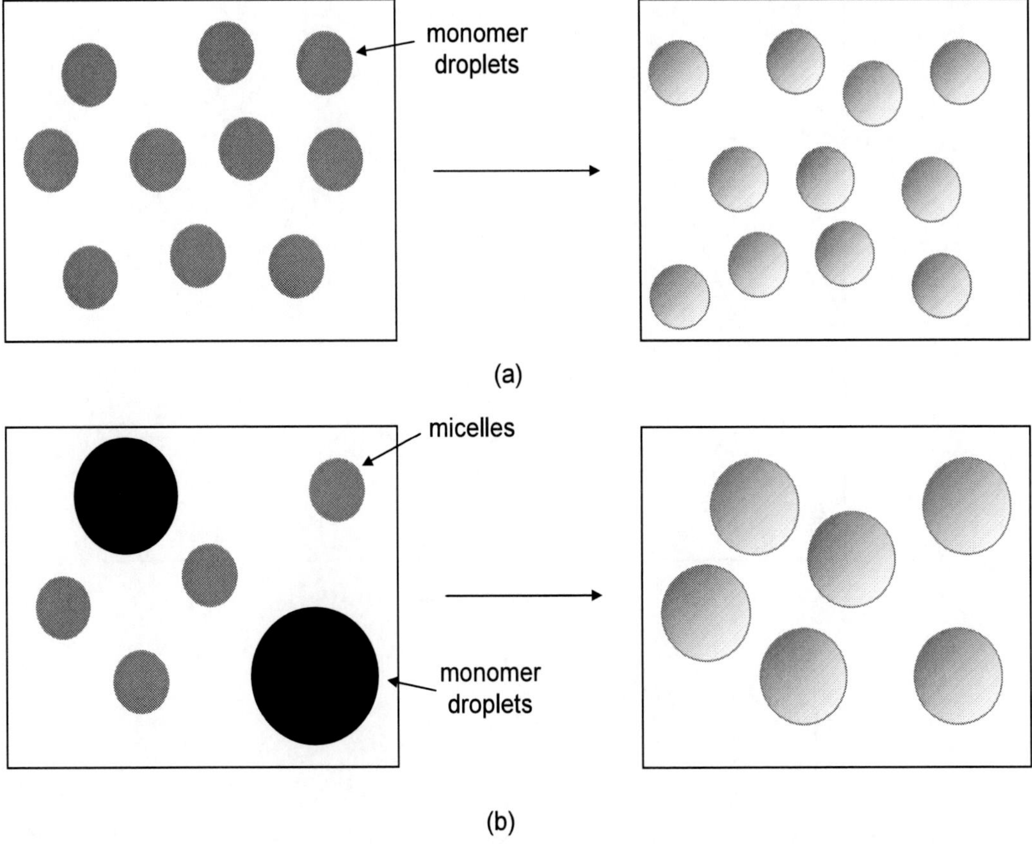

Figure 3. Another comparison between (a) miniemulsion polymerization and (b) conventional macroemulsion polymerization.

Along with miniemulsion polymerization, microemulsion polymerization has also been developed. In this special kind of polymerization, the surfactant amount is quite large, which results in the presence of all of the monomer inside the micelles. As the micelles are smaller in size and the monomer is present only in micelles, the final size of the polymer particles is also quite small (10 to 100 nm) [1,3,4]. The final latex in this case may still contain some inactive micelles that were not polymerized, as the radicals did not enter these micelles. Figure 6 shows the process of microemulsion polymerization. One should note that the rate of polymerization in this case also follows a path similar to miniemulsion polymerization, owing to the similarity in the process: in miniemulsion polymerization, the monomer droplets are polymerized, whereas in the case of microemulsion polymerization, the monomer in the micelles is polymerized. Thus, within the three different methods of emulsion polymerization, i.e., conventional emulsion polymerization, miniemulsion polymerization and microemulsion polymerization, once can achieve a wide range of particle sizes according to the requirement. Also, by operating emulsion polymerization in surfactant-free conditions, particle sizes in a much higher range can also be achieved.

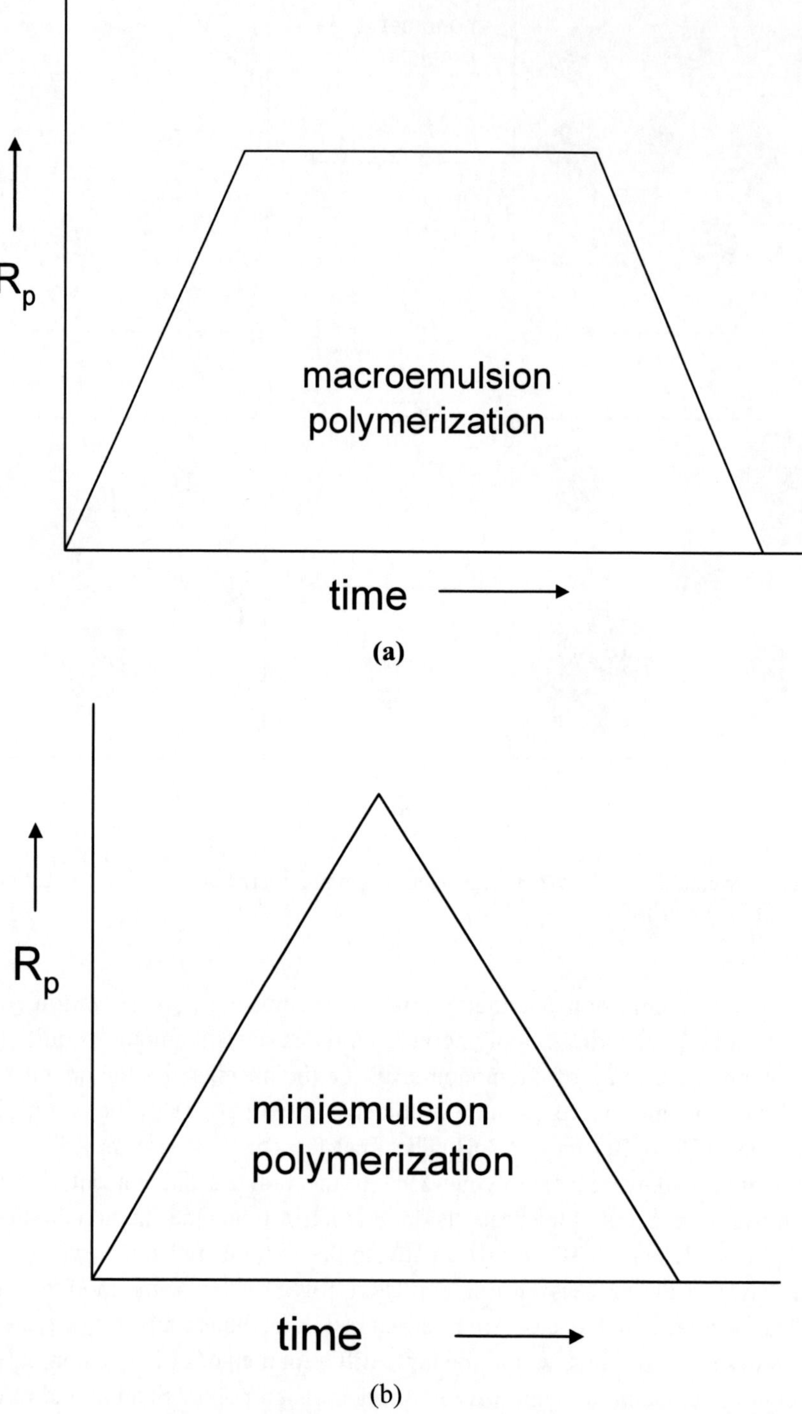

Figure 4. Kinetics of (a) macroemulsion polymerization and (b) miniemulsion polymerization.

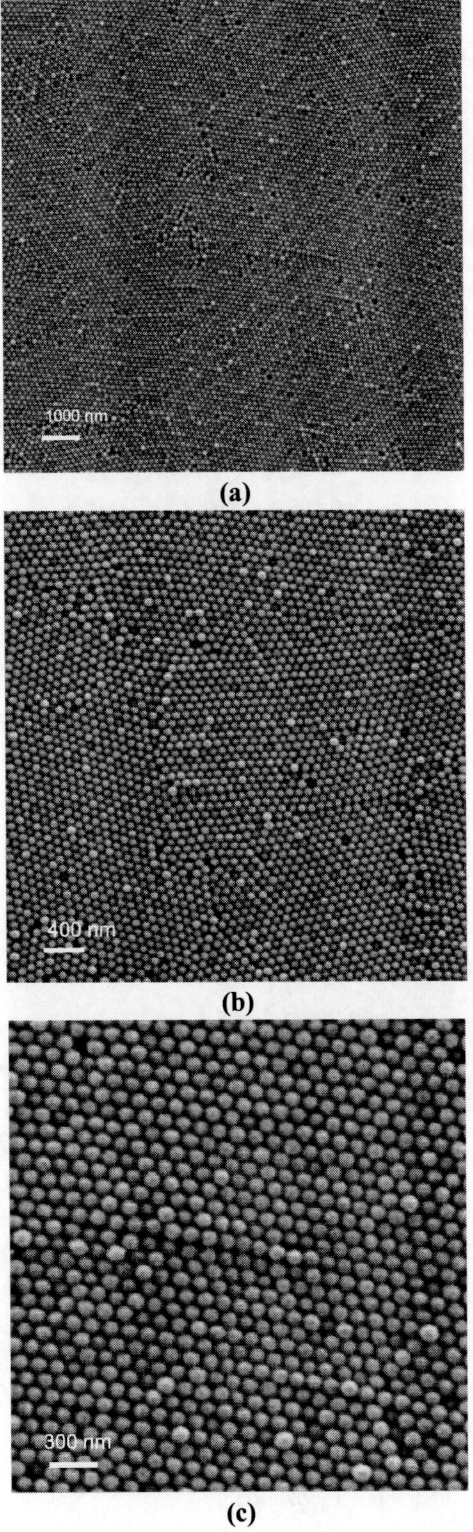

Figure 5. SEM micrographs of polystyrene latex generated by miniemulsion polymerization at different magnifications.

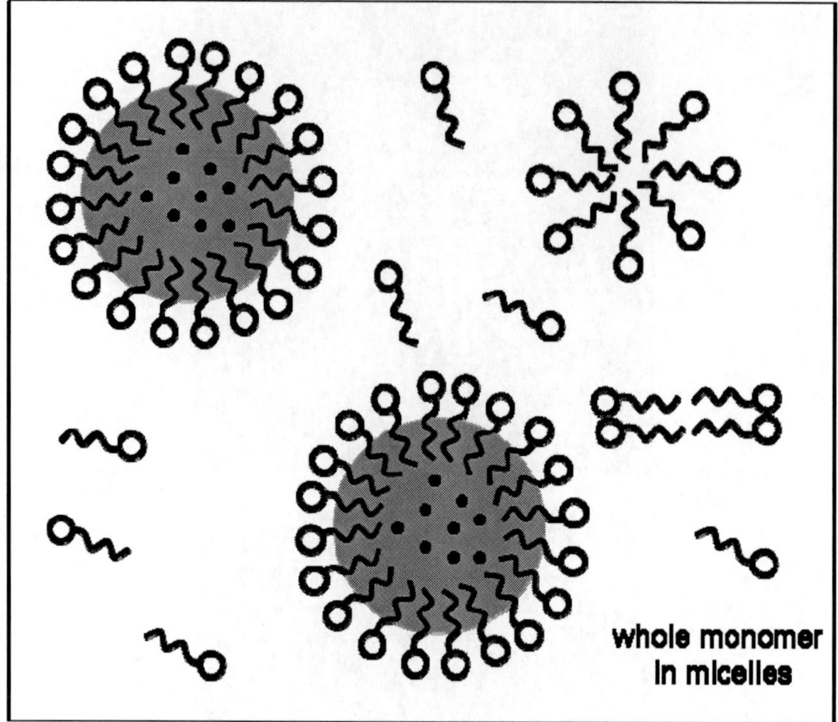

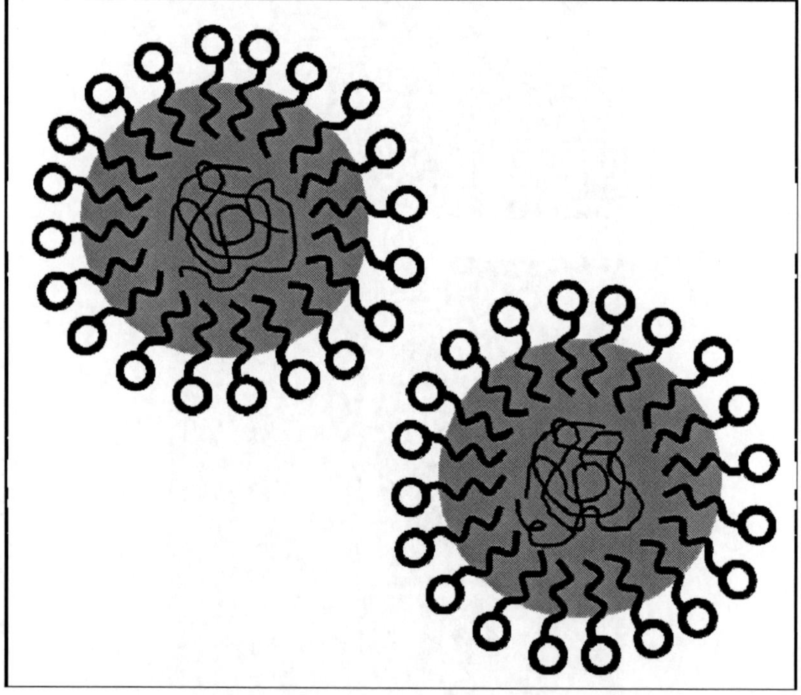

Figure 6a. Process of microemulsion polymerization.

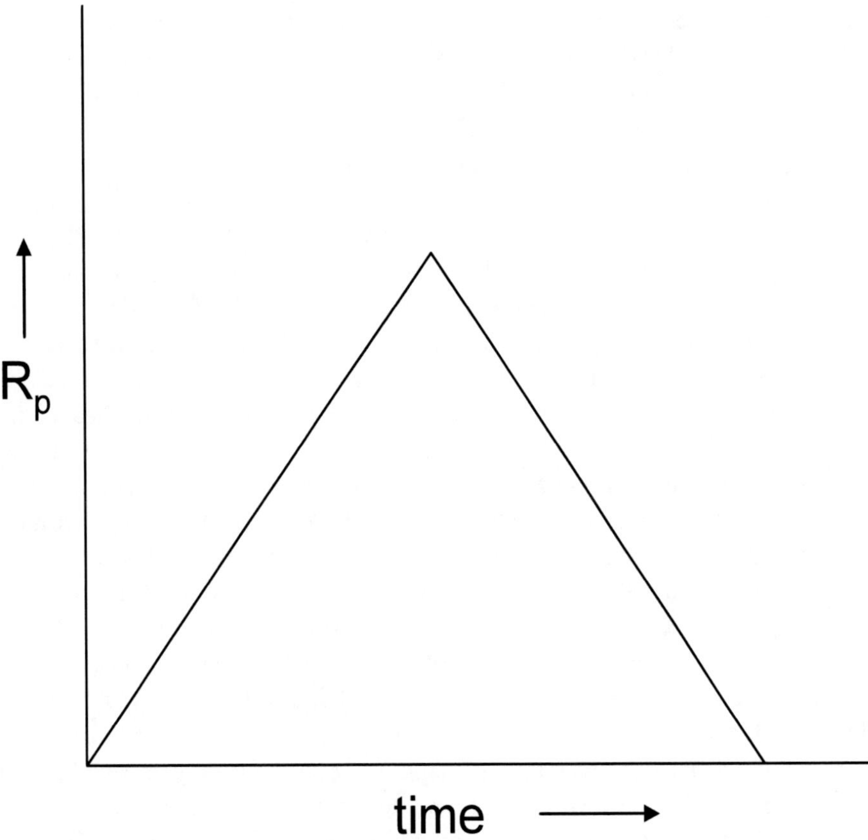

Figure 6b. Kinetics of microemulsion polymerization.

Also, in order to polymerize more hydrophilic monomers, one can use inverse miniemulsion polymerization, the process of which is compared with that of miniemulsion polymerization in Figure 7. Here, a hydrophobic reaction solvent or dispersion medium like cyclohexane is used instead of water, and the process is exactly similar to miniemulsion polymerization. Instead of a hydrophobe as a costabilizer, one must use a lipophobe such as sodium chloride, and the stabilizer is also different.

C. Non-Living Miniemulsion Polymerization

As mentioned earlier, the size of monomer droplets depends on the amount of the surfactant and costabilizer as well as the extent of mechanical shear applied to the emulsion. The costabilizer is very important, as it helps the miniemulsion retain its identity by eliminating the collapse of the small droplets to form big droplets by the Ostwald ripening process. The costabilizer, owing to its extremely hydrophobic nature, does not allow the monomer in the small droplets to diffuse out. The mechanical shear is generated by stirring, ultra-turrax or by ultrasonication. The mechanism of ultrasonication is primarily cavitation. Figure 8 represents the sonication process to generate the miniemulsions. One must be clear that the addition of a costabilizer stops the conversion of a miniemulsion into a conventional emulsion; however, the addition of a costabilizer to conventional emulsion does not automatically convert it into a miniemulsion. It is only after addition of high shearing energy that it becomes a stable miniemulsion. The costabilizer is generally required to be very hydrophobic, soluble with the monomer and with a low molecular weight. However, recent studies also used polymers as costabilizers in which a small amount of the polymer from the same monomer under polymerization is used as a costabilizer [5-7]. As these polymers are also generally water insoluble and soluble in their own monomer, they can also act as good costabilizers. This, therefore, completely eliminates the use of volatile hexadecane or other low molecular weight costabilizers in these miniemulsion polymerizations. The low molecular weight components are not desirable in the final latex, as these can easily migrate to other materials owing to their low molecular weight, thus causing health and safety concerns. In a few of such studies, the effect of hydrophobe levels and molecular weight on droplet size and size distribution of these droplets was described. The effect of the amount of surfactant on droplet size and size distributions was also quantified. It was confirmed that the addition of the monomer-soluble polymer to the miniemulsion was able to slow down the Ostwald ripening and, thus, the emulsions could be stabilized against diffusional degradation. The authors reported that by the use of polymeric costabilizer, droplet size polydispersities were near 1.023 [5-7]. Furthermore, the rate of polymerization was dependant on the amount of costabilizer used, and these miniemulsions after polymerization were observed to generate much lower polydispersities in comparison with the latexes from miniemulsions using alkanes as stabilizers. Also, by changing the amount of costabilizer, one can change the sizes of the monomer droplets. PMMA was found to be an excellent hydrophobe for MMA polymerization in these studies, and it was predicted that, based on similar lines, other systems can also be equally beneficial, such as polystyrene for styrene polymerization. It was concluded that as long as the polymer shows the two most important characteristics of a hydrophobe, i.e., water insolubility and monomer solubility, it can be used in any number of different systems. A very important advantage of the polymeric stabilizers over the other low molecular weight stabilizers was the robustness of the polymeric costabilizers in the presence of any monomer impurities or small amounts of inhibitors or retarders, etc. The miniemulsion systems with low molecular weight costabilizers tend to be sensitive to such factors, whereas the use of polymeric costabilizers makes the system almost immune to such factors or fluctuations in the system.

Much further advancement in the generation of tailor-made costabilizers have been achieved. An example is the use of monomeric costabilizers, which can also be polymerized along with the monomer during the course of polymerization, as shown in Figure 9 [8]. It would reinforce the costabilizer in the particle, thus removing any concerns regarding the diffusion of these compounds out of the particles. Another example is the use of initiators, which can additionally act as costabilizers, as shown by Schork et al. [9]. The use of lauroyl peroxide is an interesting example in which the peroxide is first used to create a stable miniemulsion followed by its decomposition to generate radicals to polymerize the miniemulsion [9]. Figure 10 shows this process for the miniemulsion polymerization of styrene. Chain transfer agents used as cosurfactants have also been reported [10].

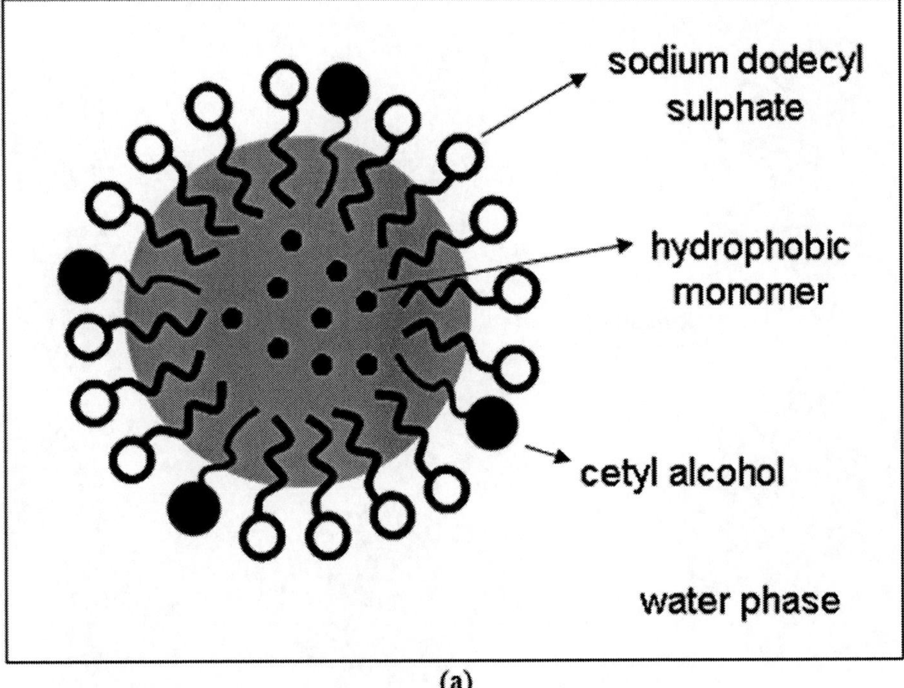

(a)

(b)

Figure 7. Comparison of (a) miniemulsion polymerization and (b) inverse miniemulsion polymerization.

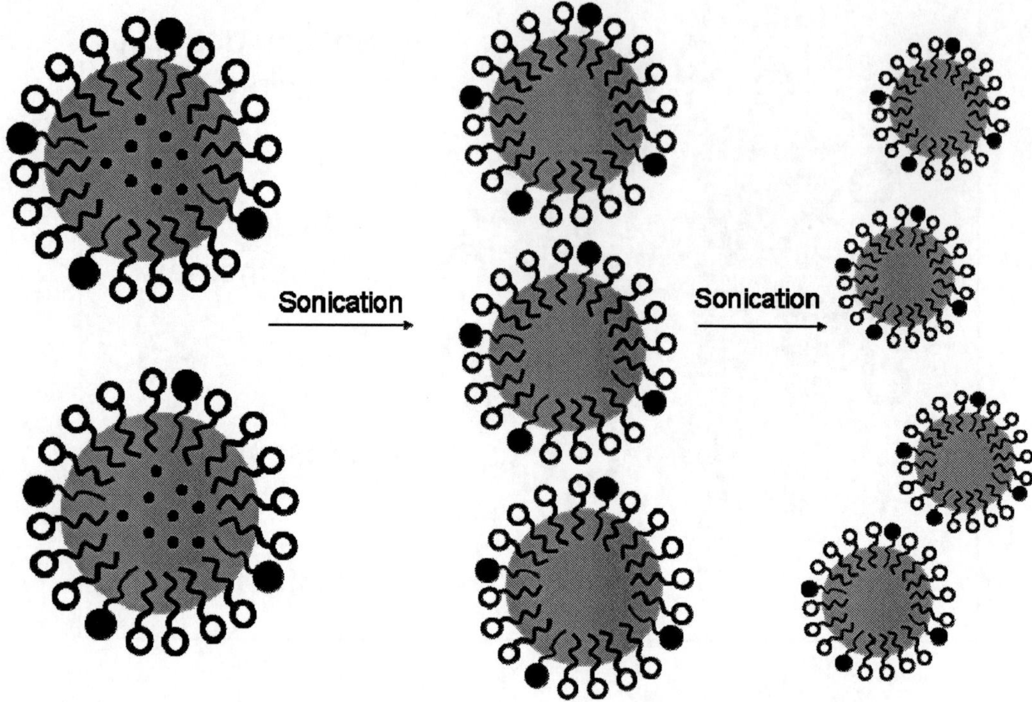

Figure 8. Use of ultrasonication for the generation of smaller monomer droplets.

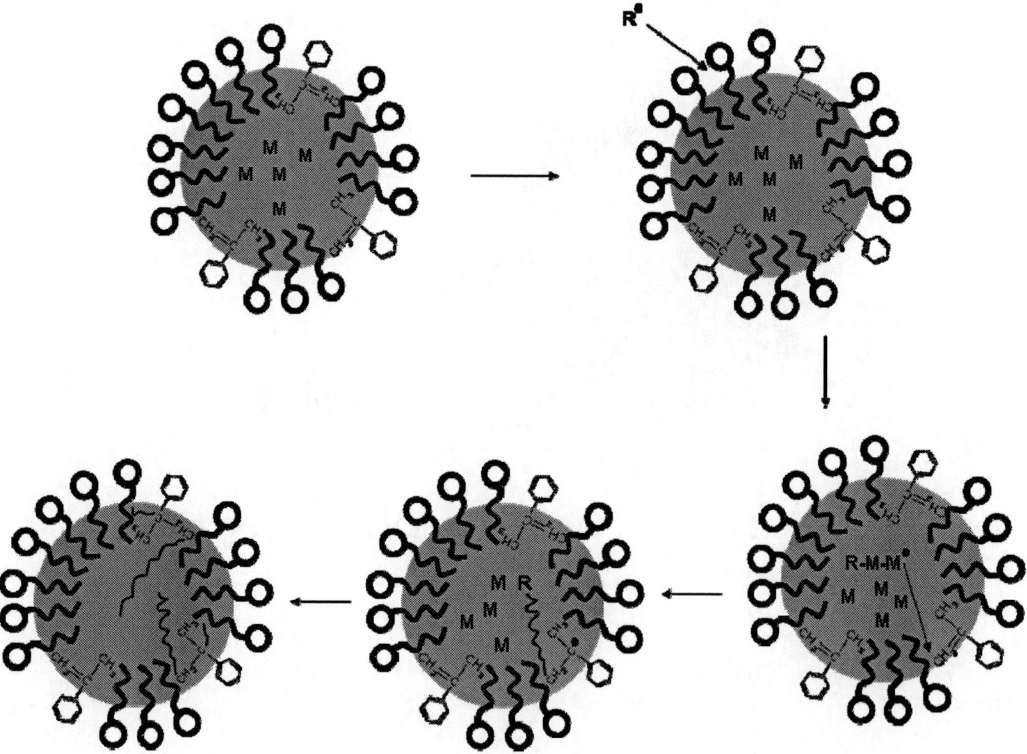

Figure 9. Use of comonomer as a costabilizer [8].

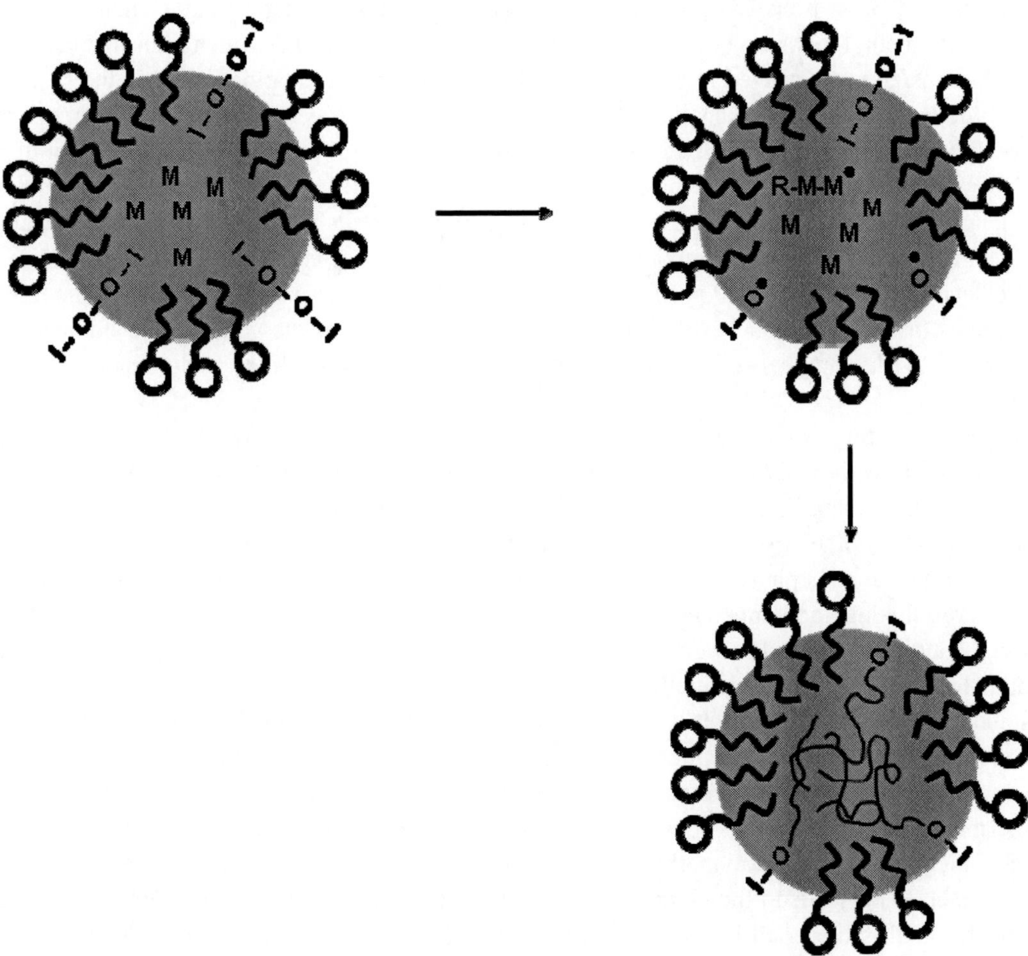

Figure 10. Use of radical initiator as costabilizer in the miniemulsion polymerization [9].

Dodecylmercaptan was used as a chain transfer agent, along with sodium lauryl sulfate as surfactant and potassium persulphate as initiator for the miniemulsion polymerization of methyl methacrylate. It was observed that the chain transfer agent was able to form stable droplets, and their sizes were comparable to the droplets formed in miniemulsions. At constant surfactant concentration, the size of the droplets was observed to decrease on increasing the concentration of costabilizer. As the surfactant concentration was below the CMC values, the droplet polymerization, therefore, dominated the particle nucleation mechanism more so than micellar or homogenous nucleation. The number of particles was observed to be dependant on the concentration of both the initiator as well as costabilizer. The rate of polymerization was observed to be dependant upon factors such as the concentration of surfactant, cosurfactant as well as initiator. The rate of polymerization is, however, independent of the concentration of surfactant if the amount of the surfactant is below CMC. One should note that for successful droplet nucleation, one should maintain the amount of surfactant below CMC, otherwise the excess surfactant would start to form micelles, which can then absorb monomer and hence initiate nucleation. It was observed by the authors that the value of chain transfer coefficient for the system of dodecylmercaptan as chain transfer agent and methyl methacrylate as monomer was in the range 0.6–0.8, which indicated that the chain transfer agent reacted less rapidly than the monomer. On the other hand, in the system with dodecylmercaptan as chain transfer agent and styrene as monomer, the chain transfer coefficient lies in the range of 15–20, indicating that it would be consumed very early in the polymerization reaction. Therefore, it would not provide stability to the forming particles for a long enough time. This, thus, leads to the loss of miniemulsion character, as the Ostwald ripening leads to the consumption of small monomer droplets to form large monomer droplets. The most important proof of droplet nucleation was observed by the size of the final particles, as its size was very close to the droplet size. Also, the ratio of the number of droplets to the number of particles was near to unity, indicating predominant droplet nucleation.

In an interesting study, a number of systems studying the polymerization of butyl methacrylate with different crosslinking monomers were reported [11]. The monomers were hydrophobic; therefore, it was beneficial to achieve their polymerization through miniemulsion, otherwise, in the conventional emulsion polymerization, their diffusion through the aqueous phase would be almost impossible, thus hindering their polymerization. The crosslinking monomers used in the study were a macromonomer crosslinker with a weight average molecular weight of 4,000 g/mol and low polydispersity, ethylene glycol dimethacrylate and aliphatic urethane acrylate comonomer. Both water soluble as well as oil soluble initiators were used. It was observed that the stable miniemulsions as well as latexes were formed with both initiator systems. As the reactions in the system involved copolymerization, it was therefore also remarked that the differences in the reactivity ratios of the monomers leads to the compositional drift in the polymer chains, with more reactive monomers tending to polymerize first, followed by the polymerization of less reactive monomers. This leads to the higher concentration of the more reactive monomers at the core, and the shell is then rich in polymer formed from a less reactive monomer. This process is sometimes not favorable, and a homogenous concentration of both polymers is required throughout the particles. This can then be achieved by controlling the addition of the monomers. The highly reactive monomers can be slowly added throughout the polymerization process to maintain an optimum ratio of the monomers in the system. It was also remarked that, based on the different water solubilities of the monomers, the kinetics of polymerization would be affected. As an example, the macro crosslinker was reported to be very hydrophobic; therefore, most of the butyl methacrylate was present on the surface of the particles owing to its higher water solubility in comparison with the other comonomer. This then formed the core of the particles with the macro crosslinker. The chains of this crosslinker were also observed to orient in a way that the less hydrophobic portion of these chains is oriented towards the surfaces of the particles that are more hydrophilic in nature. It was also observed that the site of the generation of the radicals was an important parameter, which affected the kinetics of polymerization. When a water soluble imitator was used, it produced radicals in the aqueous phase and reacted with the butyl methacrylate molecules dissolved in water. As the more hydrophobic macro crosslinker molecules were present further away in the core of the particles, this system therefore led to their polymerization in the later phase of reaction. This led to the generation of a homopolymer of butyl methacrylate to large extent on the shell of the particles.

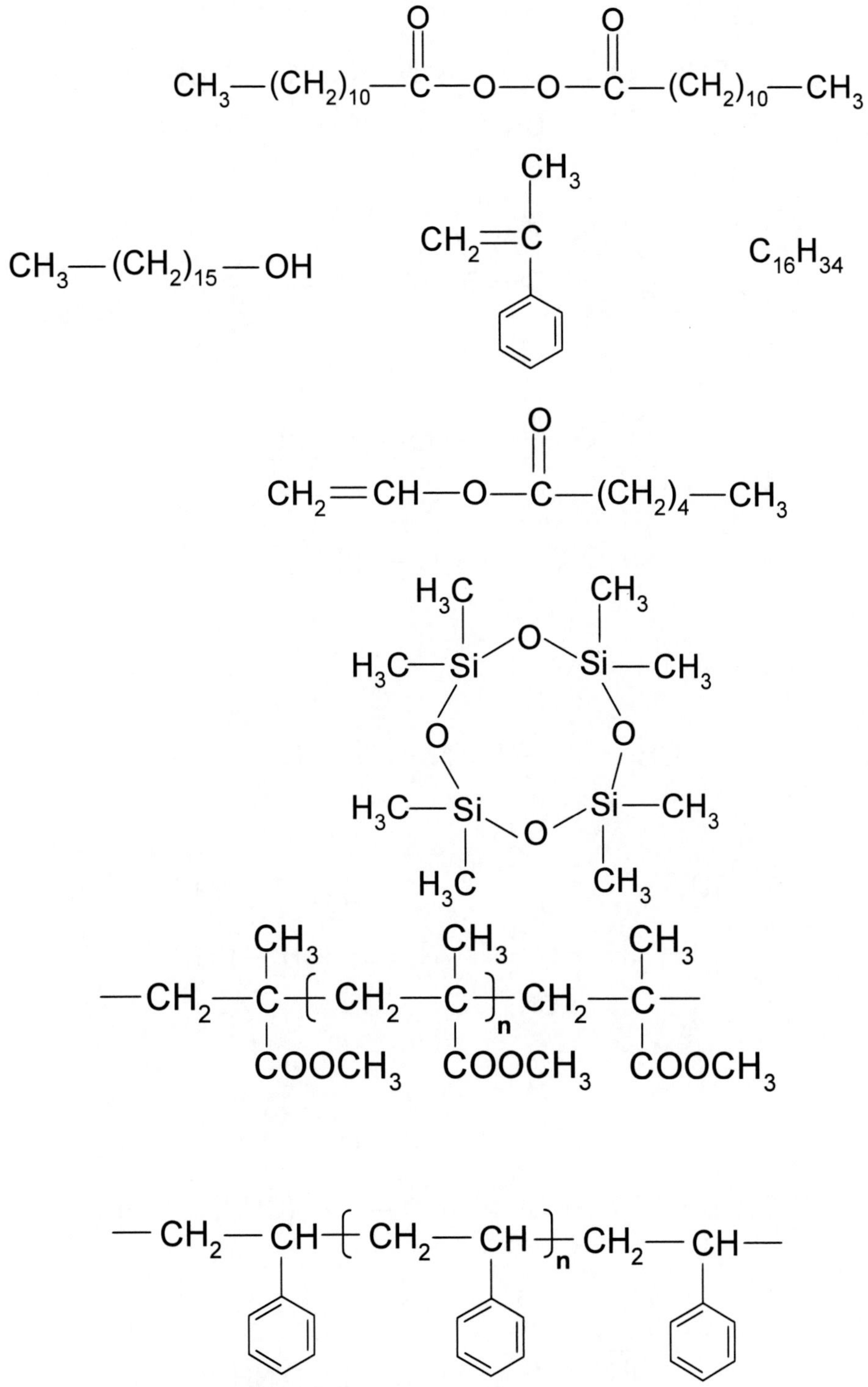

$$CH_3-(CH_2)_{10}-C(=O)-O-O-C(=O)-(CH_2)_{10}-CH_3$$

$$CH_3-(CH_2)_{15}-OH \qquad CH_2=C(CH_3)-C_6H_5 \qquad C_{16}H_{34}$$

$$CH_2=CH-O-C(=O)-(CH_2)_4-CH_3$$

$$[(H_3C)_2Si-O]_4$$

$$-CH_2-C(CH_3)(COOCH_3)-[CH_2-C(CH_3)(COOCH_3)]_n-CH_2-C(CH_3)(COOCH_3)-$$

$$-CH_2-CH(C_6H_5)-[CH_2-CH(C_6H_5)]_n-CH_2-CH(C_6H_5)-$$

Figure 11. Various stabilizers used in the miniemulsion polymerization process.

$$CH_3{-}CH_2{-}C(CH_3)(CN){-}N{=}N{-}C(CH_3)(CN){-}CH_2{-}CH_3$$

$$C_6H_5{-}C(=O){-}O{-}O{-}C(=O){-}C_6H_5$$

$$K^{+}\,{}^{-}O_3S{-}O{-}O{-}SO_3^{-}\,K^{+}$$

$$Na^{+}\,{}^{-}O_3S{-}O{-}O{-}SO_3^{-}\,Na^{+}$$

$$CH_3{-}C(CH_3)(CN){-}N{=}N{-}C(CH_3)(CN){-}CH_3$$

$$(CH_3)_2CH{-}CH_2{-}C(CH_3)(CN){-}N{=}N{-}C(CH_3)(CN){-}CH_2{-}CH(CH_3)_2$$

$$CH_3{-}(CH_2)_{10}{-}C(=O){-}O{-}O{-}C(=O){-}(CH_2)_{10}{-}CH_3$$

$$NH_4^{+}\,{}^{-}O_3S{-}O{-}O{-}SO_3^{-}\,NH_4^{+}$$

Figure 12. Various initiators used in the miniemulsion polymerization process.

Thus, a number of new advancements have been reported in recent years that allow us to completely avoid the use of conventionally used costabilizers like alkyl alcohols or long chain alkanes. Figure 11 shows a number of different costabilizers reported in the literature. A large variety of water soluble and oil soluble initiators have been reported in the literature. The structures of a few of them are presented in Figure 12.

D. Miniemulsion in Controlled Living Polymerization

As mentioned above, miniemulsion polymerization is very well suited to conducting living polymerizations such as nitroxide-mediated polymerization, atom transfer radical polymerization and reversible addition fragmentation transfer polymerization. The following are examples that show the successful use of living polymerization techniques in miniemulsion:

D.1. Nitroxide-Mediated Polymerization

The kinetics of nitroxide-mediated polymerization is quite complex when the segregation effects are considered. However, these effects come into play only when the particles are very small. As miniemulsion polymerization does not require the diffusion of generally water soluble monomers or nitroxides through the aqueous phase, the polymerization therefore becomes very straightforward in comparison with conventional emulsion polymerization in the absence of the nucleation step. Also, a variety of nitroxides are available, like TEMPO or its variations, and SG1 nitroxides. Apart from that, unimolecular systems, i.e., alkoxyamines, are also used. In many cases, nitroxide-terminated oligomers are also used, which act both as initiator as well as nitroxide to generate the dormant species that are reversibly terminated by the nitroxide.

TEMPO-mediated polymerization has generally been found to be successful for styrene. In early studies, conventional free radical initiators like KPS or benzoyl peroxide were employed along with TEMPO as nitroxide and DOWFAX 8390 or sodium dodecylbenzene sulfonate as surfactants. Good control of the polymerization reaction as well as the properties of the polymer particles could be achieved. Later on, the focus shifted to the use of TEMPO terminated polymer chains [12,13]. Figure 13 is an example of such a process, in which polystyrene terminated with TEMPO is used both as initiator as well as nitroxide. Hexadecane was used as a costabilizer, whereas DOWFAX 8390 was used as a surfactant. It was observed that by changing the amount of surfactant, different particle sizes could be obtained; however, the rate of polymerization was not affected, which is a markedly different behavior in comparison with the conventional emulsion polymerization. The TEMPO terminated oligomers of polystyrene used in this case are beneficial because they lead to the proper estimation of a number of starting chains in the system and thus can lead to better control of molecular weight. Apart from that, it was also observed that TEMPO partitions between the aqueous phase along with the monomer phase, but the TEMPO-terminated polystyrene would be present only in the monomer droplets, owing to its extensive water insolubility. The TEMPO-terminated polystyrene was synthesized by the bulk polymerization of styrene with benzoyl peroxide as initiator in the presence of TEMPO. It was also observed that the polymerization rate of miniemulsion with TEMPO-terminated polystyrene as initiator was much lower than without its addition. Increasing the concentration of initiator from 5 to 20% was observed to have no impact on the rate of polymerization. The polydispersity in the molecular weight of the polymer chains synthesized by using TEMPO-terminated polystyrene was well under control and ranged from 1.2 to 1.7, whereas the polymer chains prepared without the use of nitroxide were very polydisperse, and their polydispersities ranged from 2.5 to 3.2. The final latexes were very stable but were polydisperse in the size of the polymer particles. The livingness of the system was proven by the continuous increase in the molecular weight with conversion, and the slope of the curve between the molecular weight and conversion was observed to increase on decreasing the concentration of initiating nitroxide.

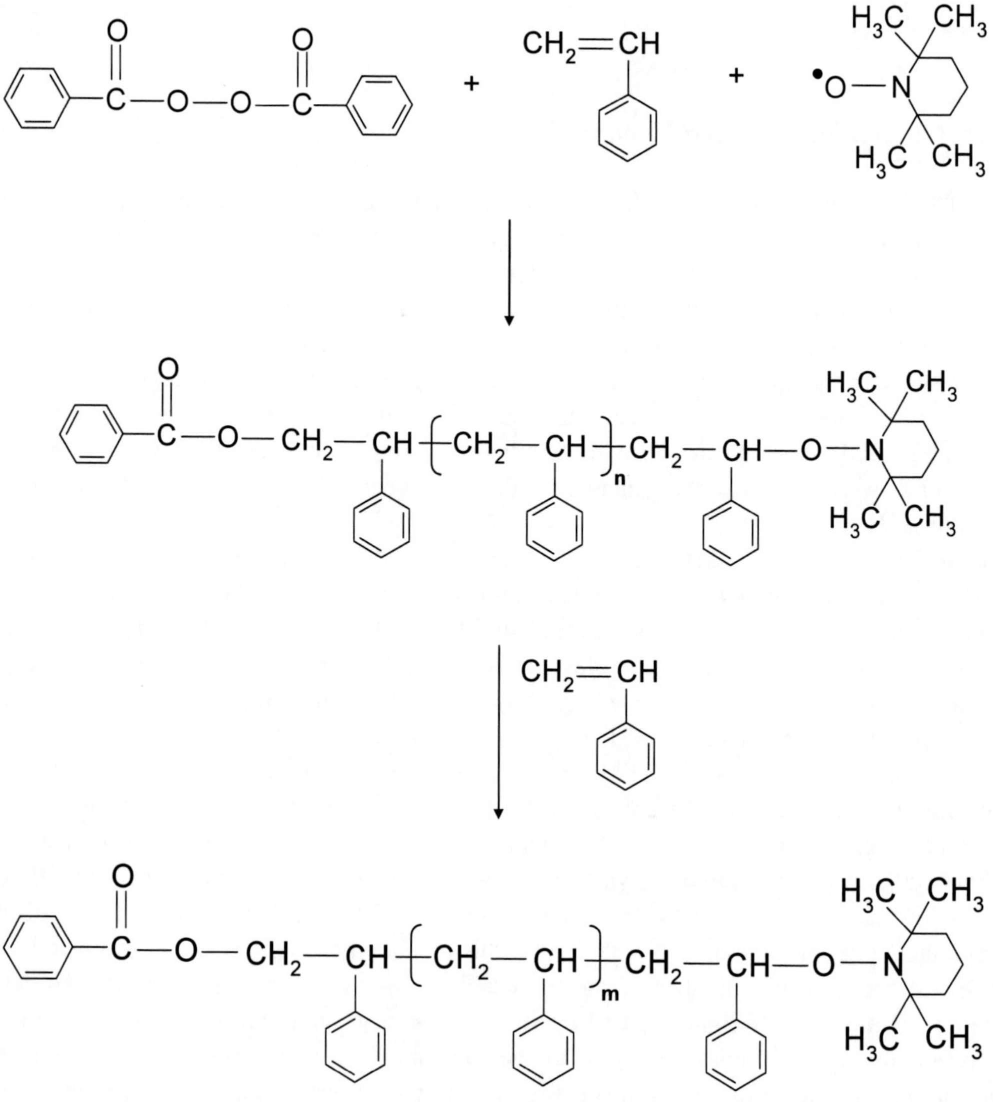

Figure 13. TEMPO-mediated polymerization of styrene.

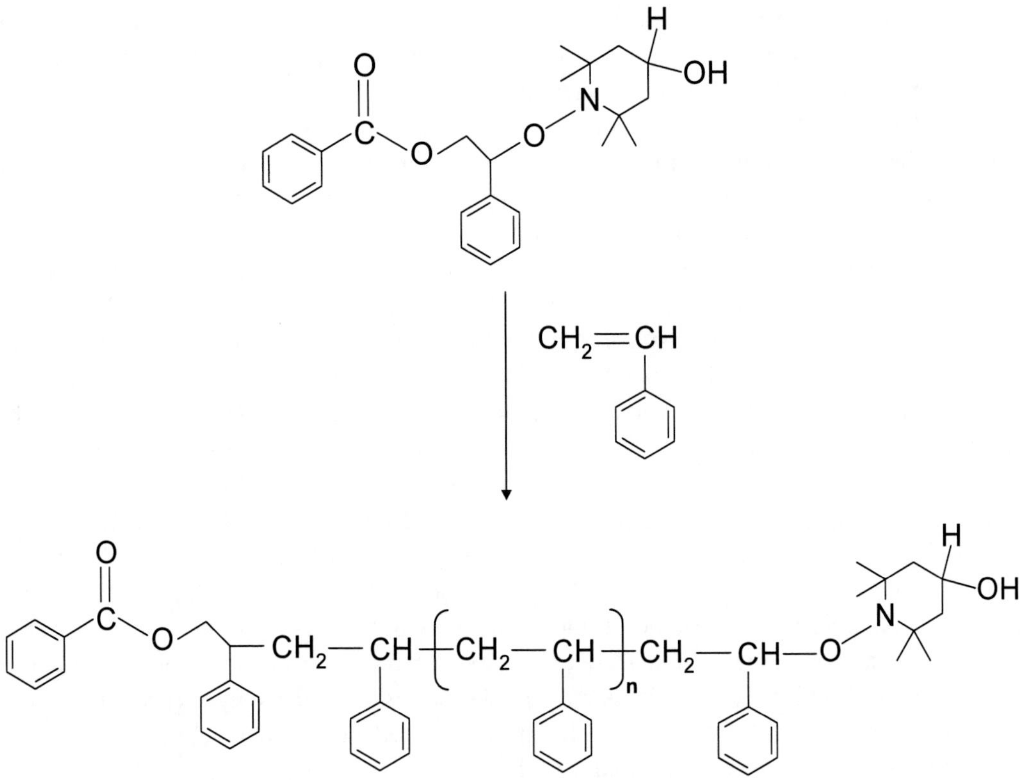

Figure 14. BST-OH nitroxide for the living polymerization of styrene [16].

Crosslinked polymer particles have also been synthesized by using miniemulsion polymerization in which a polymer capped with TEMPO was used as an initiator and tetradecane was used as hydrophobe. Dodecyl benzenesulfonic acid sodium salt was used as surfactant in these studies [14,15]. In these studies of the generation of the crosslinked polystyrene particles, a number of differences were observed in the polymerization process using the crosslinker in the miniemulsion polymerization compared to bulk or solution polymerization. The relative rate of consumption of divinylbenzene was observed to be lower in miniemulsion polymerization. This, therefore, leads to the conclusion that the number of crosslinks in the particles or crosslink density was lower in this case of particle generation by miniemulsion polymerization. It was suggested that the interface between the particle and aqueous phase played an important role in the generation of the crosslinked polymer particle network. It was suggested that the low reactivity of the vinyl groups in the miniemulsion polymerization can be useful in a commercial application in which crosslinking can be achieved in a further separate step, e.g., after the application of the particles on a particular substrate.

Apart from this, a unimolecular system has also been used, as shown in Figure 14, for BST or BST-OH alkoxyamines [16]. Styrene miniemulsion polymerizations were carried out at 135°C, and this was found to have better control over the number of chains than the two-component system. A mathematical model was also generated to predict the behavior of the nitroxide-mediated polymerization in miniemulsion using alkoxyamines and all of the factors, such as partitioning of monomer and nitroxide, particle nucleation, desorption of radicals from the particles, entry of radical inside the droplets, and polymerization reactions in the organic and aqueous phases, etc., were considered. It was observed that the experimental results agreed well with the model predictions and, thus, the model provided a better insight into the kinetics and mechanism of the polymerization. It was observed that the partioning of the nitroxides was significant in its effect on polymer kinetics in the absence of thermal initiation of the styrene. The model also predicted that the livingness of the system continued to decrease throughout the course of polymerization, owing to the termination of the radicals.

As TEMPO nitroxide is limited in the number of polymerizable monomers and requires high polymerization temperatures, other nitroxides, therefore, have also been developed. One such nitroxide that has been most successful is SG1 nitroxide, which is more flexible in the monomers that can be polymerized, and it also can be used at lower polymerization temperatures [17,18]. It was shown by Farcet et al. that butyl acrylate can be polymerized by using SG1 as nitroxide. In fact, they observed that a molar ratio of approximately 5 between the SG1 and the alkoxyamine carrying SG1 in the polymerization reaction provided much better control of the polymerization reaction. Figure 15 shows the miniemulsion polymerization of butyl acrylate [17,18]. The living nature of these chains could also be confirmed by various analytical characterizations, as one end of the polymer chains was found to contain SG1 nitroxide, indicating that the polymer chains were accurately controlled throughout the polymerization reaction. The polymerization reactions were observed to generate stable latexes and no coagulum was observed during the polymerization reaction, indicating that the initial problem of significant coagulum generation in the living polymerizations had been eliminated by the use of specialty nitroxides. However, as mentioned earlier, the particle size distribution obtained with miniemulsion polymerization was quite broad. The molecular weight distribution of the polymer chains was very well controlled, and the dispersities in the molecular weight were low. Both the alkoxyamine based on SG1 nitroxide as well as free SG1 nitroxide were used in the polymerization of butyl acrylate where polystyrene and hexadecane were used as hydrophobes. It was also observed that the initiator efficiency was less than 1 at the reaction temperature of 112°C, but the efficiency could be significantly improved when either the polymerization reaction was raised to 125°C or the macroinitiator was used, e.g., poly(butyl acrylate) end capped with SG1. The authors also reported the polymerization of styrene by using the similar miniemulsion system. Compared to the more controlled polymerization system in the case of butyl acrylate, the styrene polymerization was observed to be less controlled, owing to the low initiator efficiency because of the pronounced persistent radical effect leading to a large concentration of dead oligomer chains and high concentration of free SG1. It resulted in the too-slow growth of the polymer chains in the aqueous phase, which translated into poor entry rate of the radicals in the monomer droplets. However, it was observed that the addition of small amounts of comonomers like acrylates enhanced the entry rate of the radicals in the monomer droplets, owing to the generation of a favorable kinetic environment of copolymerization that allowed the reduction of the persistent radical effect, and the propagation rate of the radicals in the aqueous phase was enhanced. It was also reported by the authors that the pH of the reaction medium is also of great consequence in the performance of nitroxide, and it was concluded that the pH should be higher than 5 in these miniemulsion polymerizations in order to eliminate the unwanted side reactions of SG1 nitroxide.

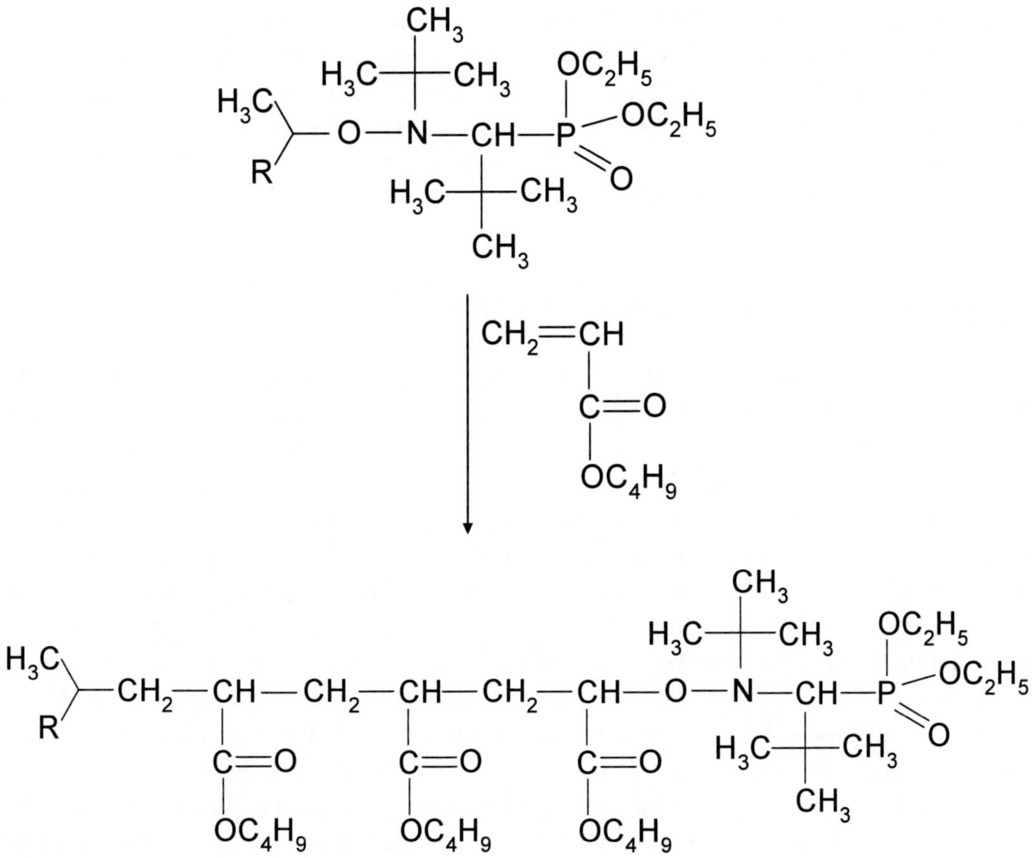

Figure 15. SG1 nitroxide-mediated polymerization of butyl acrylate [17,18].

A number of other nitroxides have also been used, e.g., as shown for the polymerization of butyl acrylate by Keoshkerian et al. [19]. In this case, the nitroxide was first reacted with styrene in the presence of benzoyl peroxide conventionally at 135°C to form polymer chains end-terminated with nitroxide [19]. This compound was subsequently used to polymerize butyl acrylate in miniemulsion polymerization. In this study, hexadecane was used as hydrophobe and sodium dodecyl benzenesulfonate was used as surfactant. The reactions were also carried out with the alkoxyamine based on TEMPO, and the comparison of the resulting molecular properties of the polymer chains was carried out with the polymer chains obtained by using the other nitroxide. The polydispersities in the molecular weight were higher when the TEMPO-based nitroxide was used, but it was confirmed that TEMPO-based alkoxyamine can also be used to achieve living polymers with acrylates, and the TEMPO system is not limited to the polymerization of styrene only. The growth of the chains in the case of TEMPO-based nitroxide system preceded faster in the beginning of polymerization, but settled to a steadier pace later.

Both monomer soluble as well as water soluble nitroxides have been used. On many occasions, water soluble nitroxides have been observed to be more successful than the monomer soluble nitroxides. One should always be careful to maintain equilibrium between the reduction of termination by increasing the concentration of nitroxide in the system (and thus by increasing the reaction time) and the increased rate of termination by propagation when the reaction times are larger. There have also been a number of theoretical studies on this subject of nitroxide-mediated polymerization to completely understand the complex kinetics. Molecular weight distributions, degree of livingness, control of polydispersity, etc., have been thereby calculated for different nitroxides. It was also suggested in these theoretical studies that by increasing the volume fraction of water in miniemulsion polymerization, the livingness of the system can be improved.

D.2. Atom Transfer Radical Polymerization

ATRP processes are to a great extent similar to stable free radical polymerization like nitroxide-mediated polymerization. As mentioned earlier, the colloidal stability is also a concern when the ATRP process is used in miniemulsion, as it tends to react with the ionic surfactant generally used in miniemulsion polymerization, thus requiring the use of non-ionic surfactants. A great deal of process development has been reported for the atom transfer radical polymerization in miniemulsion. A number of studies were reported applying the direct or forward ATRP in miniemulsion, but reverse ATRP, in which a conventional free radical polymerization initiator like AIBN can be used with the transition metal compound in its higher oxidation state, was observed to be more suitable for miniemulsion polymerization. This eliminates the use of air sensitive Cu(I) species and requires only the use of the Cu(II) species, which is more stable in air. A narrow molecular weight distribution as well as a linear increase in the molecular weight as a function of conversion was reported, and the final latexes were stable over a period of time. In one such study on reverse ATRP processes [20], a non-ionic surfactant Brij 98 was dissolved in water, mixed with a solution of monomer with $CuBr_2$/dNbpy (4,4′-di(5-nonyl)-4,4′-bipyridine) complex along with hexadecane costabilizer and sonicated to form a miniemulsion. Both water soluble as well as oil soluble initiators were used for the polymerization. When the oil soluble initiator was used, it was pre-dissolved in the monomer phase, whereas the water soluble initiator was added in the aqueous phase, and polymerization was subsequently initiated. It was observed that the polymerization rate was independent of the size and number of particles and the amount of surfactant. The shear forces were able to influence only the size of the particles and not the polymerization rate. The study also provided an interesting comparison of the kinetics of miniemulsion polymerization when oil or water soluble initiators were used. In the case of oil soluble initiator, the polymerization was observed to proceed by droplet nucleation mode as the monomer soluble initiator was already present in the droplet during the miniemulsion generation phase and, owing to its water insolubility, tends to remain in the monomer droplets. However, when water soluble initiator was used, both micellar as well as droplet nucleation were reported to take place. As the radicals are generated in aqueous phase, they react with the monomer molecules dissolved in water, and these chains can be stabilized by the absorption of surfactant on the surface to generate new polymer particles. However, these radicals also enter the monomer droplets once they reach a critical chain length and are no longer water soluble, and hence also nucleate the polymerization in monomer droplets. In another study [21], the authors also demonstrated the application of reverse atom transfer radical polymerization in miniemulsion polymerization of *n*-butyl methacrylate, in which higher resulting solid contents could be achieved and the amount of the non-ionic surfactant could be significantly reduced. A series of hydrophobic hexa-substituted tris(2-

aminoethyl)amine (TREN) complexed with $CuBr_2$ was used, and a radical deactivator and water soluble initiator, 2,2′-azobis[2-(2-imidazolin-2-yl)propane] dihydrochloride, was used for the study. The reduced amount of surfactant ensured that no micellar nucleation could take place, and the nucleation of polymer particles was carried out predominantly by the droplets. The controlled molecular weights with low polydispersities were achieved, and the latexes were stable over the period of time. The use of specific ligands and the water soluble initiator with fast decomposition rate was observed to generate well-controlled polymerization as well as colloidal stability. The authors also varied the amount of surfactant to achieve further insight into the mechanism of nucleation by micellar and droplet nucleation at different extents of surfactant concentration. At a surfactant concentration >13% by weight relative to the amount of monomer in the system, bimodal molecular weight distributions as well as bimodal particle size distribution were observed, which were assigned as the result of micellar and droplet nucleation mechanism at higher concentrations of surfactant.

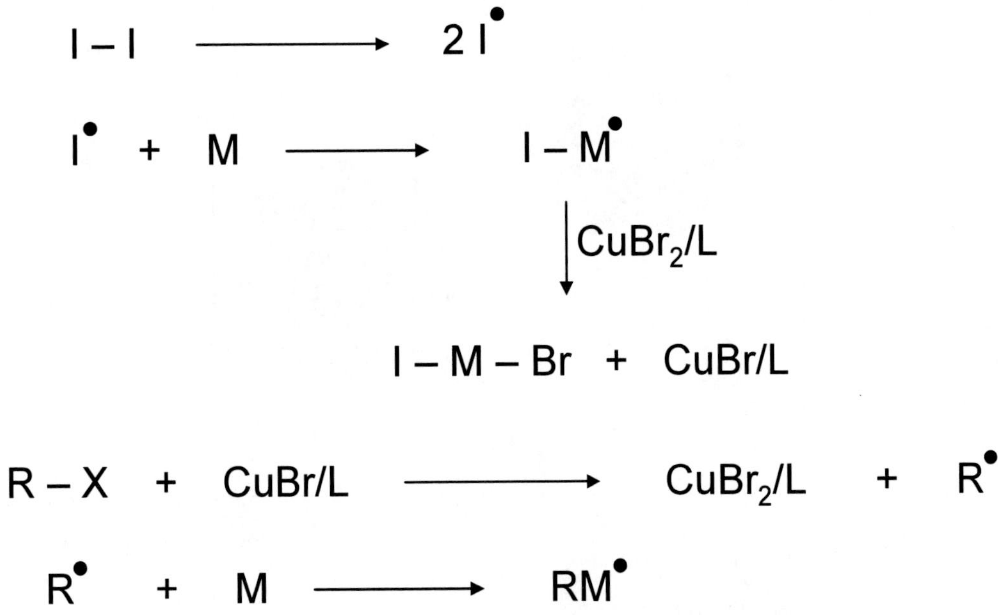

Figure 16. Kinetic process of simultaneous reverse and normal atom transfer radical polymerization.

It was also reported in the literature that a simultaneous reverse and normal initiation of the ATRP system in miniemulsion can be much more beneficial than the reverse or normal ATRP alone [22]. It was reported that such a system is able to reduce the concentration of catalyst in the system, without sacrificing the level of control over the polymerization [23]. The simultaneous reverse and normal initiation works by the addition of a small amount of conventional free radical polymerization initiator in the presence of alkyl halide, i.e., ATRP initiator. The process has been outlined in Figure 16. The catalyst concentrations required for such a process to be carried out is less by 5–8 times the amount of catalyst required for the conventional atom transfer radical polymerization process. Controlled polymerizations of acrylates, methacrylates as well as styrene were reported for these systems [24-26], in which

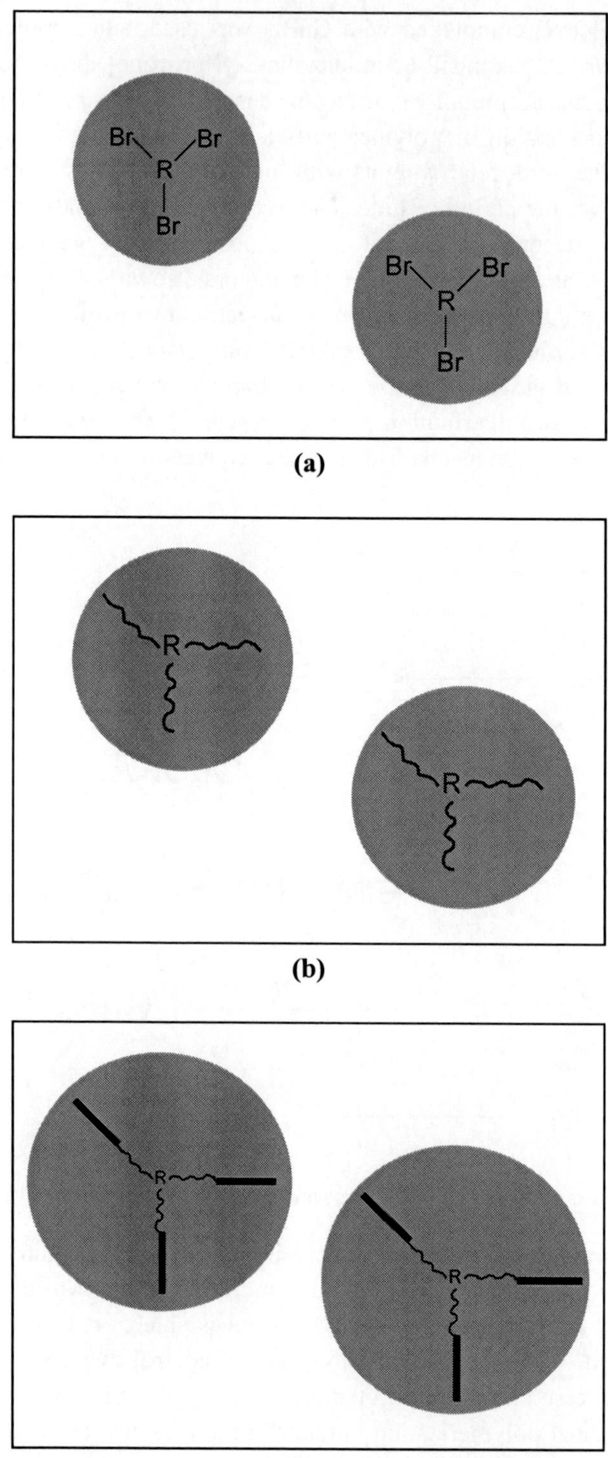

Figure 17. Simultaneous reverse and normal atom transfer radical polymerization to achieve living miniemulsion polymerization [24].

the molecular weight of the polymer chains increased linearly with conversion and polydispersities in the molecular weights of the polymer chains were observed to be less than 1.3. The scope of such a polymerization process was further broadened by the use of water soluble initiators in combination with the ATRP initiator. Although the polymerization reaction could still be successfully carried out, the control on the polymerization reaction was not as good as in the case of oil soluble initiator. One reason proposed by the authors for such a phenomenon is that the water soluble initiator produced the radicals in aqueous phase, which then react with the monomer molecules solubilized in water and form oligoradicals. These radicals, however, are very fast in reacting with each other and, thus, terminating the polymer chains before these radicals could enter the monomer droplets and achieve the living state by the end capping with ATRP initiator. This leads to the reduction in the initiation capacity. Such a system was also demonstrated to be useful in the generation of 3-arm star polymers with styrene, as shown in Figure 17 [24]. A trifunctional alkyl halide was used for the purpose, and $CuBr_2$/BPMODA was used as catalyst. The authors also confirmed that proper choice of ligands is also a very important consideration, as the ligand catalyst complex must be completely soluble in monomer and should also be insoluble in water.

The above-mentioned process of simultaneous reverse and normal initiation method of polymerization was very successful in achieving homopolymers from various monomers as well as polymers of different architectures. However, the process suffered from the inherent limitation owing to the use of free radical initiator. A fraction of the radicals generated from the free radical initiator were observed to grow into homopolymer chains only, and it was not acceptable when the final polymer was formed as a block copolymer. Therefore, to circumvent this limitation, a new technique, called activator generated by electron transfer for atom transfer radical polymerization, was developed [27]. This method completely avoids the use of free radical initiator and uses instead a reducing agent to react with Cu(II) species to generate Cu(I) species to initiate the polymerization reaction. In this particular study, ascorbic acid was used as a reducing agent, because of its water solubility. It was observed by the authors that the hydrophobic reducing agents were seen to be floating on the surface of the miniemulsions, and they were unable to enter the droplets; however, the water soluble ascorbic acid was able to dissolve in the aqueous phase and had no limitation to approach the surface or the inside of the particles. Very accurate control of the amount of reducing agent was required, as it would then translate into the remaining Cu(II) in the system, which is required to control the ATRP process.

The inverse miniemulsion process has also been reported to successfully achieve the polymerization of water soluble monomers [28,29]. Water soluble poly(oligo(ethylene glycol) monomethyl ether methacrylate) was synthesized by inverse miniemulsion polymerization following the activator generated by the atom transfer approach. Cyclohexane was used as a dispersion medium, and oil soluble sorbitan monooleate surfactant was used. Ascorbic acid was used as reducing agent and CuBr2/tris[(2-pyridyl)methyl]amine was used as catalyst. Water soluble poly(ethylene oxide)–bromoisobutyrate-based macroinitiator was used to initiate the polymerization reaction, and poly(ethylene glycol) monomethyl ether was used as costabilizer in the polymerization system. An advantage of the ATRP system over the stable free radical polymerization achieved by nitroxide-mediated polymerization is the possibility of carrying out the polymerization reaction at lower temperatures. For example, in this case of inverse miniemulsion polymerization, the reaction was carried out at 30°C, which is much lower than the processing temperatures of 120–135°C required for the nitroxides to perform efficiently. A molar ratio of 1.05 between the concentration of the initiator and the catalyst was used for the polymerization. Well-controlled molecular weights of the polymer chains were achieved, and the polydispersity in the molecular weight was less than 1.3. It was also pointed out that poly(ethylene glycol) monomethyl ether provided the latexes with colloidal stability and did not interfere in the polymerization reactions. Figure 18 shows the process of inverse miniemulsion polymerization for the generation of poly(oligo(ethylene glycol) monomethyl ether methacrylate) [29].

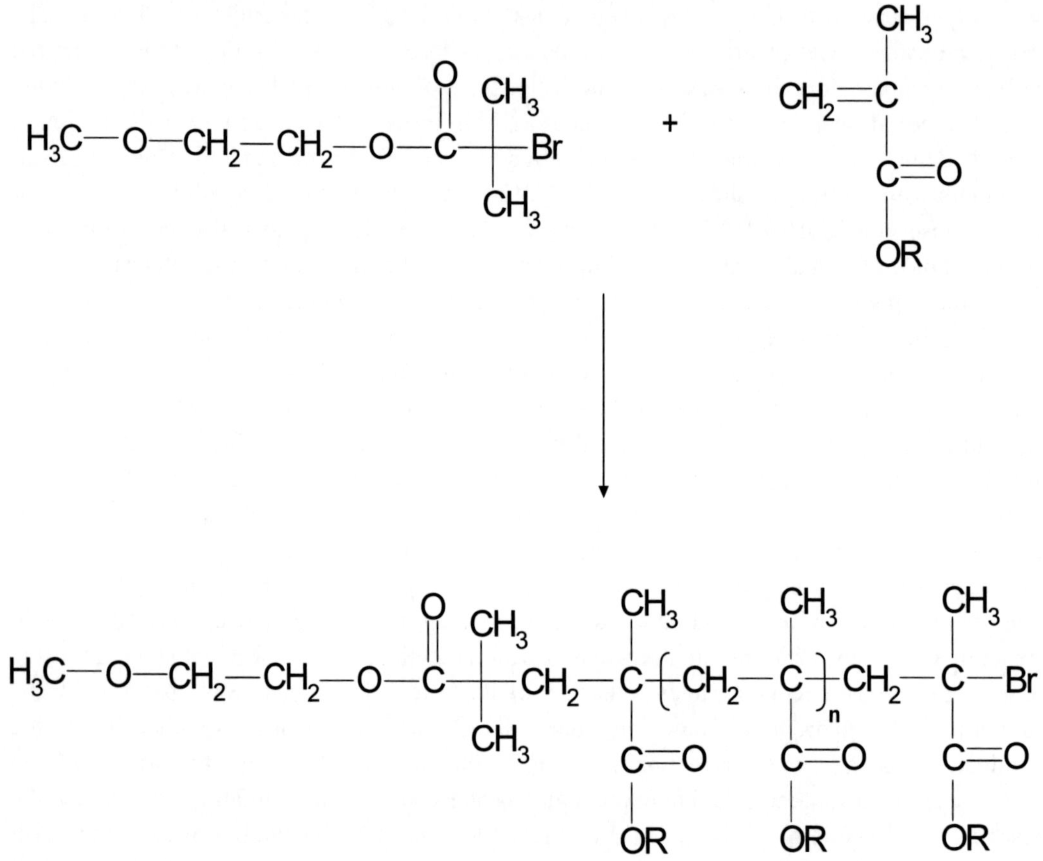

Figure 18. Inverse atom transfer radical polymerization for the generation of poly(oligo(ethylene glycol) monomethyl ether methacrylate) [29].

D.3. Radical Addition Fragmentation Transfer Polymerization

RAFT polymerization is based on the reversible transfer methodology, in comparison with the reversible termination approach used in other controlled polymerization approaches such as nitroxide-mediated polymerization or atom transfer radical polymerization. The main disadvantage of the techniques operating by reversible termination is the partitioning of the deactivating species in the aqueous phase as well as organic phase. It complicates the concentration of active and dormant species in the polymer particles. However, the techniques based on reversible transfer do not suffer from these disadvantages, as the number of free propagating radicals in these polymerization methodologies practically remains unchanged.

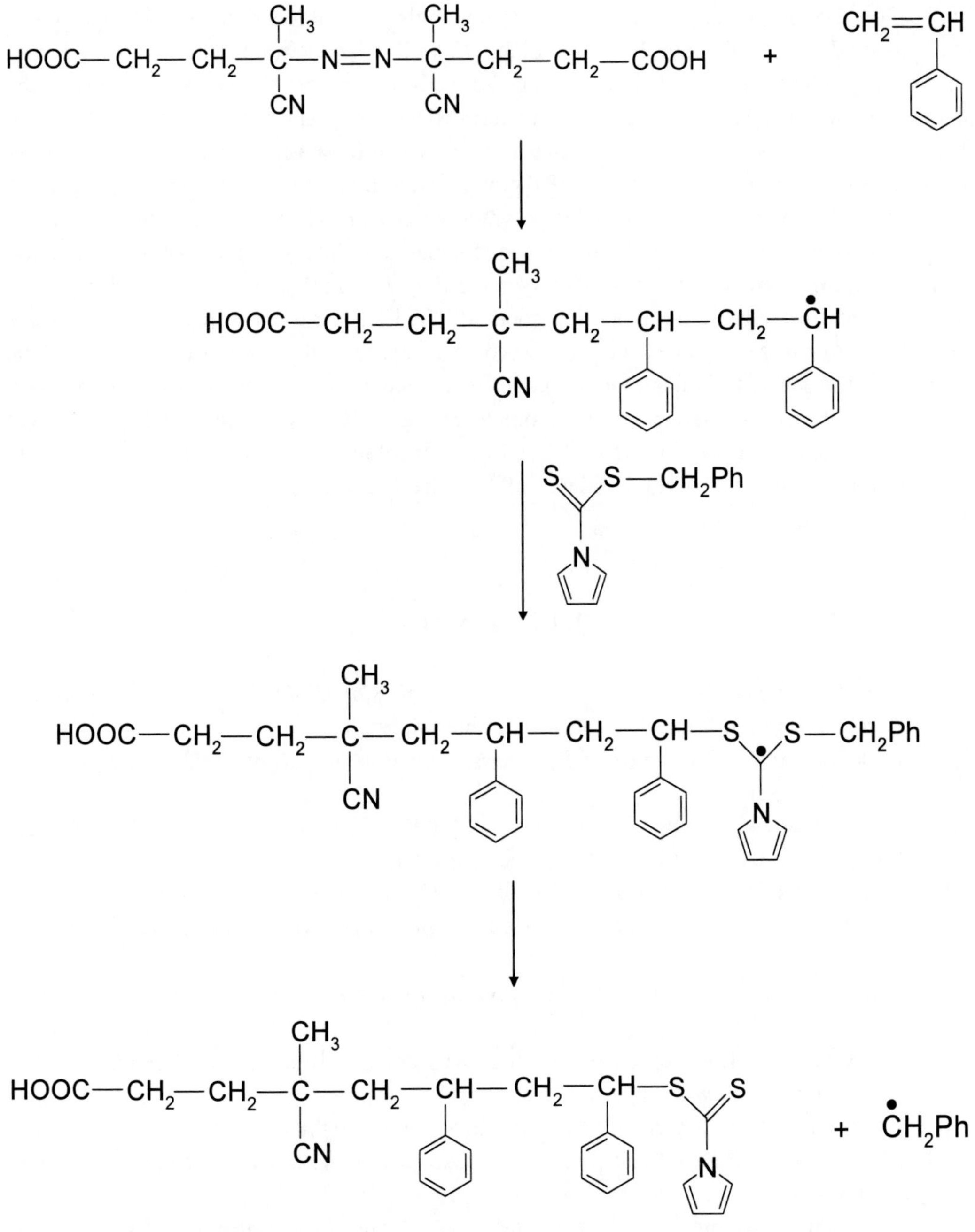

$HOOC-CH_2-CH_2-C(CH_3)(CN)-N=N-C(CH_3)(CN)-CH_2-CH_2-COOH$ + $CH_2=CH$(Ph)

↓

$HOOC-CH_2-CH_2-C(CH_3)(CN)-CH_2-CH(Ph)-CH_2-\dot{C}H(Ph)$

↓ S=C(N-pyrrolyl)–S–CH_2Ph

$HOOC-CH_2-CH_2-C(CH_3)(CN)-CH_2-CH(Ph)-CH_2-CH(Ph)-S-\dot{C}$(N-pyrrolyl)$-S-CH_2Ph$

↓

$HOOC-CH_2-CH_2-C(CH_3)(CN)-CH_2-CH(Ph)-CH_2-CH(Ph)-S-C(=S)$(N-pyrrolyl) + $\dot{C}H_2Ph$

Figure 19. RAFT miniemulsion polymerization for the polymerization of styrene [30].

In earlier studies on RAFT polymerization of styrene in miniemulsion [30], sodium dodecyl sulfate was used as surfactant and cetyl alcohol was used as cosurfactant. Water soluble initiator potassium persulphate was used, and phenylethyl dithiobenzoate was employed as RAFT agent. Controlled polymerization observed with stable latexes was reported. Figure 19 details the process of polymerization. A similar study was reported by Butté et al. [31]. Inverse RAFT polymerization has also been reported to polymerize water soluble polymers like poly(acrylamide) [32,33]. Cyclohexane was used as reaction medium, $MgSO_4$ was used as costabilizer and 2-(2-carboxyethylsulfanylthiocarbonylsulfanyl)propionic acid was used as RAFT agent. Both oil soluble as well as water soluble initiators were used for the study. The water soluble initiator 4,4′-azobis(4-cyanovaleric acid) was dissolved in the aqueous phase, whereas the oil soluble initiator AIBN was premixed with the monomer so that it was directly present in the monomer droplets when these droplets were formed during the process of homogenization. The RAFT polymerization could also be carried out at lower temperatures than the nitroxide-mediated polymerization runs. Here, a reaction temperature of 60°C was used, which is still higher when compared to the room temperature reaction temperatures of atom transfer radical polymerization. Polymer latexes generated by using both water and oil soluble initiators had good colloidal stability and were stable over a few days. The polymerization rates were observed to decrease with reaction time. The authors reported that by using the oil soluble initiator the molecular weight achieved was higher than the weight achieved by using a water soluble initiator. However, the polydispersity was higher when the polymerization was achieved with oil soluble initiator. Thus, it was observed that the water solubility of the initiator affects the kinetics of polymerization and final outcome of polymer latexes.

References

[1] P. A. Lovell and M. S. El-Aasser (1997). *Emulsion Polymerization and Emulsion Polymers*, Editors, John Wiley and Sons Limited, England.

[2] G. Odian (2004). *Principles of Polymerization, Fourth Edition*, John Wiley & Sons, Inc., New Jersey.

[3] K. Matyjaszewski and T. P. Davis (2002). Editors, *Handbook of Radical Polymerization*, John Wiley & Sons, Inc., New Jersey.

[4] K. Landfester (2001). *Macromolecular Rapid Communications*, 22, 896-936.

[5] J. L. Reimers and F. J. Schork (1996). *Journal of Applied Polymer Science*, 60, 251-262.

[6] J. Reimers and F. J. Schork (1996). *Journal of Applied Polymer Science*, 59, 1833-1841.

[7] F. J. Schork, Y. Luo, W. Smulders, J. P. Russum, A. Butté, and K. Fontenot (2005). *Advances in Polymer Science*, 175, 129-255.

[8] J. L. Reimers, F. J. Schork (1996). *Polymer Reaction Engineering*, 4, 135.

[9] J. L. Reimers and F. J. Schork (1997). *Industrial & Engineering Chemistry Research*, 36, 1085-1087.

[10] D. Mouran, J. Reimers and F. J. Schork (1996). *Journal of Polymer Science, Part A: Polymer Chemistry*, 34, 1073-1081.

[11] H. M. Ghazaly, E. S. Daniels, V. L. Dimonie, A. Klein, and M. S. El-Aasser (2001). *Journal of Applied Polymer Science*, 81, 1721-1730.

[12] G. Pan, E. D. Sudol, V. L. Dimonie, and M. S. El-Aasser (2002). *Macromolecules*, 35, 6915-6919.

[13] G. Pan, E. D. Sudol, V. L. Dimonie, and M. S. El-Aasser (2001). *Macromolecules*, 34, 481-488.
[14] M. N. Alam, P. B. Zetterlund, and M. Okubo (2006). *Macromolecular Chemistry and Physics,* 2006, 207, 1732-1741.
[15] P. B. Zetterlund, M. D. Alam, H. Minami, and M. Okubo (2005) *Macromolecular Rapid Communications*, 26, 955-960.
[16] J. W. Ma, J. A. Smith, K. B. McAuley, M. F.Cunningham, B. Keoshkerian, and M. K. Georges (2003). *Chemical Engineering Science*, 58, 1163-1176.
[17] C. Farcet, J. Nicolas and B. Charleux (2002). *Journal of Polymer Science, Part A: Polymer Chemistry*, 40, 4410-4420.
[18] J. Nicolas, B. Charleux, O. Guerret, and S. Magnet (2004). *Macromolecules*, 37, 4453-4463.
[19] B. Keoshkerian, A. B. Szkurhan, and M. K. Georges (2001). *Macromolecules*, 34, 6531-6532.
[20] K. Matyajaszewski, J. Qiu, N. V. Tsarevsky, and B. Charleux (2000). *Journal of Polymer Science, Part A: Polymer Chemistry*, 38, 4724-4734.
[21] M. Li and K. Matyajaszewski (2003). *Macromolecules*, 36, 6028-6035.
[22] M. Li, K. and Matyjaszewski (2003). *Journal of Polymer Science, Part A: Polymer Chemistry*, 41, 3606-3614.
[23] J. Gromada and K. Matyajaszewski (2001). *Macromolecules*, 34, 7664-7671.
[24] M. Li, K. Min and K. Matyajaszewski (2004). *Macromolecules*, 37, 2106-2112.
[25] M. Li, N. M. Jahed, K. Min, and K. Matyajaszewski (2004). *Macromolecules*, 37, 2434-2441.
[26] K. Min, M. Li and K. Matyajaszewski (2005). *Journal of Polymer Science, Part A: Polymer Chemistry*, 2005, 43, 3616-3622.
[27] K. Min, H. Gao and K. Matyajaszewski (2005). *Journal of American Chemical Society*, 127, 3825-3830.
[28] J. K. Oh, F. Perineau and K. Matyajaszewski (2006). *Macromolecules*, 39, 8003-8010.
[29] J. K. Oh, C. Tang, H. Gao, N. V. Tsarevsky, and K. Matyajaszewski (2006). *Journal of American Chemical Society*, 128, 5578-5584.
[30] G. Moad, J. Chiefari, Y. K. Chong, J. Krstina, R. T. A. Mayadunne, A. Postma, E. Rizzardo, and S. H. Thang (2000). *Polymer International*, 49, 993-1001.
[31] A. Butté, G. Storti and M. Morbidelli (2001). *Macromolecules*, 34, 5885-5896.
[32] G. Qi, C. W. Jones and F. J. Schork (2007). *Macromolecular Rapid Communications*, 28, 1010-1016.
[33] M. F. Cunningham (2008). *Progress in Polymer Science*, 33, 365-398.

Chapter 6

Generation of Copolymer or Core Shell Particles by Conventional Polymerization

A. Introduction

It is commonly observed that emulsion polymerization is rarely used to generate homopolymer particles, as the commercial applications of the latex particles generally require more than one functionality or property from them, which are generally fulfilled by synthesizing copolymer latex particles. These copolymer particles can have different morphologies, e.g., random copolymer particles, block copolymer particles, or grafted particles in which polymer chains from a monomer are grafted onto the polymer chains generated from a different monomer, etc. Generation of special morphologies, like block copolymer or grafted copolymer particles, is not straightforwardly achieved by conventional emulsion polymerization, as emulsion polymerization leads to uncontrolled free radical polymerization of the monomers with one another and generates random copolymer particles unless the monomer feeding protocols and the monomer reactivity ratios are not considered. Figure 1 is a representation of these types of polymer chains. The gradient copolymer particles contain the copolymer chains in which the monomer concentration changes along the radius of the particles; i.e., in the case of binary polymerization, polymer chains at the core are rich in one monomer and the concentration of this monomer decreases as one follows the radius outwards. The concentration of the other monomer increases, therefore, along this direction. This gradient in the concentration of the monomers along the radius of the particles is the result of different reactivity ratios of the monomers. The more reactive monomer tends to react faster, thus causing the less reactive monomer to lag behind. This fast polymerization of the more reactive monomer leads to its concentration at the core of the particles, and the less reactive monomer starts to form the shell of the particles as its polymerization is delayed owing to the low reactivity ratio. Figure 2 also provides some demonstrations of the polymer particles of different morphologies, like core shell grafted copolymer particles, core shell hydrophilic shell and hydrophobic core copolymer particles, as well as core shell particles of different surface characteristics, including strawberry topography, orange peel topography, moon crater type topography, etc. The example of hydrophilic shell and hydrophobic core copolymer particles is the classical styrene-N-isopropylacrylamide system, in which the hydrophilic poly(acrylamide) chains keep drifting out to the surface of the particles during the course of polymerization owing to their hydrophilic nature, whereas the polystyrene chains, owing to their hydrophobic nature, concentrate at the core of the particles.

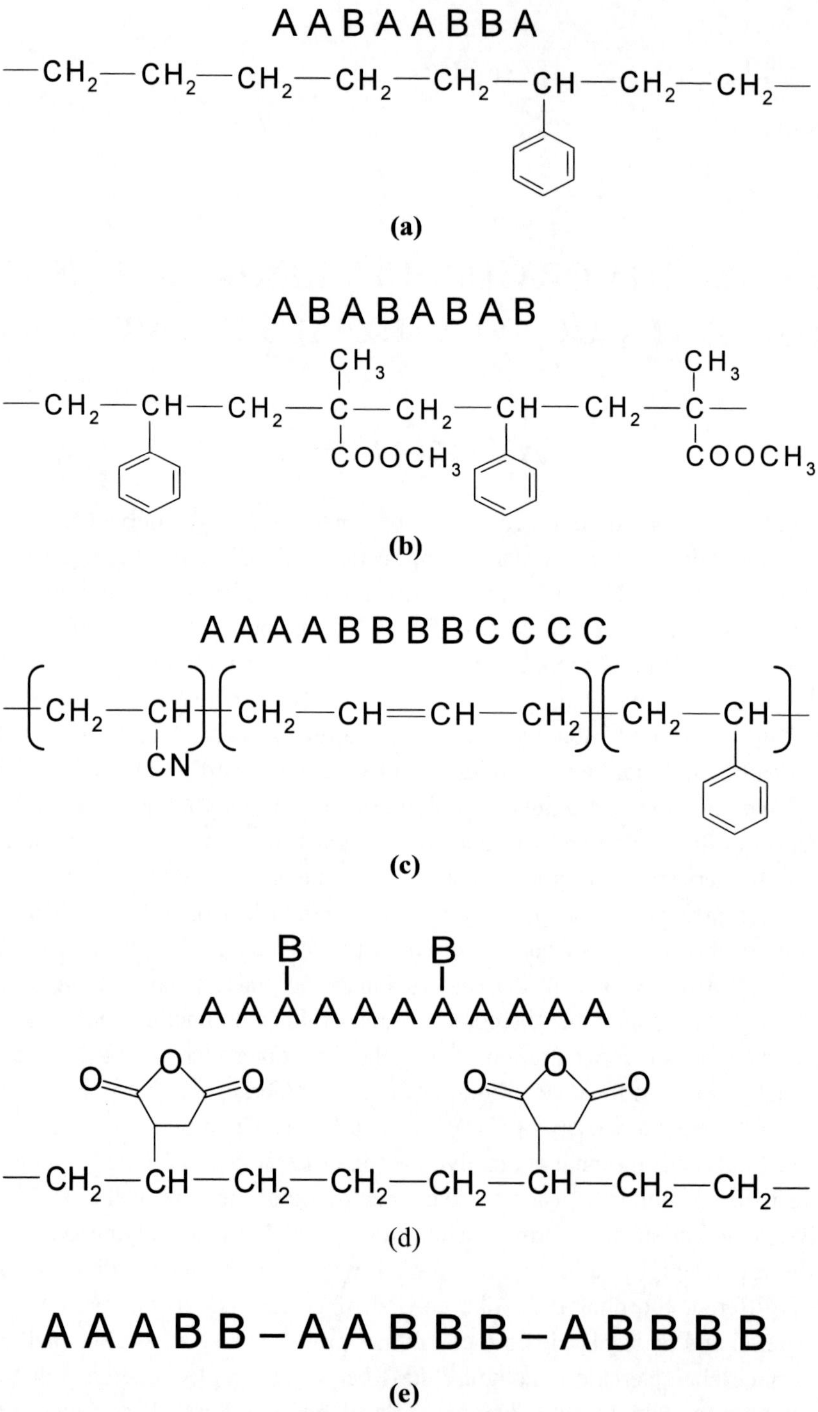

A A B A A B B A

$—CH_2—CH_2—CH_2—CH_2—CH_2—CH(C_6H_5)—CH_2—CH_2—$

(a)

A B A B A B A B

$—CH_2—CH(C_6H_5)—CH_2—C(CH_3)(COOCH_3)—CH_2—CH(C_6H_5)—CH_2—C(CH_3)(COOCH_3)—$

(b)

A A A A B B B B C C C C

$—[CH_2—CH(CN)]—[CH_2—CH{=}CH—CH_2]—[CH_2—CH(C_6H_5)]—$

(c)

B B
A A A A A A A A A A A

$—CH_2—CH—CH_2—CH_2—CH_2—CH—CH_2—CH_2—$ (each CH bearing a succinic anhydride ring: O, O, O)

(d)

A A A B B – A A B B B – A B B B B

(e)

Figure 1. Different copolymer morphologies: (a) random copolymer, (b) alternating copolymer, (c) block copolymer, (d) graft copolymer and (e) gradient copolymer.

B. Compositional Drift in Emulsion Polymerization

As mentioned above, binary or ternary monomer copolymerization in emulsion has a characteristic of generating compositional drift, i.e., monomer concentration at different points along the radius of the diameter may be quite different from each other and these concentrations thus, can also be different from the initial feed ratios. The compositional drift is caused by the different reactivity ratios of the monomers in the emulsion polymerization process as well as by the partioning of the monomers in the aqueous and organic phase. This partioning leads to different monomer ratios in the growing polymer particles compared to the initial ratios fed to the system. Monomer partioning is the result of different water solubilities of the monomers—e.g., a more hydrophilic monomer would be partitioned more in the aqueous phase than the organic phase, whereas an opposite behavior would be demonstrated by the hydrophobic monomer. One must also be careful that, apart from the differences in the chemical behavior of the monomers, the differences in the methodologies by which the emulsion polymerization is carried out, e.g., the feeding of the monomers to the system, also significantly affects the final particle properties like size and morphology, thus, these system changes also provide the opportunity to tune the polymer particles according to need. However, this control in the chemical composition as well as surface morphology of the particles cannot be achieved by batch addition of the monomers, owing to the above-mentioned differences in the partitioning of the monomers. Semibatch addition of the monomers is preferred over the batch addition, as the semibatch addition allows accurate control over the chemical composition of the copolymer particles.

C. Semibatch or Semicontinuous Emulsion Polymerization

As mentioned above, the semicontinuous mode of addition of monomers throughout the course of emulsion polymerization allows one to control the chemical composition of the copolymer chains, thus allowing the composition of monomers in the polymer chains to be the same as the initial monomer ratios. The semibatch addition of monomers can be achieved by further two methods, namely, flooded addition and starved addition of monomers [5]. Flooded addition of monomers corresponds to monomer addition protocol whereby the monomer addition rate is faster than the polymerization rate of the monomers. The methodology of flooded addition of monomers to the polymerizing system may be beneficial in many applications of structured latex generation, but this method of monomer addition can also cause a severe amount of secondary nucleation of polymer particles owing to the excess of monomers present in the system. As the addition of monomers occurs at a much faster rate than their consumption by polymerization, this therefore leads to the accumulation of excess monomers in the system, which does not enrich the polymer particles, thus forming a separate monomer phase. As the radicals are constantly generated from the initiator molecules, they then initiate these monomer molecules to generate an additional family of particles, the sizes of which are smaller than the primary particles. Figure 3 is an example of extensive secondary nucleation in the polymerization process. The secondary nucleation can also be achieved in the batch polymerization system to generate core shell morphologies when the addition of shell-forming monomers to the preformed seed particles is carried out, as it is generally observed that the fully polymerized particles are somehow less willing to accept the other monomers to form the shell and, thus, these shell forming monomers tend to form secondary particles in the system. However, this phenomenon of secondary nucleation can also be used to one's advantage and, in fact, in many cases the secondary nucleation is deliberately generated. The smaller sized particles along with the larger sized particles allow one to pack more particles into the latex; e.g., in such combinations of small and large particles, latexes with higher solid contents can be produced as the smaller particles settle into the interstitial voids between the larger particles. This, thus, eases the transportation of a large amount of material owing to the higher extents of solids in latexes. This type of morphology in particle sizes is also beneficial in the control of the rheology of many systems. In some cases, even three different particle size families are generated to achieve these commercial advantages.

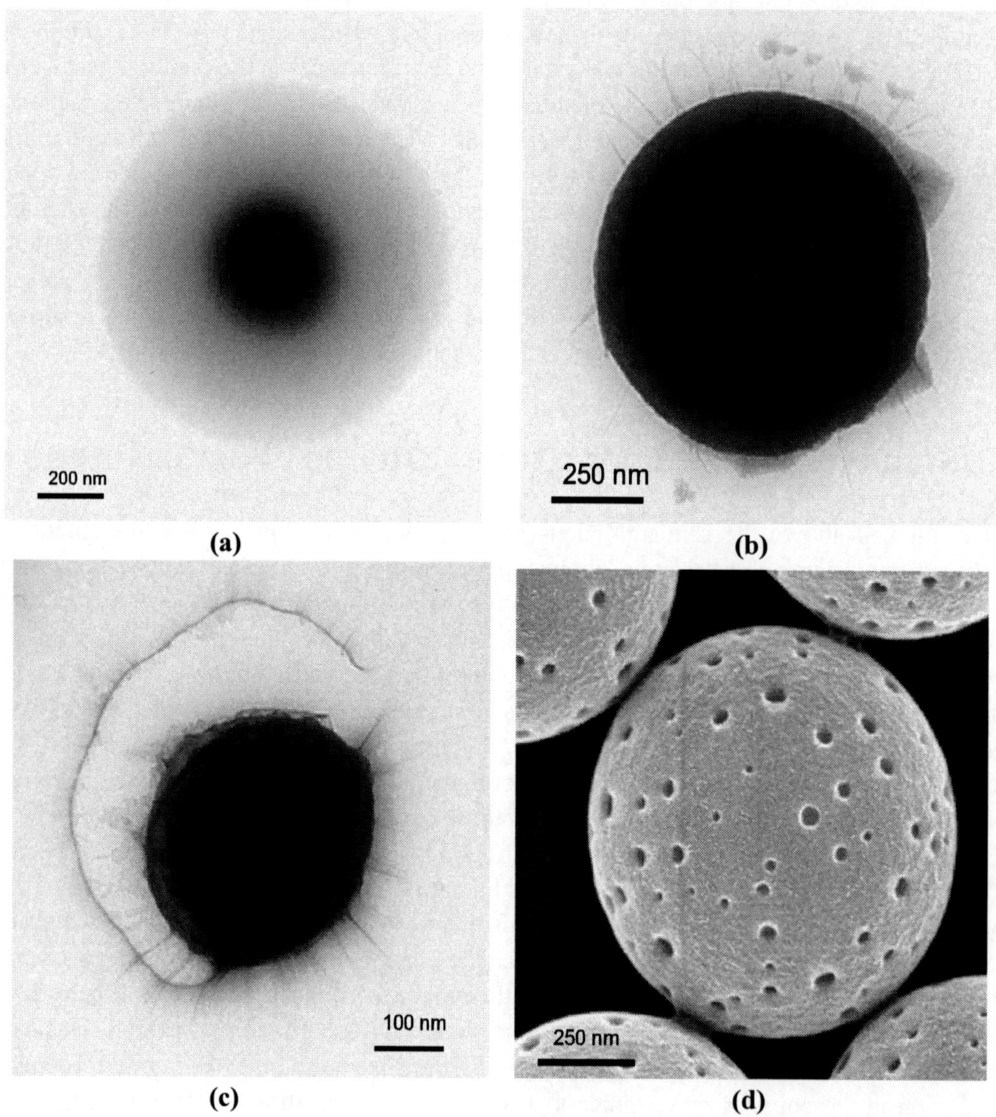

Figure 2. (Continues on next page.)

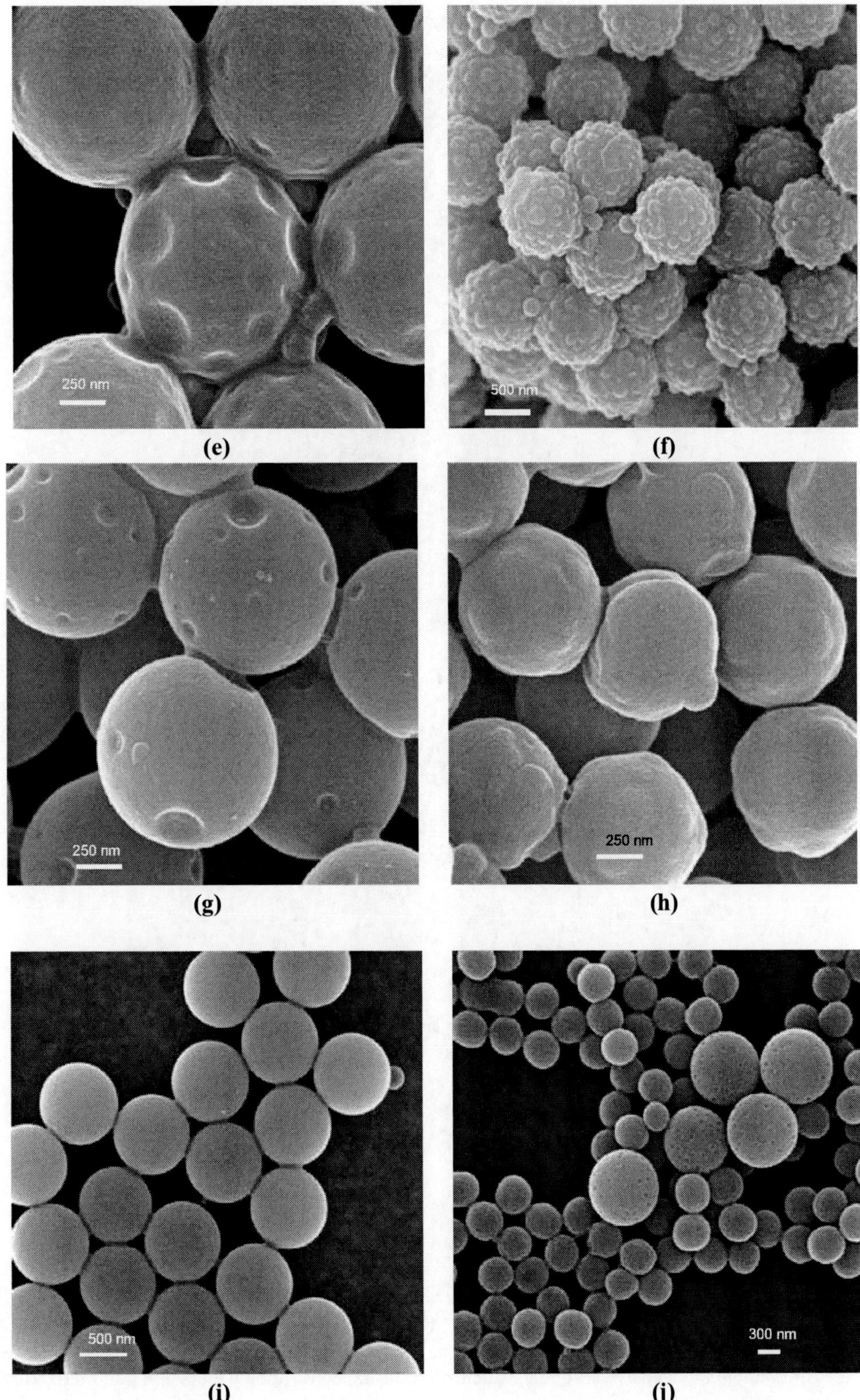

Figure 2. Different morphologies of copolymer particles generated by conventional emulsion polymerization [1-4].

Figure 3. Example of extensive secondary nucleation during conventional emulsion polymerization [1].

The second methodology of monomer addition is starved addition, in which the monomer addition rate is slower than the monomer polymerization rate, and this allows the chemical composition of the polymer chains to be equal to the ratio of monomers in the feed. The addition of the monomers can also be automatically controlled by using the reactors in which the monomers can be added at a constant rate. One can also program the addition of the monomers, based on the conversion of the monomers in the polymerization, to attain a very accurate chemical composition of the copolymer chains. Variable feed rate can also be achieved in these systems to generate core shell particles. Apart from that, the semicontinuous processes allow one to change the crosslinking pattern in the copolymer particles by controlling the feed concentrations. As an example, some applications require the use of particles with a crosslinked core and soft shell. A crosslinked core is required for strength, whereas a soft shell may be beneficial in merging the particles with each other, for example in surface coating applications. Many systems also require the opposite morphology, i.e., soft core and hard shell, which again can be generated by using the controlled addition of crosslinker at the later stages of polymerization. As mentioned, the use of semibatch polymerization allows one to control the copolymer composition and morphology; however, this process also allows the achievement of better process safety in the industrial preparations of the latexes on a large scale. These processes are safer than the batch polymerization process, as the monomer amount inside the reactor is always limited to the extent that it is polymerized faster, and as the whole of the monomer to be polymerized is not present in the reactor, it allows better control of heat dissipation and elimination of the risk for uncontrolled polymerization.

Although the use of semibatch conditions allows one to achieve better control over the copolymer compositions, it can also negatively impact the molecular weight distribution of the polymer chains. It has been generally observed that these conditions result in broader molecular weight distributions and a significant amount of chain transfer to the polymer, owing to the presence of a higher amount of polymer in the system. Secondary nucleation may be desired in a few systems; however, in most of the cases, its presence is not wanted and eliminating the presence of secondary nucleation is a challenge. The starved conditions of monomer addition allow one to achieve latexes without secondary nucleation, but one must also be careful to control the amount of surfactant in the system or charges on the surface. This condition is more critical when, owing to the changes in surface composition, the hydrophilicity or hydrophobicity of the particles' surface changes. As an example of copolymerization of a more hydrophobic and more hydrophilic monomer, initially the surface of the particles may be hydrophobic; however, as the chains rich in hydrophilic monomer diffuse out of the particles on the surface, the nature of the panicles' surface changes to become more hydrophilic. This change in the surface character from hydrophobic to hydrophilic may allow the release of surfactant from the surface of particles. This would then result in non-entry of the additional hydrophobic monomer added to the system into these particles with hydrophilic surfaces. This would, therefore, result in the generation of secondary particles in the system, indicating that the system must be carefully controlled in order to achieve the optimum particles.

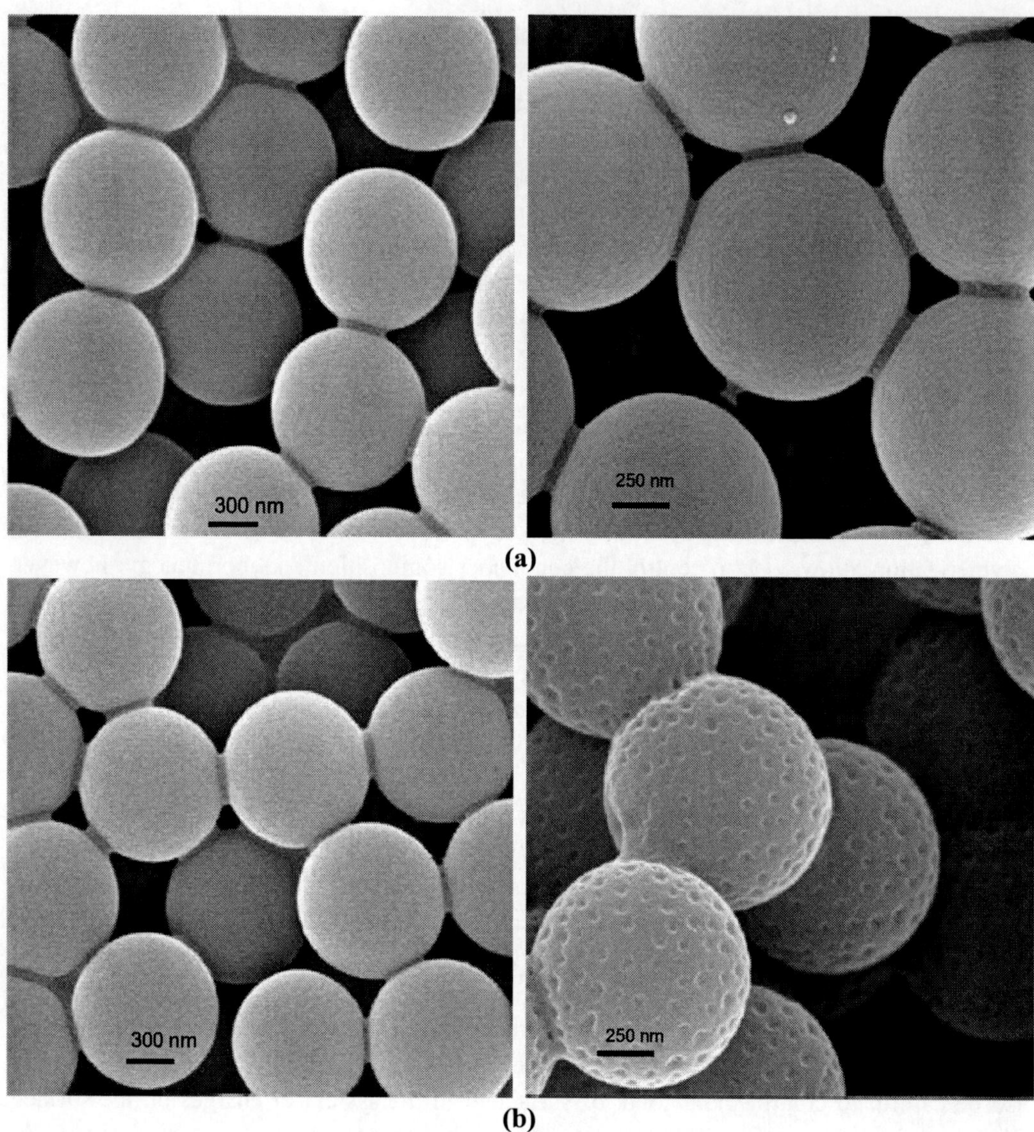

Figure 4. Surface functionalized particles when fully polymerized polystyrene particles (on the left) are modified with (a) styrene and (b) functionalizing monomer on the surface (on the right).

D. Considerations in the Generation of Structured Latexes with Emulsion Polymerization

Batch polymerization of copolymer particles of polystyrene and poly(methyl methacrylate) were reported without the use of initiators [6]. These copolymer particles of poly(methyl methacrylate-*co*-styrene) were prepared by thermally initiated emulsion copolymerization. The first step included the generation of seed particles of polystyrene and poly(methyl methacrylate) which were then dialyzed, and in the second step were swollen with a second batch of monomers and were subsequently polymerized. The seed polymer particles were prepared by conventional emulsion polymerization using sodium dodecyl sulfate emulsifier and ammonium persulphate free radical initiator. As no initiator was used during the polymerization process, the authors observed a broad polydispersity in the size of the polymer particles. But the authors mentioned that this polydispersity in the particle sizes is beneficial in generating latexes with high solid contents. The size distribution of polymer particles was observed to improve by using the seed particles in the emulsion copolymerization method. It was also observed that as the chain ends do not contain the initiator components at one of their ends—as is generally the case when the initiators are used in the emulsion polymerization—the particles therefore were generated with different morphologies in comparison with the conventional process. It was observed that totally different particle morphologies, such hemispherical, sandwich-like, core-shell, and inverted core-shell particle morphologies, were obtained depending on the polymerization conditions. It was concluded that the different monomer to polymer ratio and the hydrophilicity of the monomer affected the final structural outcome of the polymer particles. The authors also compared the particle morphologies generated from the thermal initiation process as well from the conventional emulsion polymerization process in which potassium persulphate was used as an initiator. It was observed that the incorporation of the initiator fragments to one end of the chains allows the polystyrene chains to become more hydrophilic, thus changing the surface nature of the polymer particles.

In an interesting study, poly(methyl methacrylate) seed was used to generate the copolymer particles of poly(methyl methacrylate) with polystyrene [7]. Both batch as well as semibatch modes of monomer addition were employed for the study. Oil soluble initiators, such as azobisisobutyronitrile, as well as water soluble initiators, including potassium persulphate, were used. It was observed that by using the oil soluble initiators, an inverted core shell morphology of the particles was obtained in which the polystyrene chains were present in the core of the particles and the poly(methyl methacrylate) covered the particles owing to its hydrophilicity. The authors mentioned that the generation of the particle morphology, in the case of hydrophilic hydrophobic polymer pair systems, depended upon hydrophilicity, stage ratio, molecular weight, viscosity and polymerization methods. An interesting result was reported when using the water soluble initiator for polymerization. In this case, the morphology was less affected by the hydrophilicity of the polymers, but was more affected by the initiator concentration and polymerization temperature. Depending on the amount of initiator added to the system, the polystyrene rich phase either sandwiched the poly(methyl methacrylate) phase or completely covered it to form a core shell morphology. These changes in the morphology of the particles while using the water soluble initiators for the emulsion polymerization process were concluded to be the result of a varying amount of the presence of $SO4^{\bullet -}$ groups present on the surface which, being hydrophilic in nature, allows the polystyrene phase to become more hydrophilic and, thus, stay on the surface of the composite particles, thereby burying the more hydrophilic poly(methyl methacrylate) phase in the core of the particles. These observations indicated that by simply using the different initiators one can completely change the morphology of the particles and, hence, can tune it according to the requirement.

It was also described in the reported study [8] that although particles may be polymerized in two subsequent phases it is still possible that a core-shell morphology is not formed. As a number of parameters affect the final particle morphology, a few of them described above, a number of other morphologies may occur. Thus, based on thermodynamic constraints, it was proven that particles with a hydrophilic core and hydrophobic shell would be inherently difficult to achieve. It was described that, based on a two-stage latex formation, the possible interfaces are polymer 1/water, polymer 2/water and polymer1/polymer 2. According to the authors, as described in the following equation, in order for the thermodynamically favorable system to form, the value of ΔG should be minimal:

$$\Delta G = \Sigma(\gamma A)_i$$

where

G is the Gibbs' free energy,
A is the area of the interface, and
γ is the interfacial tension at the interface.

It was further explained that for the hydrophilic monomer 2 and hydrophobic monomer 1, a core shell arrangement would be generated, since polymer 1/water interface would be eliminated and the tension of the polymer 2/water interface would be small owing to the hydrophilicity of the polymer. The behavior of the system was described by the following equation:

$$\Delta G = (\gamma A)_{polymer1/polymer2} + (\gamma A)_{polymer2/water} - (\gamma A)_{polymer1/water}$$

Thus, it was predicted that the morphology of the resulting particles would be inverted when polymer 1 becomes more hydrophilic than polymer 2.

In another study on the generation of functional and structured latexes [9], poly(*N*-isopropylacrylamide) (PNIPAAM) microgel particles were used as a matrix for the polymerization of a hydrophobic monomer, *N*-methylpyrrole (MPy). PNIPAAM microgel particles were used to absorb a small amount of the hydrophobic monomer which was subsequently polymerized to yield a second polymer phase of embedded poly(*N*-methylpyrrole) (PMPy) particles. It was observed under microscope that the composite particles had a "raspberry-like" morphology. It was also quantitatively observed that the PMPy polymer phase occupied the majority of the particle volume.

PNIPAAM microgel was prepared by surfactant-free emulsion polymerization using ammonium persulphate as an initiator. It was observed that under the described reaction conditions no coagulation was observed. Two different types of embedded PMPy polymer particles were evident: small diameter particles (~15 nm) and larger particles (~60 nm). It was also observed by the authors that the larger PMPy particles were located relatively close to the periphery of the microgel particles. It was further remarked that the resulting raspberry-like morphology on the particles' surface confirmed the notion that PMPy prefers to polymerize on itself rather than at an unoccupied site in the PNIPAAM network. Although stable latexes were achieved, in certain special cases, however, coagulation of the system was observed. When the amount of initiator was reduced, the stability of the system was totally disturbed, indicating that the presence of a critical amount of ionic groups from the initiator was necessary in the microgels in order to generate stable composite particles. The authors also were successful in the description of the polymerization process of MPy. MPy was observed to be present inside the microgel particles under orthokinetic conditions via the solubilized MPy molecules. As the initiator was subsequently added, the water-soluble oxidant diffused towards the MPy monomer molecules present within the PNIPAAM microgel particles and led to the oxidative polymerization of MPy primarily within the microgel particle leading to the generation of cationic PMPy oligomers. The authors confirmed that the PMPy particles subsequently grew by a nucleation and growth process. PMPy was observed to incorporate anions from the local environment, such as SO_4^{2-} and $-OSO_3^-$ groups from PNIPAAM microgel, thus facilitating anchoring of the PMPy nuclei to the PNIPAAM network.

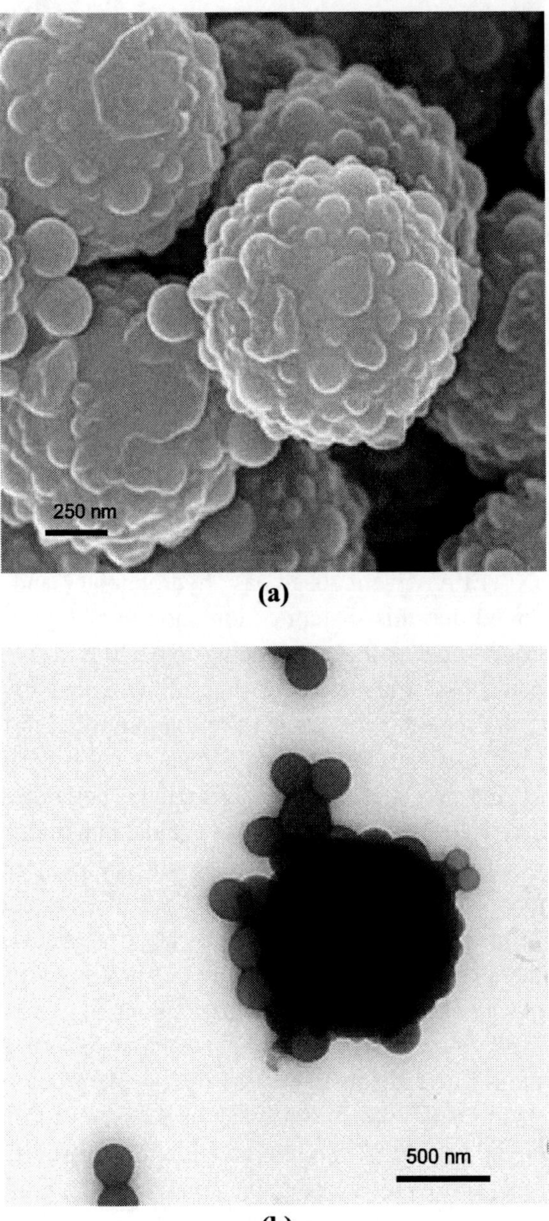

Figure 5. Functionalized polystyrene particles where styrene was added before the functionalizing monomer in the shell forming batch.

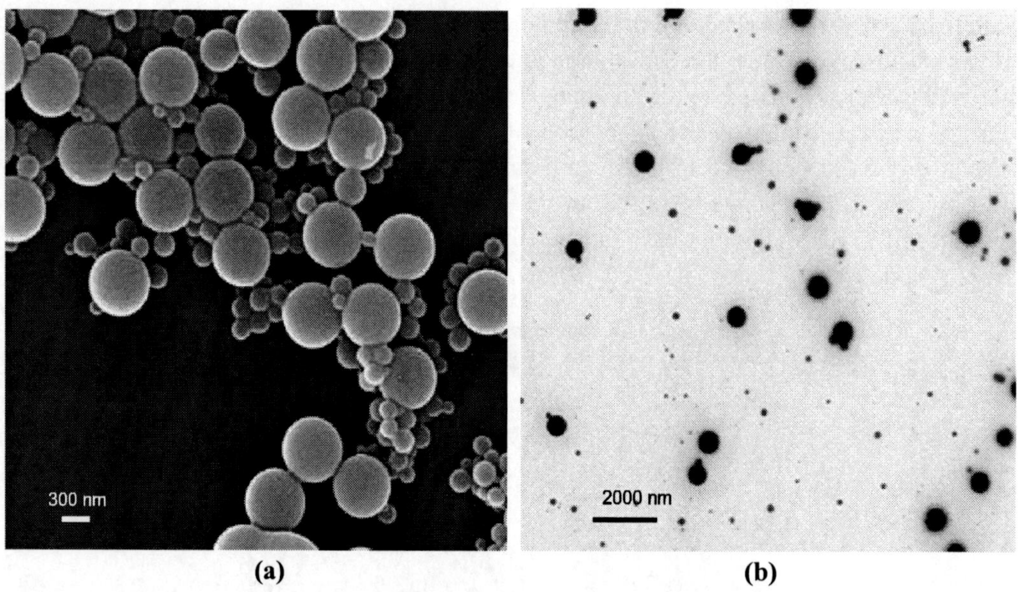

Figure 6. Functionalized polystyrene particles where styrene and the functionalizing monomer in the shell forming batch were added together.

A "topology-controlled" process has been reported which was shown to combine the particle morphology with the design consideration to achieve free radical polymerization in emulsion in such a way that allows the grafting of a polymer onto the seed of the other polymer [10]. A redox initiation system was used, which consisted of a hydrophilic reducing agent and a hydrophobic oxidizing agent; e.g., cumene hydrogen peroxide (CHP) and tetraethylenepentamine (TEPA) were used as hydrophobic and hydrophilic species, respectively. It was claimed that this system of initiation would generate radicals mainly at the polymer/water interface due to the partitioning of CHP and TEPA primarily into the organic phase and aqueous phases, respectively, thus, allowing one to generate block or graft copolymer according to the requirement. To confirm the proposed process, a layer of water-soluble dimethylaminoethyl methacrylate (DMAEMA) polymer was chemically grafted to the surface of hydrophobic polymer latexes like polyisoprene, polybutadiene and polystyrene. The surface functionalized latexes were stable in the acidic conditions, in comparison to the unmodified latexes, which were coagulated just by the addition of few drops of dilute hydrochloric acid solution. These findings also confirmed that the hydrophilic polymer was not physically adsorbed on the surface of the particle, but was chemically bound to the seed latex particles. The authors concluded that the amount of grafting on the surface of the latex particles was a function of the latex particle size distributions. It was pointed out that for such a system to be successful, one must carefully control the loci of radical generation and subsequent polymerization. The radicals generated during the process must be accessible to the monomer, which in this study was controlled in such a way that the majority of the radicals were formed on the surface of the hydrophobic latex particles. It was also observed that the nature of the seed latexes also affects the amount of the grafting of the hydrophilic polymer onto the surface. As polystyrene latex particles are below their glass transition temperature at room temperature, they therefore exhibit the glassy state, which did not allow the generated radicals on the surface to diffuse inside the particles where the concentration of the hydrophilic monomer was much lower. Thus, this allowed the concentration of the radicals on the surface to be high, allowing higher grafting densities. On the other hand, as other latexes used in the study, e.g., polybutadiene or polyisoprene, are rubbery in nature and have more porosity in the chain structure in comparison with to the polystyrene chains, they therefore are unable to hold all of the generated radicals onto the surface, thus causing a decrease in the concentration of radicals on the surface, which translates into decreased grafting density of the hydrophilic polymer.

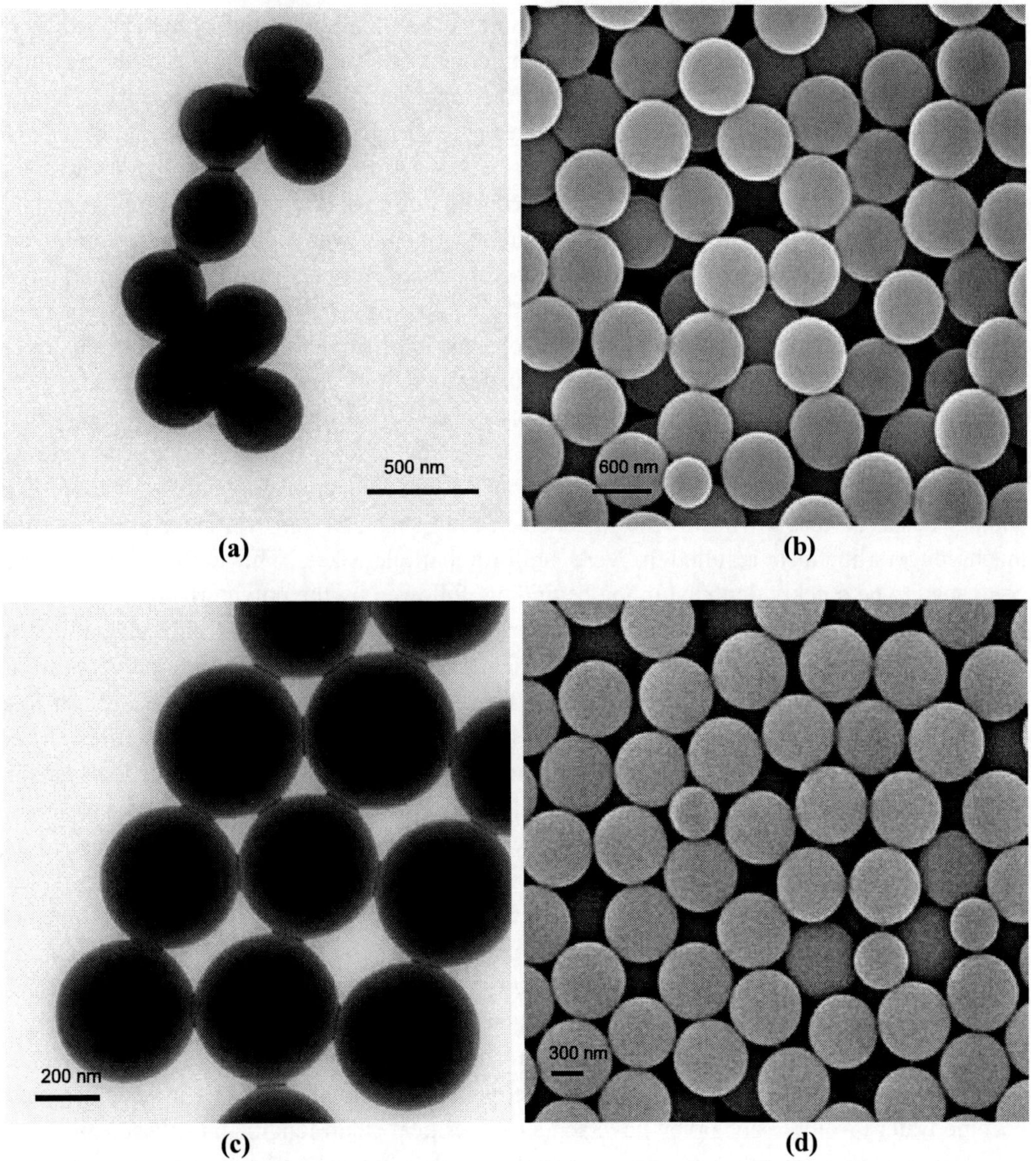

Figure 7. Shot addition of shell forming monomers when only functionalizing monomer was added (b) and when a mixture of styrene and functionalizing monomer was polymerized (d); (a) and (c) represent the seed particles.

Surfactant free emulsion copolymerization was reported in another study of the generation of copolymers of styrene with ionic comonomers by using potassium persulphate as initiator [11]. Emulsion polymerizations carried out in the absence of conventional surfactants have been reported to generate very homogenized latexes, but the particle sizes are bigger than the emulsion polymerization with surfactant, the polymerization rate is slow and the solid fractions obtainable is roughly 10%. The sulphate groups present at the end of the chains from the free radical initiators have negative charges and provide stability to the particles. Ionic comonomers used in the study were the sodium salt of p-sodium styrenesulfonic acid (NaSS) and the sodium salt of 2-sulfoethyl methacrylate (NaSEM). In the case of NaSEM, a greater extent of this comonomer resulted in lower sizes of the resulting particles and, as expected, the increasing the concentration of the initiator was also responsible for the reduction in the particle size. The particle size was, however, enhanced by increasing the ionic strength in the aqueous solution. With NaSS as ionic comonomer, the particle sizes decreased with increasing the concentration of the monomer, similar to that observed for NaSEM. When the amount of styrene monomer was increased in the system, the particle sizes were observed to increase. Four stages of conversion in the polymerization process were observed. The rates were low in the beginning, which were observed to increase at 6% conversion. The rate then remained constant until 60% conversion, followed by an acceleration phase and subsequently leveling off around 85% conversion. The particles were observed to grow faster in the beginning; however the rate slowed down after 6% conversion. The nucleation stage was observed to be over at 6% conversion, which is significantly lower in value in comparison with to the conventional emulsified emulsion polymerization. The authors observed that, owing to the faster nucleation stage, the growth stage was therefore much longer, but also resulted in very uniform particle sizes. The ionic monomer was confirmed to be quickly depleted in the beginning, followed by the polymerization of styrene. As mentioned above, the emulsion polymerization without surfactant is carried out to generate larger particles with much better particle size distributions, and these particles are of significant use when one requires the processes in which an exact amount of surface ions must be known. The colloidal considerations in the case of emulsion polymerization carried without surfactant play a much more significant role in comparison with emulsified systems. As a result, the morphology of the particles is extremely sensitive to the processing conditions. The colloidal stability, diffusion of the monomers from the aqueous phase to the organic phase, mechanism of particle nucleation, number of particles, etc., are a few important parameters that are different between the emulsified and non-emulsified systems. The nucleation in the case of emulsified systems takes place by the micellar nucleation mode, where the polymerization of the monomer present in the micelles is initiated by the radicals entering these micelles, thus generating polymer particles. However, as no surfactant is present in the system in the case of non-emulsified polymerization process, the nucleation proceeds by the homogenous nucleation mode, where the growing radicals in aqueous phase become water insoluble once they have achieved a critical chain length and subsequently start to collapse with each other to reach a stable polymer particle that is still enriched with monomer. More monomer diffuses from the monomer droplets and the particles grow in size during the course of polymerization. Owing to the stability considerations, the number of particles is much smaller in the case of the non-emulsified emulsion polymerization compared with the emulsified case. Owing to the commercial importance of these particles, it is, therefore, of importance to study these particles and optimize the conditions in which the required polymer properties can be achieved. It can also be possible to coat these particles with functional shells to achieve further control of surface properties. In one such study of the surface functionalization of the particles generated by surfactant free emulsion polymerization, the particles were covered with a thin shell of an initiator to initiate polymers by controlled living polymerization of atom transfer radical polymerization [1,12,13]. The water soluble free radical initiator potassium persulphate was used, as its radical fractions also function towards providing colloidal stability to the particles. This technique to immobilize the atom transfer radical polymerization initiator on the surface of the particles provides a very special approach to graft polymer chains or polymer brushes from the surface of the particles. Apart from that as the particles do not contain any surfactant on the surface, the atom transfer radical polymerization can be successfully carried out, as the ATRP agents are known to react with the ionic surfactants used in the emulsified polymerizations. The ATRP initiator in this study was a dual functional molecule. It has on one end a methacrylic or acrylic group, which in the first instance can be homopolymerized or copolymerized with other monomers such as styrene on the surface of the particles. The other end of the molecule contained the ATRP initiator site, which remains intact in the first instance of copolymerization to form the shell on the particles, but can later be activated by the addition of suitable reagents. The amount of the initiator can also be controlled on the surface by changing the monomer ratios in the batch added during the formation of the shell, and this, thus, represents the amount of grafting of polymer on the particles that can thus be controlled. The process to immobilize the ATRP initiator was achieved by using the above-described batch and semibatch processes. In the batch process, the polymer seed is formed first with the desired characteristics, and the functionalizing monomer (molecule with one end as a methacrylic or acrylic group and the other end as ATRP initiation moiety) is polymerized afterwards on the surface. In the semi-batch mode of particle functionalization method, a single step process can be achieved by adding the functionalizing monomer to the seed particles before full polymerization is achieved. The studies reported that the various surface morphologies were achieved in these two modes of polymer formation. To fully analyze the whole system, the impact of reaction conditions, such as the weight ratio of the functionalizing monomer to the seed polymer, preswelling of the seed particles with the functionalizing monomer, mode of addition of the monomer to the seed, effect of crosslinker, surfactant, etc., was analyzed. Figure 4 shows the functionalized particles when the fully polymerized seed particles were covered with a shell by using the batch polymerization mode. In the first case as shown in Figure 4a, pure styrene was used as the shell forming material, whereas in the second case, as shown in Figure 4b, pure functionalizing monomer was used to form a shell on the polystyrene seed particles. These trials with homopolymer shell generation were of importance in order to obtain the information on the behavior of the monomers in the system independent of the other monomers. It is clear from the results that both monomers generated quite different particle morphologies on the surface. The seed as well as core-shell particles generated from polystyrene seed had a smooth appearance, and the core-shell particles were also free from any secondary nucleation, indicating that the styrene added in the second step was able to diffuse onto the surface of the seed particles. On the other hand, the morphology of the polymer shell generated from the polymer of the functionalizing monomer was quite rough and patchy, indicating a mismatch of compatibility between the seed and shell-forming monomer, and the inclusion of this functionalizing monomer on the surface of the seed particles may not be thermodynamically initiated, but may be only kinetically inspired due to the stirring of the latex. However, no secondary nucleation was also observed in this case, indicating that the polymer formed from the functionalizing monomer may not be stable enough to form its own particles. However, from the NMR studies on the dried latex particles dissolved in deutrated tetrahydrofuran, it was evident that the amount of functionalizing monomer groups present in the particles was lower than the amount of functionalizing monomer added initially during the polymerization reaction, indicating that some amount of polymer formed from functionalizing monomer was still in the aqueous phase, which was confirmed by the presence of some coagulum present in the system. It can also be possible that the functionalizing monomer was not fully polymerized, as the polymerization in the absence of surfactant is also slow. An interesting observation was reported by the authors when the two shell forming monomers were added to the system. In the first case, styrene was added first, followed by the addition of the functionalizing monomer, whereas in the second case, these monomers were added together. In both the cases, extensive secondary nucleation was observed, as shown in Figures 5 and 6.

It is also interesting to note that by changing the mode of addition of monomers to the system, the final morphology of the particles obtained was totally different, as Figures 5 and 6 clearly show. There are a number of factors that can affect the final outcome of the polymerization reaction in this case: compatibility of the functionalizing monomer with the core, compatibility of the copolymer chains of styrene and functionalizing monomer with polymer chains, competition between the diffusion of these chains on the seed surface as well as rate of growth of the secondary particles to become stable, sticking of the copolymer particles on the surface of the seed particles because of kinetic forces, etc. Apart from that, the change in compatibility of the copolymer chains with the seed as a function of changing monomer ratio is also important to note, as in the case of early addition of styrene to the system wherein the styrene diffuses to enrich the polymer particles, therefore decreasing its effective ratio with functionalizing monomer, which changes the chemical composition of the copolymer chains.

The above results describe the polymerization trials in the batch mode in which the seed particles were fully polymerized and washed and subsequently covered with a thin shell. In this section, the semibatch addition of shell-forming monomers to seed particles at the stage when the seed is 70% polymerized is explained. One would observe from these results that, as indicated above, changes in one or more parameters in the polymerization producedure changes the morphology of the resulting particle altogether. Figures 7 and 8 illustrate this process. In this case, both methodologies of fast (flooded) and delayed (starved) addition of the shell-forming monomers were analyzed, and their impact on the particle properties were quantified. In this case also, pure monomer as well as a combination of monomers were used to form the shell on the seed particles. The shot or fast addition of the monomers (functionalizing monomer in Figure 7b and combination of styrene and functionalizing monomer in Figure 7d) led to the generation of very smooth particles without the generation of the secondary nucleation, whereas Figures 7a and 7c represent the corresponding polymer particles before the addition of shell-forming monomers. The NMR studies indicated that the amount of monomers present in the particles was less than the amount of the monomer added initially, indicating that the monomers may not have been fully polymerized. However, as the functionalizing monomer molecules could be immobilized chemically on the surface, it signaled a homogenous distribution of the ATRP initiator on the surface. As shown in Figure 8, the delayed or starved addition of monomers to the surface generated a different morphology with a large amount of secondary nucleation in the case when only functionalizing monomer was added to the system (Figures 8a and 8b) and relatively smaller amount of secondary nucleation when the functionalizing monomer was added along with styrene (Figures 8c and 8d).

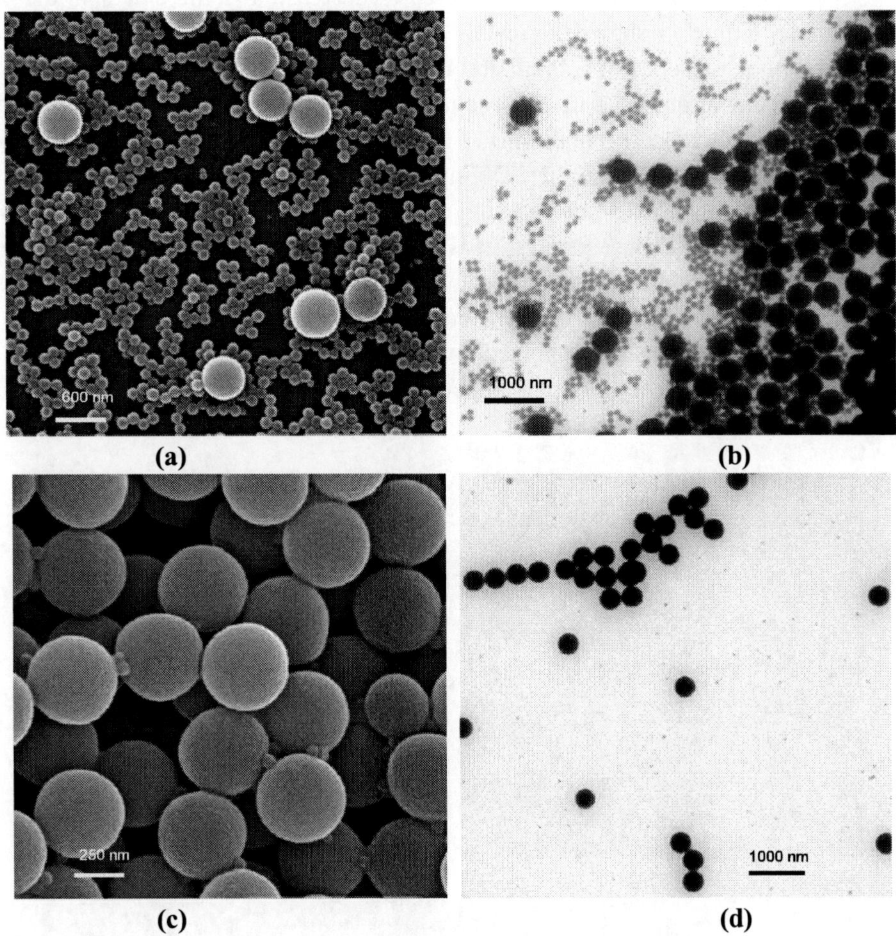

Figure 8. Starved addition of shell forming monomers when only functionalizing monomer was added (a, b) and when a mixture of styrene and functionalizing monomer was polymerized (c, d).

A very interesting comparison can thus be made between the particles generated with and without surfactant. In Figure 9, one can see the batch emulsion polymerization of the generation of the shell on the polystyrene seed particles by using styrene and functionalizing monomer. As one can see, the particles formed are completely different in morphology and, owing to the use of surfactant, the generated particles in emulsified polymerization were also much smaller. Figure 10 shows the trials with all different possibilities of particle generation using emulsified conditions, i.e., by using batch or semi-batch conditions, by using one monomer or a combination of monomers or by using fast or slow addition methodologies for the monomers. In all cases, the emulsified polymerization did not suffer from the presence of secondary nucleation, indicating that the particle morphologies in surfactant-free polymerization are much more sensitive to the changes in process conditions. Another important result was obtained when a small amount of crosslinker divinylbenzene was added to the system. Irrespective of the use of surfactant in the system, the polymerizations carried out with crosslinker resulted in total elimination of the secondary nucleation, though the particles' surfaces were observed to be much rougher in comparison to the particles without crosslinker. The particle sizes were also observed to decrease to some extent in the case of added crosslinker. Figure 11 shows the results of the particles in the surfactant-free emulsion polymerization in the presence of a small amount of crosslinker. Irrespective of any mode of polymer particle generation, the particles seem to be more or less similar, thus less affected by the changes in the process conditions. Figure 10 also shows the emulsified particles generated in the presence of crosslinker. Divinylbenzene enhances the colloidal instability of the forming particles and reduces the swelling of the polystyrene seed particles by the second batch of monomers, which subsequently leads to the formation of a large number of nuclei which, being colloidally unstable, continue to coalesce with the polystyrene particles. The effect of increasing the amount of shell-forming monomers on the morphology of the functionalized particles was also analyzed. As shown in Figure 12, the morphology also changed significantly when the amount of functionalizing monomer was increased.

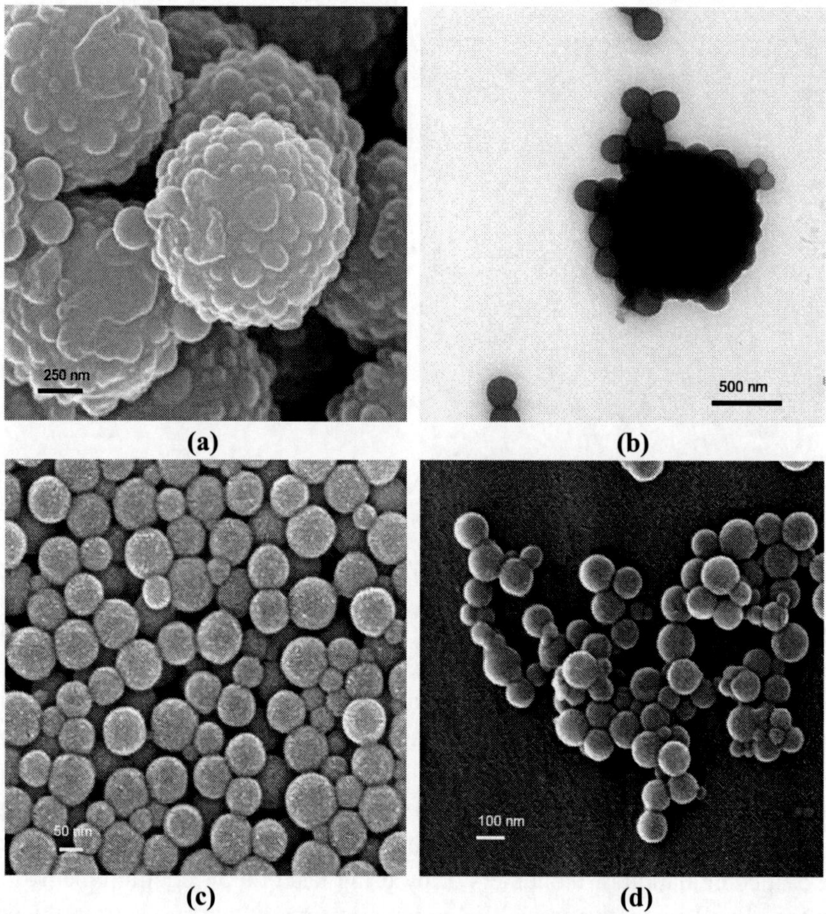

Figure 9. Comparison of particles generated without and with surfactant: (a,b) particles generated without surfactant and functionalized with a copolymer of styrene and functionalizing monomer, (c) pure polystyrene particles generated by the use of surfactant and (d) particles of (c) functionalized with a copolymer of styrene and functionalizing monomer.

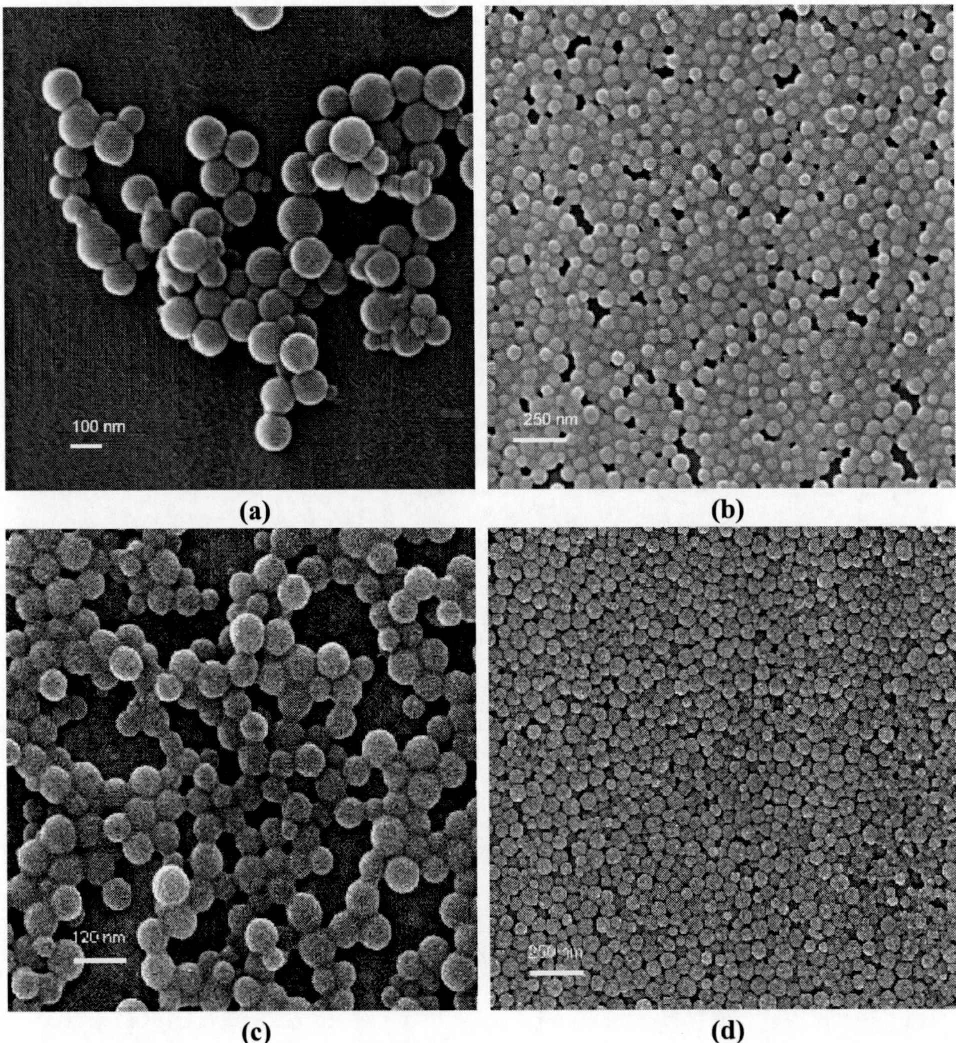

Figure 10. Various trials of particle generation and subsequent surface functionalization by the use of emulsified conditions and changing the process methodologies, i.e., shot addition, starved addition, batch or semibatch generation of particles, etc.

In an interesting study on the generation of hydrophilic and hydrophobic composite particles, the copolymerization of styrene with water soluble N-isopropylacrylamide (NIPAAM) was carried out in batch as well as semibatch conditions. The particles generated by batch conditions are shown in Figure 13, whereas the particles generated in semibatch conditions are shown in Figure 14. In this case, the evolution of particle morphology as a function of conversion or time is also shown. Here the crosslinker was also used, which was also observed to affect the final particle morphology. The effect of increasing the amount of NIPAAM in the copolymer particles as well as increasing the amount of crosslinker is also shown in Figure 15. By increasing the amount of hydrophilic monomer in the shell, a more hydrophilic surface of the particle could be observed, which may also lead to a certain stickiness among the particles (Figures 15a and 15d). However, increasing the amount of crosslinker also leads to more smooth particles on the surface (Figures 15b and 15c).

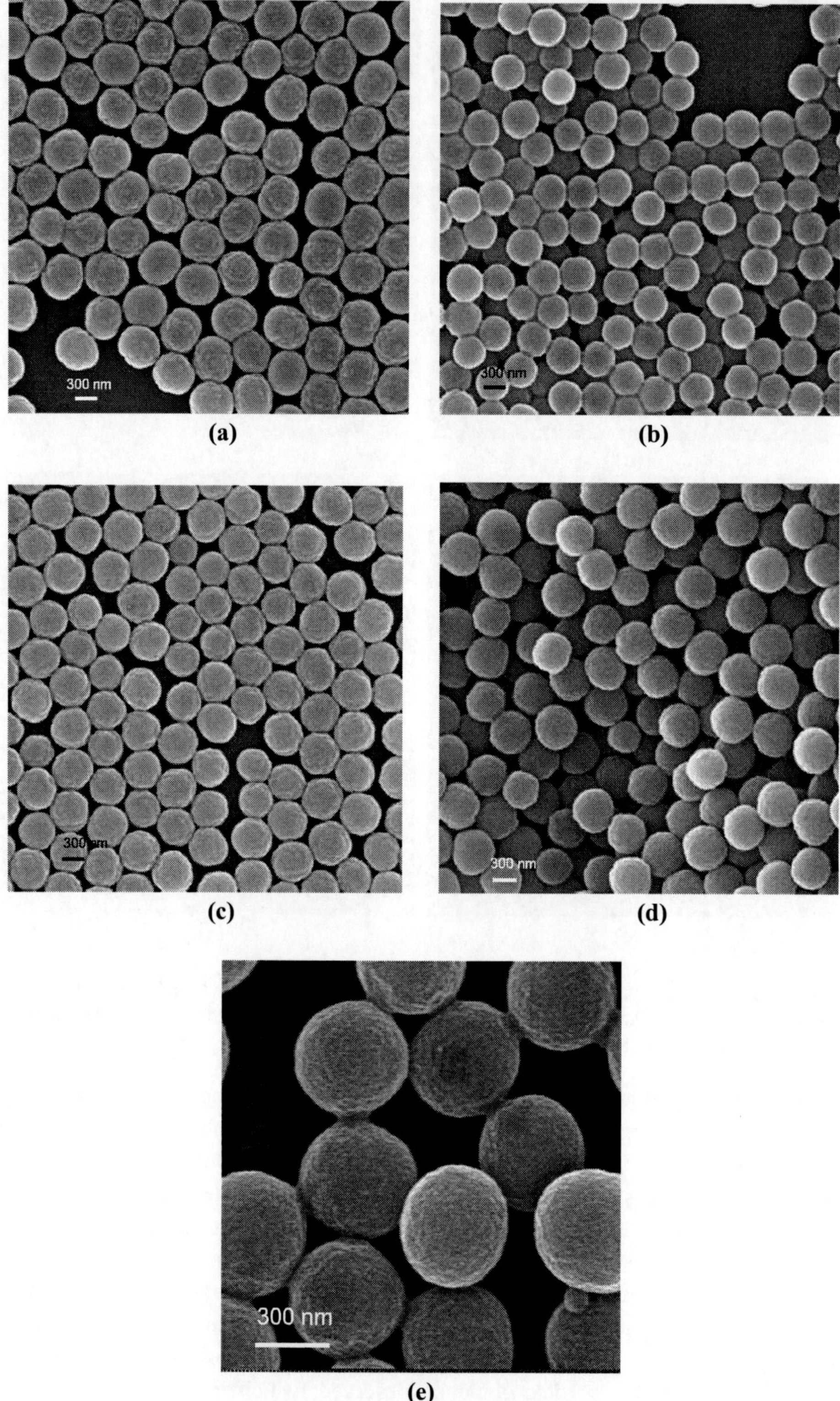

Figure 11. Various morphologies of surface functionalized particles generated by the use of a small amount of crosslinker and using different reaction methodologies.

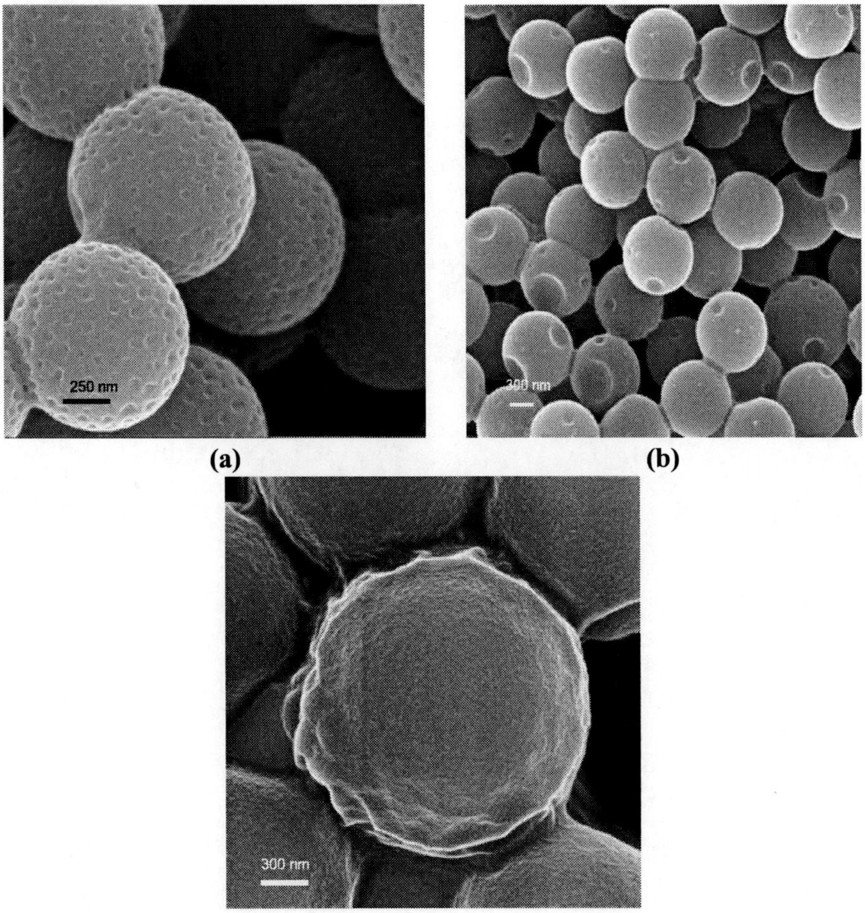

(a) (b)

(c)

Figure 12. Morphology generation on the surface of the particles when the amount of functionalizing monomer is increased.

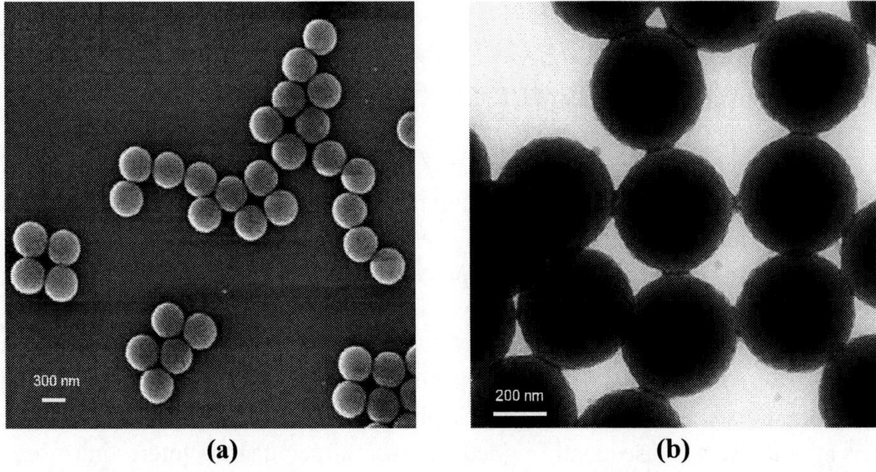

(a) (b)

Figure 13. Polystyrene-co-poly(N-isopropylacrylamide) copolymer particles generated by batch polymerization.

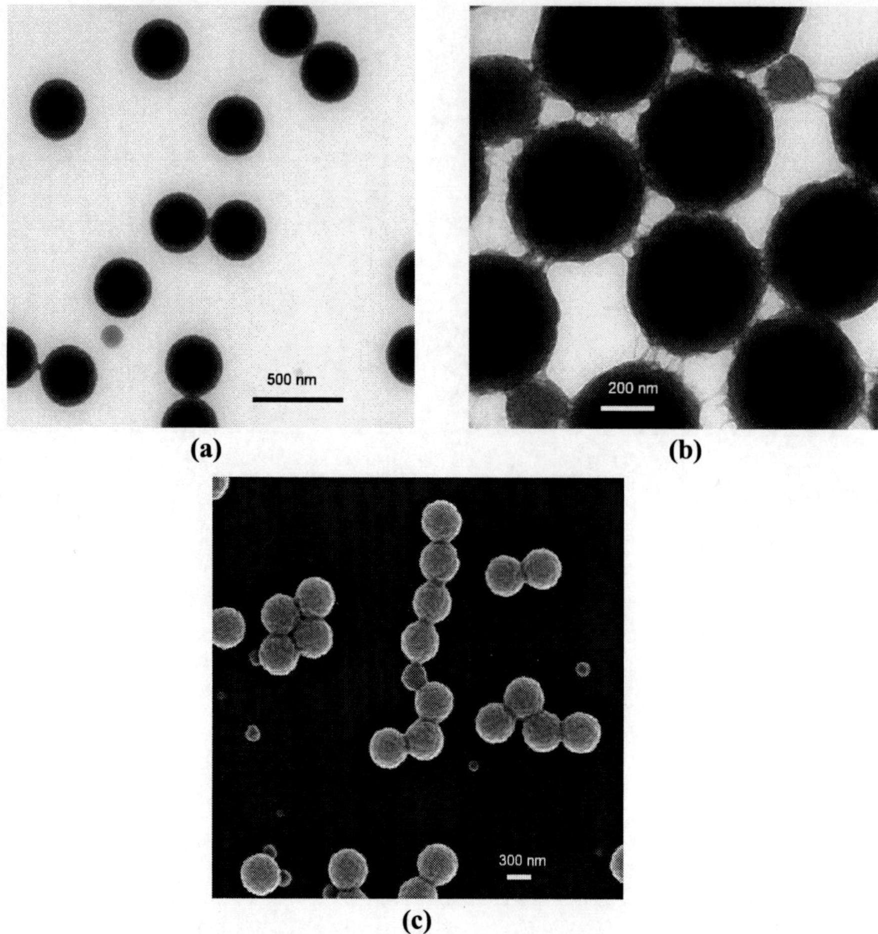

Figure 14. Polystyrene-co-poly(N-isopropylacrylamide) copolymer particles generated by semibatch polymerization, (a) before and (b,c) after the full polymerization.

E. Considerations in the Generation of Structured Latexes with Miniemulsion Polymerization

An interesting study on the copolymerization of vinyl acetate and butyl acrylate by using the batch miniemulsion and macroemulsion polymerization methods was reported, and the particle properties as well as the kinetics of the polymerization reaction owing to these two different polymerization methods were compared [14]. Sodium hexadecyl sulphate was used as surfactant and hexadecane was used as cosurfactant. Miniemulsions were generated in two different ways. In the first case, hexadecane was mixed with the organic phase, i.e., monomer mixture, which was then mixed with the aqueous phase containing surfactant. In the second case, hexadecane was added to the aqueous phase first and the system was sonified. The monomer mixture was subsequently added to this miniemulsion. Interesting observations were reported regarding the amount of adsorption of surfactant onto the monomer droplets. It was observed that at different concentrations of sodium hexadecyl sulphate, the presence of a small amount of hexadecane resulted in a significant increase in the amount of sodium hexadecyl sulphate adsorbed on the surface of the droplets. A further increase in the amount of hexadecane, however, resulted only in a slight increase in the amount of adsorption. It was also observed that as hexadecane to sodium hexadecyl sulphate ratio and initial concentration of sodium hexadecyl sulphate was increased, the droplet size of monomer mixture was reduced and the emulsion stability was enhanced. Induction period was observed to be present in the miniemulsion polymerization trials, which was attributed to the presence of oxygen-containing air bubbles in the system which were formed during the extensive emulsification process. As an alternative explanation for the presence of induction period in the polymerization process, it was remarked that the increased surface resistivity to the entry of the radicals generated in the aqueous phase could be responsible, which occurs due to the presence of a thick layer of hexadecane-sodium hexadecyl sulphate at the surface of monomer droplets.

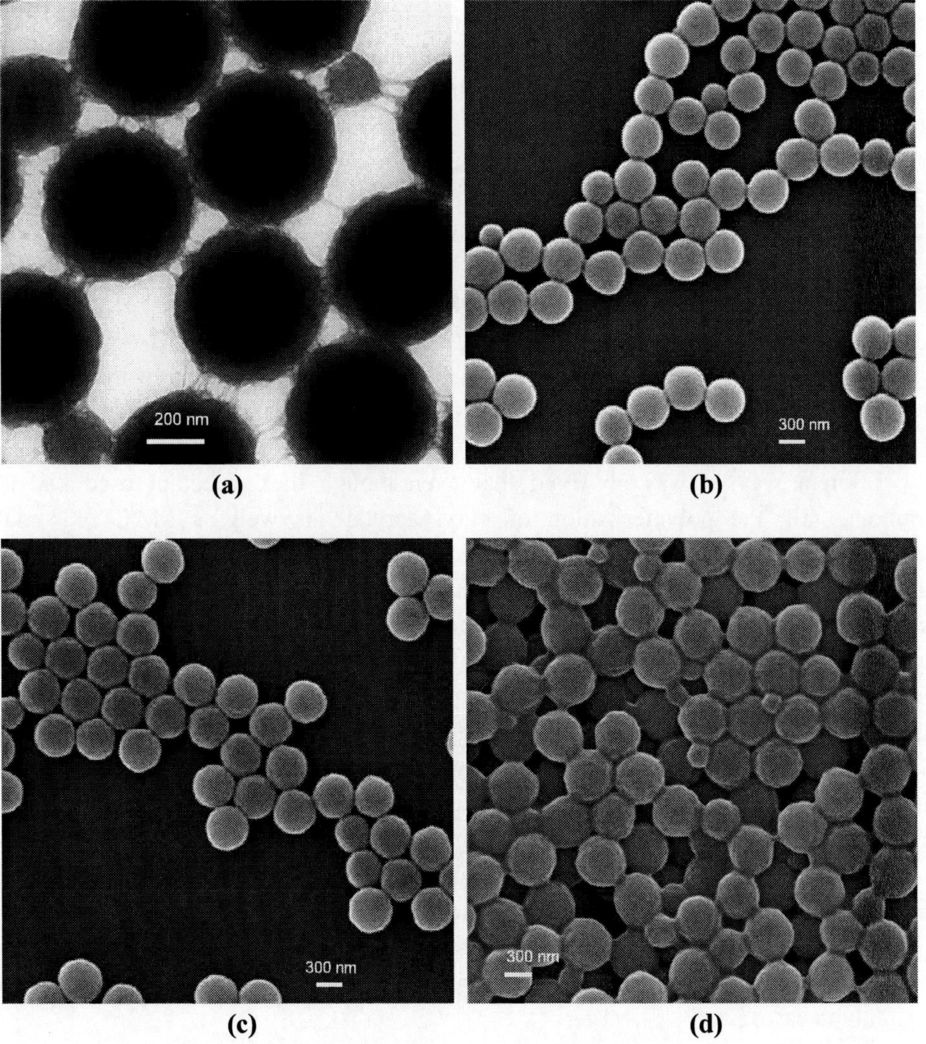

Figure 15. Various permutations of polystyrene-co-poly(N-isopropylacrylamide) copolymer particles generated by increasing the amount of hydrophilic polymer in the particles (a,d) and copolymer particles generated by increasing the amount of crosslinker in the particles (b, c).

The chemical compositions of the copolymer chains generated by conventional emulsion and miniemulsion polymerization were significantly different. The polymer chains synthesized by miniemulsion process had lower contents of vinyl acetate units up to 70% conversion compared to the chains formed in emulsion polymerization process, thus confirming that the copolymerization achieved by the two methods have different kinetic behavior and leads to the generation of chemically different copolymer chains. One may wonder that the addition of hexadecane to the system brings about such significant changes. It was concluded by the authors that hexadecane helped in the formation of stable emulsions with very small droplet size. It was also observed to hinder the rate of interparticle monomer transport during the polymerization and it was also observed to act as swelling promoter in the monomer-polymer system present in the particles.

Copolymerization of butyl acrylate and 2-(methacryloyoxy)ethyl trimethyl ammonium chloride (MAETAC), which forms a hydrophobic and hydrophilic monomer pair, was achieved using interfacial redox initiator system [15]. This system included cumene hydroperoxide and tetraethylenepentamine as initiators, hexadecane as cosurfactant and Triton X-405 as surfactant. As butyl acrylate monomer is hydrophobic, it therefore stayed in the droplets, whereas owing to hydrophilicity, 2-(methacryloyoxy)ethyl trimethyl ammonium chloride significantly partitions in the aqueous phase. Similarly, the redox initiator system also forms a hydrophobic and hydrophilic pair system. Cumene hydroperoxide, owing to its hydrophobicity, remains significantly in the monomer droplets, whereas hydrophilic tetraethylenepentamine is significantly present in the aqueous phase. Interestingly, the authors observed that when a water soluble initiator was used to polymerize the system in conventional emulsion polymerization mode, only polymerization of MAETAC monomer could be achieved, and butyl acrylate remained unpolymerized. It clearly indicates that no radicals could enter the micelles to initiate the polymerization of hydrophobic monomer when a surfactant with a long hydrophobic tail was used. But as the redox system polymerizes in a totally different way, it was observed that even though the surfactant used had a long hydrophobic tail, the polymerization of butyl acrylate as well as MAETAC could be successfully achieved. This owes to the different initiation mechanism in the case of the redox initiator system; the radicals are generated at the aqueous organic interface and, thus, the radicals need not penetrate through the surfactant layer in order to initiate the polymer particles. It was also suggested that, owing to the water solubility of MAETAC monomer, there are two places in which it can be polymerized viz. the aqueous phase, where the amount of butyl acrylate is not so great, thus generating polymer chains which have almost none or a much lower amount of butyl acrylate. The polymerization of MAETAC can also take place at the interface of organic phase with aqueous phase and, in this case of polymerization, the formed copolymer chains also include the butyl acrylate component. It was also observed that, in the case of miniemulsion polymerization, the homogenous nucleation occurred at low MAETAC concentration, but this was not the case at higher concentrations of MAETAC. Findings showed that although in both cases the kinetics of polymerization was significantly different, 18% of the MAETAC was polymerized in the aqueous phase in the final resulting latexes in both cases.

In another study, terpolymerization of butyl acrylate, methyl methacrylate and vinyl acetate were reported in semicontinuous mode [16]. Hexadecane was used as cosurfactant and potassium persulphate was used as initiator. In the beginning of the polymerization, it was observed that the terpolymer was richer in methyl methacrylate and butyl acrylate and it contained almost no vinyl acetate. This indicated that the reactivity ratios of the monomers are very different. It was also observed that the homogeneity of the terpolymer improved when the instantaneous conversion increased as a result of the increase in the initiator concentration. The polymerization rate in the miniemulsion polymerization decreased as the concentration of the hexadecane cosurfactant increased. At the beginning of the process when the monomers accumulate in the reactor, it was observed that the instantaneous conversion was sensitive to the polymerization conditions, but the sensitivity was observed to significantly decrease as the polymerization proceeded further. It was concluded that the combination of anionic as well as nonionic emulsifiers was required to achieve stability in the system, and the anionic surfactant on its own was not sufficient to impart colloidal stability.

REFERENCES

[1] V. Mittal, N. B. Matsko, A. Butté, and M. Morbidelli (2007). *Polymer*, 48, 2806-2817.

[2] V. Mittal, N. B. Matsko, A. Butté, and M. Morbidelli (2007). *European Polymer Journal*, 43, 4868-4881.

[3] V. Mittal, N. B. Matsko, A. Butté, and M. Morbidelli (2008). *Macromolecular Materials and Engineering*, 293, 491-502.

[4] V. Mittal, N. B. Matsko, A. Butté, and M. Morbidelli (2008). *Macromolecular Reaction Engineering*, 2, 215-221.

[5] K. Matyjaszewski and T. P. Davis (2002). Editors, Handbook of Radical Polymerization, John Wiley & Sons, Inc., New Jersey.

[6] Y. Z. Du, G. H. Ma, H. M. Ni, M. Nagai, and S. Omi (2002). *Journal of Applied Polymer Science*, 84, 1737-1748.

[7] I. Cho and K. W. Lee (1985). *Journal of Applied Polymer Science,* 30, 1903-1926.

[8] S.Lee and A. Rudin (1992). *Journal of Polymer Science, Part A: Polymer Chemistry,* 30, 865-871.

[9] J. Mrkic and B. R. Saunders (2000). *Journal of Colloid and Interface Science,* 222, 75-82.

[10] D. J. Lamb, J. F. Anstey, C. M. Fellows, M. J. Monteiro, and R. J. Gilbert (2001). *Biomacromolecules*, 2, 518-525.

[11] M. D. Juang and I. M. Krieger (1976). *Journal of Polymer Science, Polymer Chemistry Edition*, 14, 2089-2107.

[12] J. N. Kizhakkedathu, R. Norris-Jones and D. E. Brooks (2004). *Macromolecules,* 37, 734-743.

[13] J. N. Kizhakkedathu, A. Takacs-Cox, and D. E. Brooks (2002). *Macromolecules*, 35, 4247-4257.

[14] M. J. Unzue and J. M. Asua (1993). *Journal of Applied Polymer Science*, 49, 81-90.

[15] Y. Luo and F. J. Schork (2001). *Journal of Polymer Science, Part A: Polymer Chemistry*, 39, 2696-2709.

[16] J. Delgado, M. S. El-Aasser and J. W. Vanderhoff (1986). *Journal of Polymer Science, Part A: Polymer Chemistry*, 24, 861-874.

Chapter 7

GENERATION OF COPOLYMER OR CORE SHELL PARTICLES BY CONTROLLED LIVING POLYMERIZATION

A. INTRODUCTION

Living polymerization techniques have an advantage over conventional polymerization techniques, as they allow molecular weight and distribution to be accurately controlled. Apart from that, these techniques allow one to achieve special morphologies like block copolymers, star copolymers, polymers with gradient concentrations in the chains, etc. These morphologies are not easily controlled in the conventional polymerizations, as the average age of the radicals is very short. However, the age of the radicals is prolonged owing to the reversible termination or reversible transfer methods of living polymerization, which allow one to thus achieve special polymer architectures. In the case of polymerization in the heterogeneous phases, i.e., emulsion and miniemulsion polymerization, the use of living polymerization techniques is very beneficial, as it provides additional functionality to the polymer particles and allows the special functionalization of the particles. These methods can be used to generate core shell block copolymer particles, particles with grafted polymers on the surface of the seed, particles with a gradient of monomer concentration as a function of the radius, particles with gradient of crosslinking from the core to the surface, etc. The possibilities allowed by using these living methods in the aqueous phase are even more interesting, as they do not require unwanted changes in the polymerization system. Although the remainder of living agents or residual from these living polymerization methods is unwanted, as they can impart color or smell to the product and may preclude them from being used for food contact applications, a significant amount of research, however, has been carried out in this direction to clean the latexes from these agents or to convert them into more friendly species. The various techniques of living polymerization can be used to generate the structured latexes like nitroxide mediated polymerization (NMP), atom transfer radical polymerization (ATRP) and radical addition fragmentation chain transfer (RAFT) polymerization. The first two techniques of nitroxide-mediated polymerization and atom transfer radical polymerization function by reversible chain termination by using special agents, whereas the technique of radical addition fragmentation chain transfer polymerization is based on reversible transfer of the polymer chains to provide living character to the polymer chains. A large number of studies have reported on these systems, and these living agents have been significantly fine tuned to provide optimum results in specific systems. A number of nitroxides have been developed over the years, like TEMPO (2,2,6,6-tetramethyl-1-piperidinyloxy) and the more advanced and more generally applicable SG1 (*N-tert*-butyl-*N*-(1-diethylphosphono-2,2-dimethylpropyl). Atom transfer radical polymerization has also been further developed to carry out inverse ATRP as well as a combination of direct and reverse ATRP to complement each other and, hence, to reduce the amount of initiator required in the system. Apart from that, these techniques can also be run as inverse emulsion or mini-emulsion polymerizations to copolymerize more hydrophilic monomers.

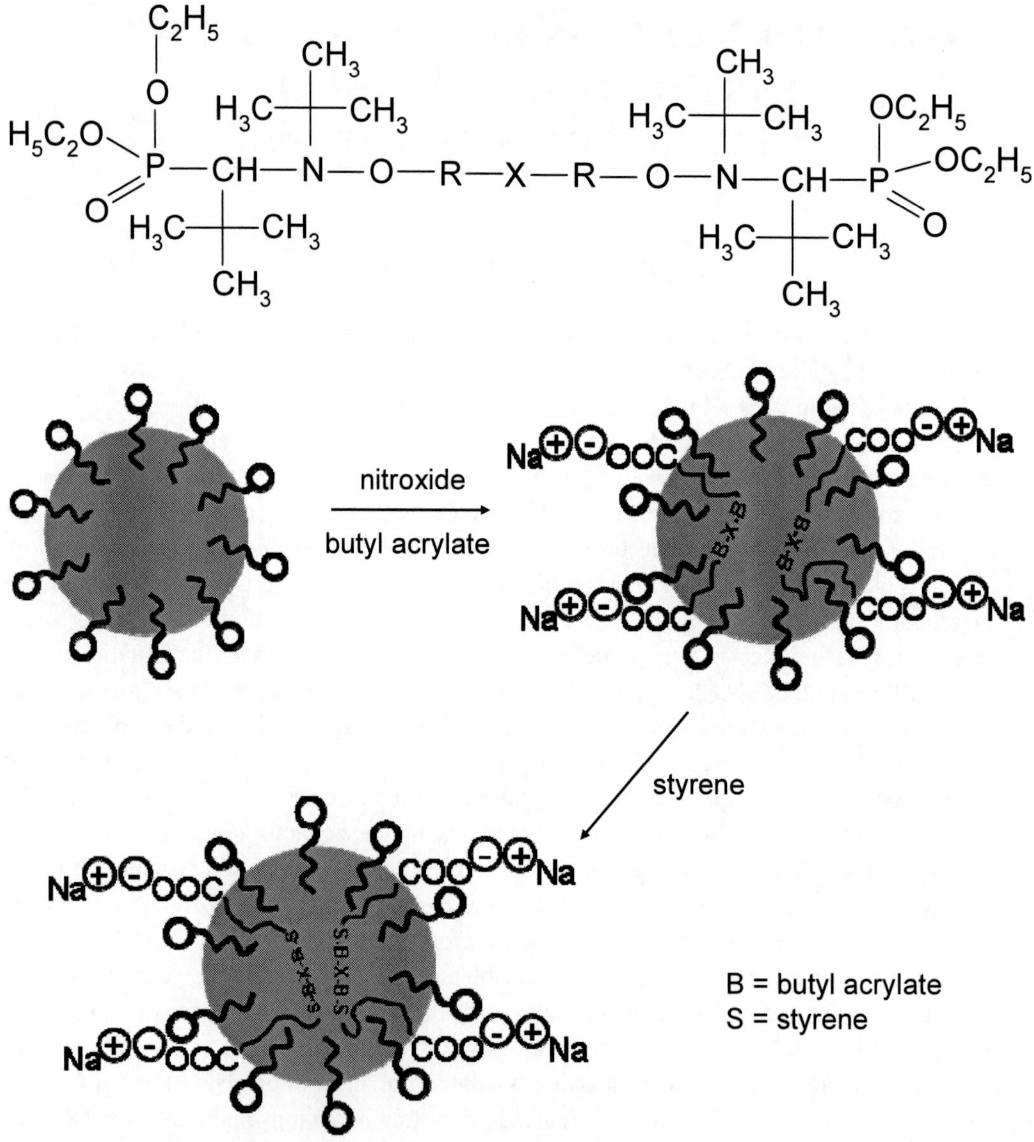

Figure 1. Bifunctional alkoxyamine used for the generation of block copolymer of butyl acrylate and styrene [2]. Sodium salt of above mentioned alkoxyamine was used.

B. Considerations in the Generation of Structured Latexes with Living Emulsion Polymerization

B.1. Nitroxide Mediated Polymerization

Nitroxide mediated polymerization has proven to be a successful technique for the generation of special macromolecules. The technique has been successfully employed for the synthesis of functional architectures in the chains in polymer particles. In one such study to generate triblock copolymers from monomers butyl acrylate and styrene, water soluble SG1 based bifunctional alkoxyamine (sodium salt of alkoxyamine was used owing to water solubility) was employed [1,2]. Dowfax 8390 was used as a surfactant in this polymerization. A seeded polymerization process, in which seed particles are generated whereby the polymer chains have low molecular weight and the solid content is also low, was used. The seed is then added with a batch of monomer to grow the particles and, this way, the generation of monomer droplets is avoided and, as a result, the initiation of polymer particles by the entry of radicals in the monomer droplets is completely eliminated. After the first batch of the monomer is polymerized, the other monomer is added, which then forms the block of the other polymer chains, thus forming copolymer particles. When a bifunctional alkoxyamine is used, one can use the two functional ends of this alkoxyamine to generate triblock copolymers. Thus, in order to generate polystyrene-b-poly(butyl acrylate)-b-polystyrene triblock copolymer particles, a seed was first generated from butyl acrylate particles. The seed was further swollen with butyl acrylate to form a central poly(butyl acrylate) block in the emulsion particles. The particles were then added with styrene to form two blocks of styrene around the central poly(butyl acrylate) block to form the triblock copolymer. Figures 1 and 2 explain in detail the process of triblock synthesis. It was reported that the molecular weights of the polymer chains increased linearly with the monomer conversion, and the polydispersity was also under control with a value of 1.64 at 85% conversion. However, it was observed that, to some extent, high molecular weight shoulders were observed in the molecular weight analysis, indicating that some penta blocks may also have been formed, leading to the increase in the molecular weight of the polymer chains. However, it was observed that no homopolymer poly(butyl acrylate) was observed, indicating that the copolymer block could be successfully initiated with styrene. However, it is generally difficult to achieve complete conversion of the first batch of monomer before the second batch of monomer is subsequently added. Also, at the high conversion of the monomers, the irreversible termination reactions increase, leading to deterioration of the molecular weight generating polydispersity in the chains; therefore, one should be careful to stop the polymerization reaction at conversions below 100%. But, this then leads to the contamination of the second block of polymer attached to the first batch, as the monomer from the first batch is not completely polymerized. This was also observed by the authors in this study in which the polystyrene blocks were to a small extent polluted by the presence of butyl acrylate when the styrene batch was added before the full conversion of butyl acrylate was achieved in order to reduce the termination reactions happening at higher monomer conversions.

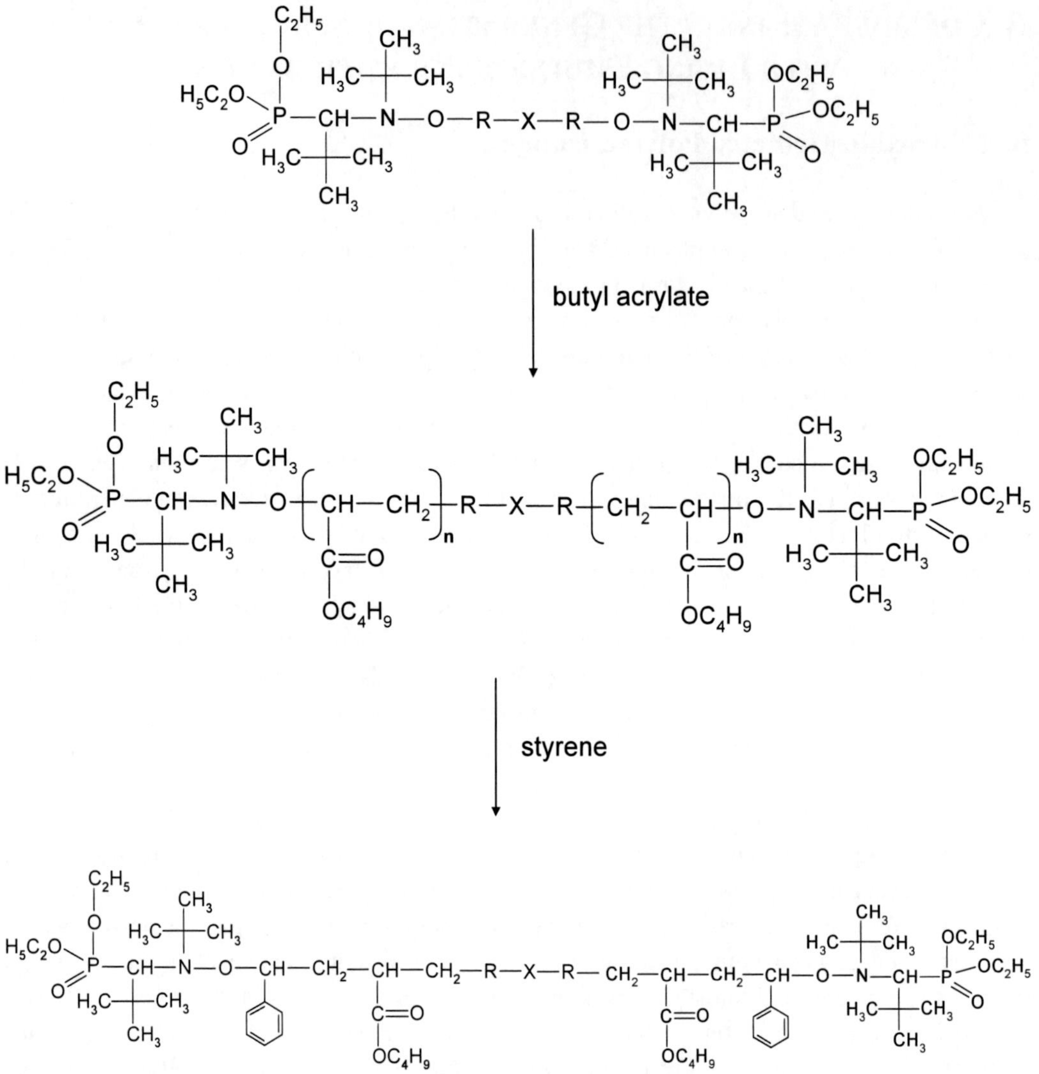

Figure 2. Polymerization reactions in the process of generation of block copolymer of styrene and butyl acrylate by using bifunctional alkoxyamine [2].

B.2. Atom Transfer Radical Polymerization

Initial trials to achieve controlled living emulsion polymerization by using the atom transfer radical polymerization method suffered from the generation of coagulum, colloidal instability and the reaction of the catalyst with the anionic surfactant sodium dodecyl sulphate, conventionally used in the emulsion polymerization processes [3,4]. But with time, significant advances have been achieved in the initiator, catalyst and surfactant systems, and also by the use of seeded polymerization techniques, controlled polymerization can be efficiently carried out by ATRP method.

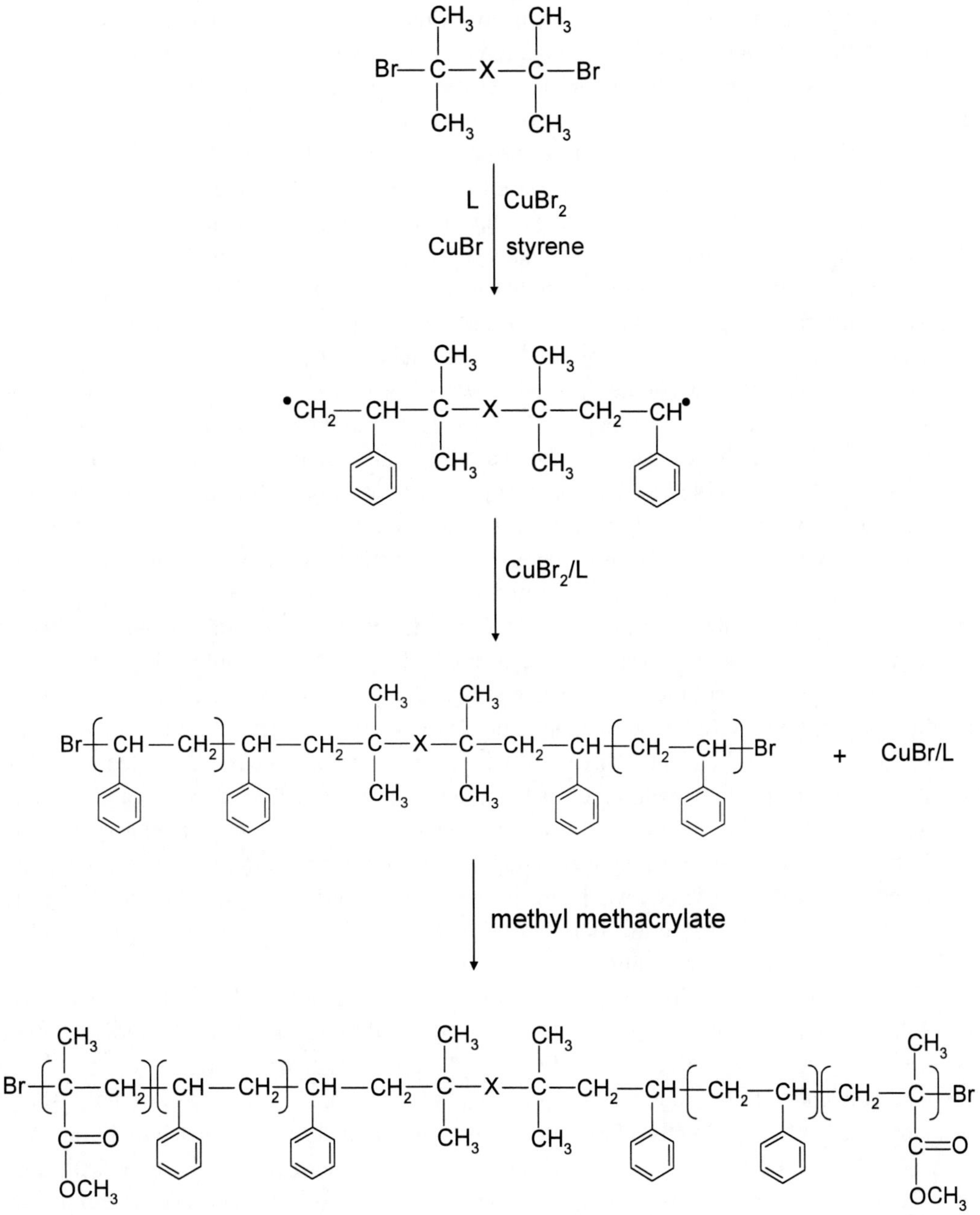

Figure 3. Polymerization process of the generation of triblock copolymer particles of poly(methyl methacrylate)-b-poly(styrene)-b- poly(methyl methacrylate) using ATRP.

In one such study, block copolymerization of 2-ethylhexyl methacrylate and methyl methacrylate was carried out in emulsion using atom transfer radical polymerization [5]. Bifunctional initiator of 1,4-butylene glycol di(2-bromoisobutyrate) was used in combination with copper bromide/4,4′-dinonyl-2,′-bipyridyl catalyst and polyoxyethylene sorbitan monooleate as surfactant. First, a block of 2-ethylhexyl methacrylate was generated in the emulsion particles, and these particles were subsequently added with methyl methacrylate to form the triblock copolymer owing to the use of bifunctional initiator. Figure 3 illustrates the process of the generation of triblock copolymer particles of poly(methyl methacrylate)-b-poly(styrene)-b- poly(methyl methacrylate). As mentioned above, for the nitroxide mediated polymerization case, the authors here also pointed out the difficulties of achieving a clean second block of polymers, as it is not beneficial to achieve full conversion of the monomer forming the first batch, owing to the increased polymer termination and transfer reactions at the high conversion. The reactivity ratios play a significant role in the generation of the chemical composition of the copolymer chains when the monomers are added together, as the more reactive monomer would polymerize faster, thus leading to the high concentration in the chains initially formed, followed by the increased concentration of the less reactive monomer at the end of the polymerization reaction. As a residual amount of 2-ethylhexyl methacrylate was present in the system when methyl methacrylate was added, it was therefore of interest to know that how this residual amount would be distributed in the second block of poly(methyl methacrylate). The authors calculated the reactivity ratios of two monomers as 0.903 and 0.930 at 70°C, which indicates that the second batch of poly(methyl methacrylate) would have random distribution of 2-ethylhexyl methacrylate units. It was observed that the molecular weight of the polymer chains increased linearly with methyl methacrylate conversion. There was no coagulum formed in the system, and the polydispersity in the molecular weight was observed to be low at 1.26. The partitioning studies of the catalyst in the organic and aqueous phase were also performed. Nonyl side chains in the bipyridine were observed to improve dissolution of the catalyst ligand complex in the organic phase, as in the ethylhexyl group in 2-ethylhexyl methacrylate. However, when methyl methacrylate started to form at the ends of the chains, the partitioning coefficient of the catalyst started to change and the balance began to shift in favor of the aqueous phase. As the catalyst ligand complex would have better affinity for the polymer formed from 2-ethylhexyl methacrylate than poly(methyl methacrylate), its concentration therefore in the poly(methyl methacrylate) blocks could be expected to be low. Thus, during the polymerization of methyl methacrylate, the catalyst was not present in the vicinity of the loci of polymerization, thus, leading to loss of polymer molecular weight control.

In another study on the use of the seeded polymerization approach, block copolymers of poly(i-butyl methacrylate) and polystyrene were prepared by using ethyl 2-bromoisobutyrate as initiator and CuBr/4,4′-dinonyl-2,2′-dipyridinyl as catalyst ligand complex [6]. Tween 80 (polyoxyethylene sorbitan monooleate) was used as surfactant. First, a seed of poly(i-butyl methacrylate) end-capped with ATRP initiator was prepared, to which a batch of styrene was then added to form a block copolymer. The rate of polymerization of styrene with atom transfer radical reagents was much slower than that of i-butyl methacrylate; therefore, the reaction was carried out at an elevated temperature of 70°C. However, it also increased the partioning of $CuBr_2$ in the aqueous phase. By using the nuclear magnetic resonance, the molar ratio of styrene to i-butyl methacrylate was observed to be very close to the theoretical values, indicating that the polymerization reactions form the block copolymer chains well under control. The molecular weight distribution was also observed to be low at 1.2. After the generation of polystyrene block, the molecular weight of the polymer chains increased from 23,100 to 30,100, and molecular weight distribution was further improved to 1.1.

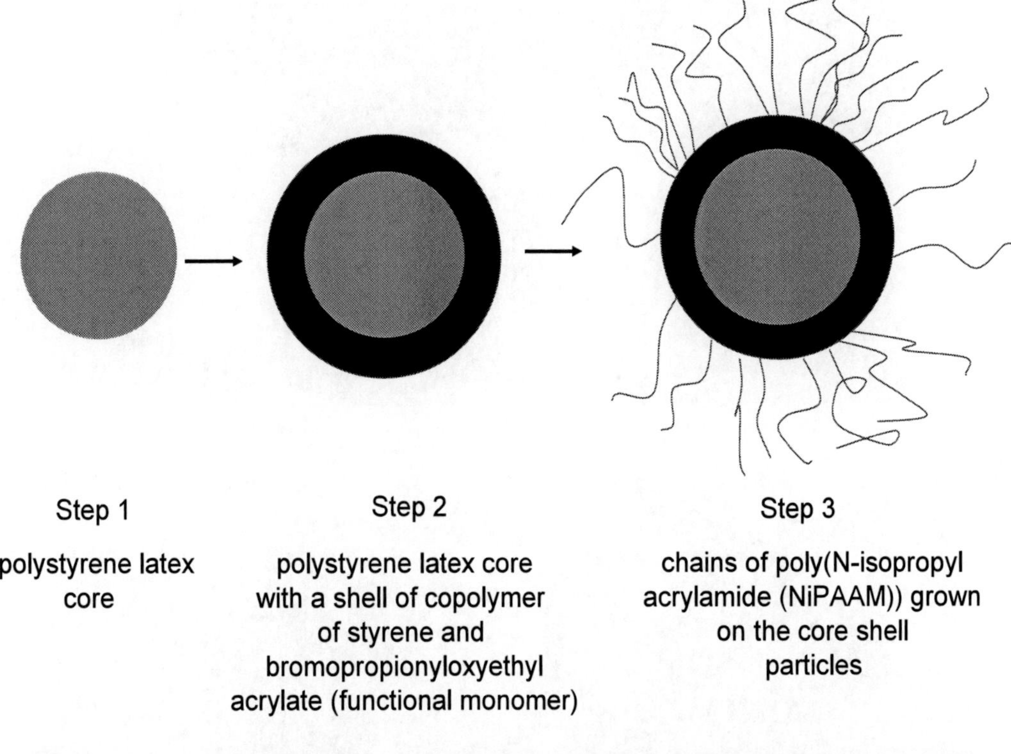

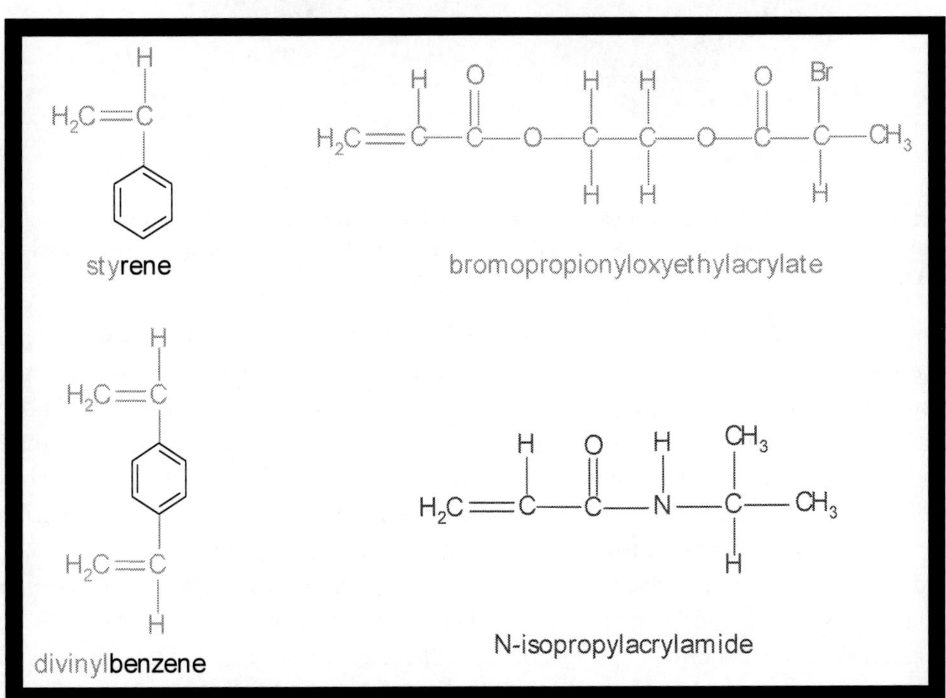

Figure 4. Scheme of generation of core shell morphology of particles by ATRP and the chemical structures of various monomers used in the process [7-9].

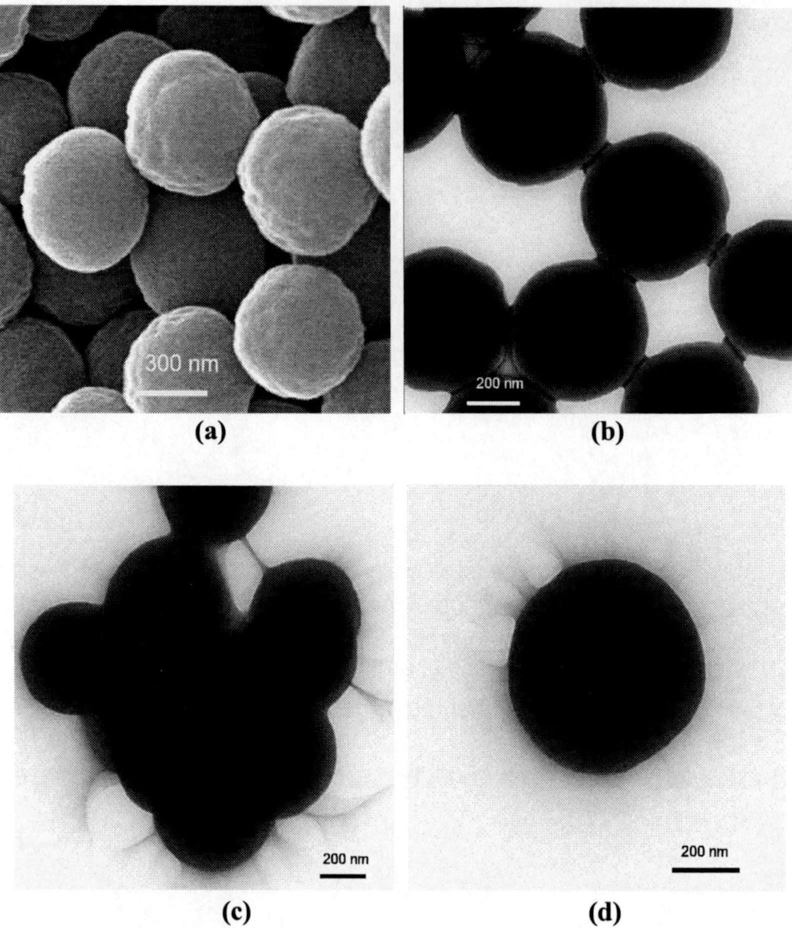

(a) (b)
(c) (d)

Figure 5. (a,b) Latex particles functionalized with a copolymer layer carrying ATRP initiator and (c,d) the grafted brushes of poly(N-isopropylacrylamide) from the surface of the functionalized particles using ATRP.

Special core shell morphology could be achieved by using atom transfer radical polymerization in emulsion carried out in the absence of surfactant [7-9]. In this specific morphology, a seed was generated first, which was covered with a thin shell of polymer containing ATRP initiator moieties. These moieties could then be used to graft polymer chains by the atom transfer radical polymerization method to achieve grafting of the polymer brushes from the surface of the polymer particles. As atom transfer radical polymerization isaffected by the ionic surfactants used during emulsion polymerization, ATRP was therefore studied without the use of surfactant. Though the emulsion polymerization in the absence of surfactants generates larger particles (higher than 500 nm in average diameter), the larger diameter particles are beneficial in terms of their characterization, e.g., to track changes in the surface morphology by microscopy. Figure 4 is a representation of the process of grafting from the surface of the particles. In this case, a seed of polystyrene is formed and is covered subsequently with a shell of copolymer of styrene and another acrylic monomer which on its other end contains the ATRP initiating site. The core and shell can be formed either as a batch process of two well-defined separate steps; or, it is also possible to generate the core shell structure in semi-batch conditions with a combination of two steps, where the shell-forming monomers are added to the seed when the seed is polymerized to, say, 70%. In the third step in Figure 4, grafting of poly(N-isopropylacrylamide) is achieved by using atom transfer radical polymerization. The benefit of atom transfer radical polymerization in generating grafts from the particles' surface is the ability to control the molecular weight and chain lengths of the grafts grown from the surface. The grafting density can also be varied by varying the amount of ATRP initiator in the shell. As the polymer particles may be used for applications in which high pressure environments are used, e.g., chromatography columns, the crosslinking of the particles therefore may also be carried out by the addition of crosslinker in seed and shell forming monomers. The aqueous ATRP has been used in such studies for the grafting processes, and the advantage of this technique in such systems is its applicability at room temperature compared to the much higher temperatures required for nitroxide mediated polymerizations. Although the applicability of aqueous ATRP for grafting reactions from various surfaces was reported to be less efficient, it can still be good enough for the requirements of the grafts on the polymer particle surfaces. Though the presence of ATRP agent at the ends of the polymer chains may not be absolutely acceptable in many of the applications, many techniques, however, have dealt with post polymerization processes to convert these groups into more acceptable groups and, in the case of graft polymerization from the surface, these groups are easily accessible for these post polymerization processes and are not embedded deep inside the polymer particles.

Figures 5 and 6 show the microscopical evaluation of the system described in Figure 4. The grafted polymer layer of poly(N-isopropylacrylamide) around the latex particles is visible as a thin mass and, depending on the adsorption of the particles on the grids used during transmission electron microscopy, can be seen as brushes sprouting from the surface of the particle or as a continuous layer covering the surface. Although the ATRP in bulk is very efficient in controlling the molecular weight by delaying or eliminating the termination reactions, as mentioned above, one must be mindful that it may not be efficient to that extent for surface grafting reactions, as most of the polymer formed by atom transfer radical polymerization was found away from the surface, i.e., unbound to the surface. It is also possible that these chains were formed bound to the surface, but later got snatched from the surface, or the chain transfer reactions with the species present in the solution can lead to the transfer of the propagating radical from the surface of the particles to the aqueous solution. It is also possible that some ATRP initiator may be present unbound to the particles' surface which is residual from the step of generating a shell on the seed particles, which can then initiate chains of pure poly(N-isopropylacrylamide) in aqueous solution forming large amount of solution polymer. However, the grafts achieved in the process are still very controlled in molecular weight and polydispersity. The particles shown in Figure 5 were prepared by polymerization of styrene to full conversion and, in the next step, generation of a thin shell was achieved by the polymerization of styrene and another monomer with other end carrying the ATRP initiating moiety. The particles shown in Figure 6 were prepared by fully polymerized seed monomer of styrene, but polymerized in the presence of crosslinker divinyl benzene. The seed particles were then covered in the next step with a thin shell of the copolymer of styrene, divinyl benzene and the third monomer carrying the ATRP initiator at its chain end. Figure 7 also underlines the effect of cleaning the latex as well as staining the latex during microscopic investigations. The grafted latex particles, when present in the state in which they were prepared, seemed to have thick layers of poly(N-isopropylacrylamide), as seen in the first micrograph of Figure 7. However, after cleaning thoroughly by multiple centrifugation and resuspension cycles, the layer seems to have become quite thin, although it is only qualitative observation and the image in the microscope can also be affected by other factors. However, both the images were taken without the use of a staining agent to compare the effect of cleaning. Most probably, the unbound chains that are formed in the aqueous solution, due to a number of factors described above, are adsorbed physically on the surface of the particles or are merely entangled in the chains bound covalently to the surface of the particles, which get washed off when an excessive cleaning operation is undertaken. Staining the polymer particles after their adsorption on the TEM grid was helpful in understanding the layer of the polymer grafted around them. Negative staining using uranyl acetate was used in order to improve the contrast of the polymer layer as compared to the surroundings. After staining the background, the boundary of the layer grafted around the polymer particles become somewhat more visible. Though the contrast by using the staining agent improves, the staining agent also acts to shrink the polymer brushes; therefore, it is not possible to achieve any quantitative results after using staining agent.

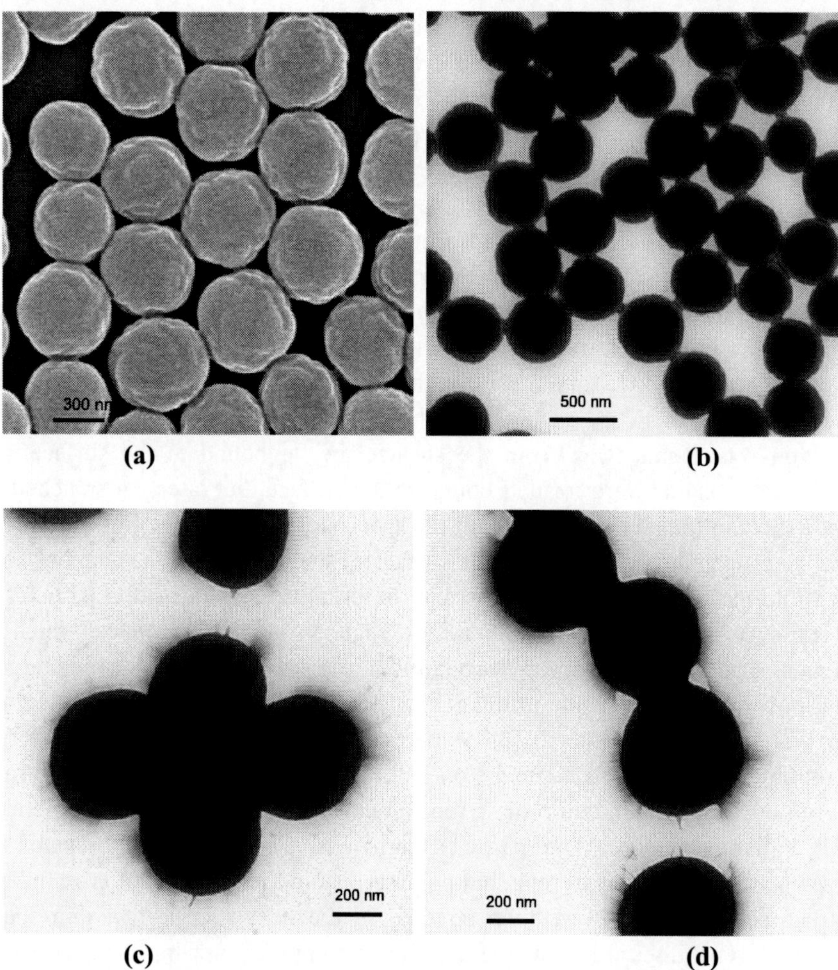

Figure 6. (a,b) Latex particles functionalized with a copolymer layer carrying ATRP initiator and (c,d) the grafted brushes of poly(N-isopropylacrylamide) from the surface of the functionalized particles using ATRP.

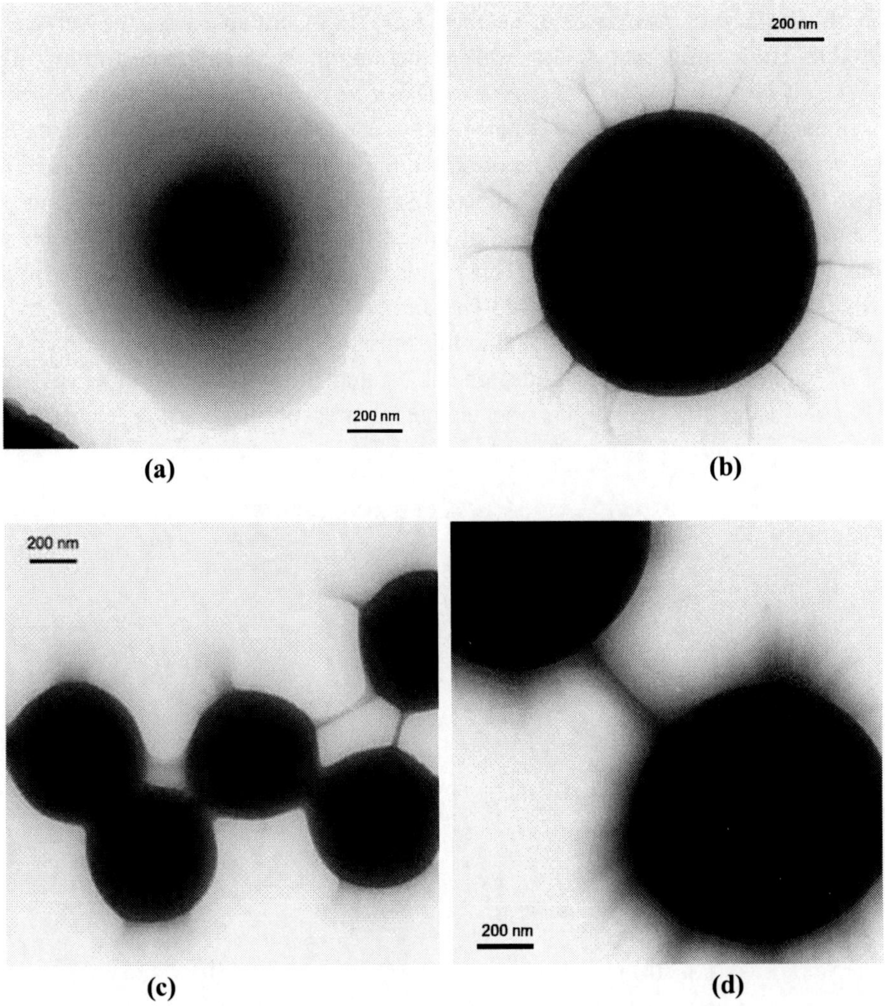

Figure 7. Effect of washing and staining the latex particles: particles (a) before and (b) after the washing, particles (c) before and (d) after staining with uranyl acetate.

As mentioned, the use of ATRP leads to controlled grafting on the surface of the polymer particles which can either be controlled by changing the density of the ATRP initiator in the shell or by changing the amount of the monomer during polymerization. By keeping the amount of initiator the same in the shell, the amount of monomer taken during grafting reaction was varied and its effect on the resulting polymer brushes was analyzed. Increasing the amount of monomer in the reaction mixture was responsible for increasing the amount of grafts that could be achieved around the polymer particles, as shown in Figure 8. An interesting set of experiments dealing with the analysis of the effect of temperature on the grafted poly(N-isopropylacrylamide) chains was also reported. As it is a thermally responsive polymer, i.e., it is hydrophobic above its critical solution temperature of 32°C and is hydrophilic below it, TEM grids therefore were prepared at 40°C, whereby the polymer chains would be hydrophobic and, thus, in the collapsed state of conformation on the polymer particles. This effect was clearly seen in the micrographs, as shown in Figure 9, for the different amount of grafts present on the surface. In any case, one could not see polymer grafts on the surface of the particle, as they were fully collapsed on the surface of the particles, thus confirming that ATRP was successful in achieving the grafting from the surface. As salt also affects the polymer brushes of poly(N-isopropylacrylamide) the same way as temperature, another set of experiments were reported in which salt solution was mixed to the latex and the particles were observed under the microscope. The grafted layer as a result of the salt was observed to become more compact and different in morphology compared to the layer seen without the use of salt, as shown in Figure 10 for different amount of polymer grafts bound to the surface. Thus, it also confirmed the grafted nature of polymer chains which can be achieved by ATRP. The authors of the study also reported the use of electron energy loss spectroscopy coupled to transmission electron microscopy to confirm the presence of poly(N-isopropylacrylamide) chains around the polymer particles, as shown in Figure 11. By using this technique, the elemental presence of various elements can be detected.

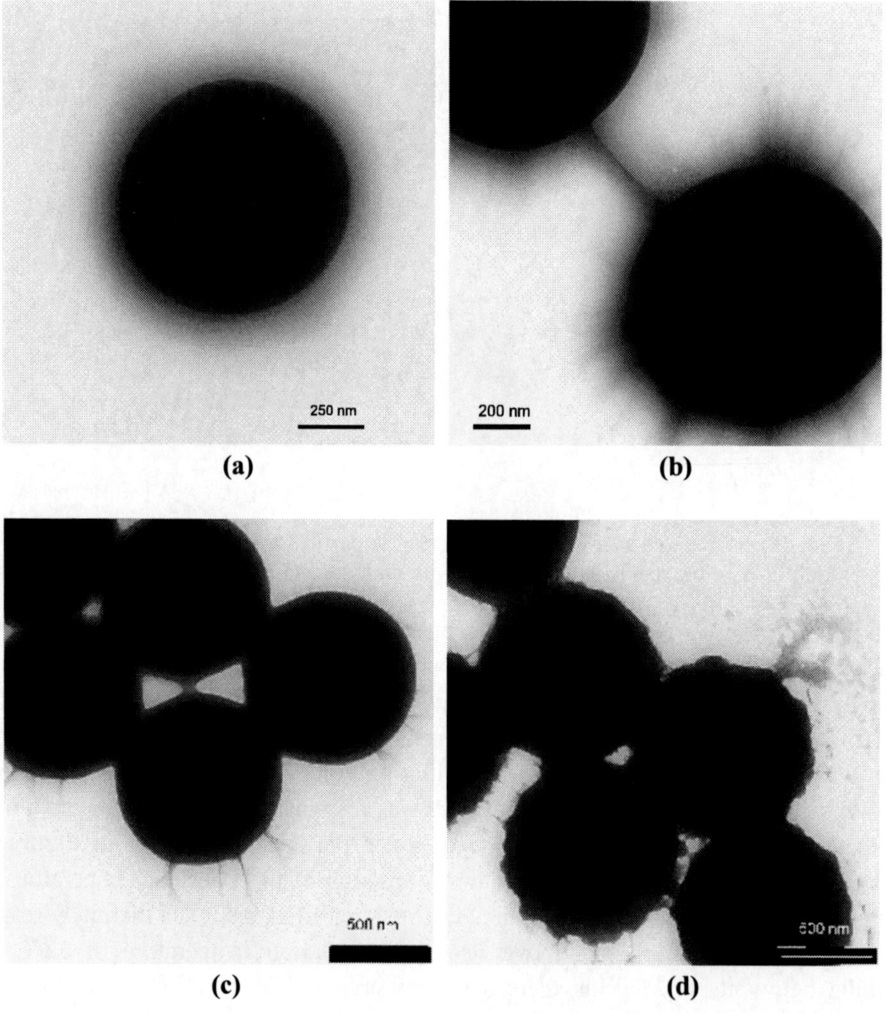

Figure 8. (a,b,d) Effect of increasing the amount of initial amount of poly(N-isopropylacrylamide) monomer in the system taken after staining. Figures 8c was the particles as shown in 8d but are observed without the use of staining agent.

As the exact thickness of the polymer brushes around the polymer particles could not be ascertained in TEM owing to less contrast of the poly(N-isopropylacrylamide) chains with the surroundings, it was of interest to quantify the exact amount of polymer grafted on the particles' surface in order to relate it to the efficiency of the ATRP process. Cryo SEM performed on the grafted particles indicated that, in fact, a very thick layer of poly(N-isopropylacrylamide) brushes was grafted on the surface of the particles, as shown in Figure 12. Free polymer was also found to be present between the latex particles. Although it confirms the high efficiency of ATRP in achieving the grafting, it was only a qualitative observation. However, many reported studies in literature have shown the cleavage of the grafted polymer chains from the surface, and the molecular weight of the chains was quantitatively related to the ATRP process [10,11].

Most of the above-mentioned studies on the grafting of the polymer brushes were performed on emulsion polymerization particles generated without the use of surfactant. The resulting particles were bigger in size (~500–800 nm); therefore, it was of interest to confirm that these grafting reactions can also be carried out on the surface of small particles. The small particles can only be achieved by the use of surfactants; therefore, either the nonionic surfactant should be used to generate these particles or the ionic surfactant should be removed from the surface of the particles by extensive cleaning. In one such study, the crosslinked polystyrene seed particles were synthesized by using sodium dodecyl sulphate as surfactant, and a shell was similarly generated on the surface of these particles by forming the copolymer in which one monomer contained a terminal ATRP moiety. The latex after shell generation was subjected to extensive washing cycles of centrifugation and redispersion in the aqueous medium, and an ATRP reaction was carried out on this latex to graft poly(N-isopropylacrylamide) chains from the surface [12]. Figure 13 shows the particles before and after the grafting. It is clear that the particles were grafted with thick layers of poly(N-isopropylacrylamide), thus indicating a successful ATRP process. As the polymer particles before the ATRP reaction may still have had some of the ionic surfactant adsorbed on the surface, the overall polymerization reaction may have been affected; however, the generated grafts seemed to be enough to change the functionality of the latex particles on the surface. The generated grafts were also quantified as a function of temperature in dynamic light scattering, as the size of the particles became different at different temperatures owing to different extents of swelling of the grafted brushes on the surface. In Figure 14, one can see that as temperature falls below 32°C, the size of the brushes increases and is maximum nearing 200 nm at 10°C. Heating beyond 32°C, the size of the brushes is practically reduced to one-tenth value at 20°C, indicating that ATRP was successful in grafting the small particles.

The particles grafted with poly(N-isopropylacrylamide) could be visualized as floating in water with a swollen mass grafted around them. The extent of swelling of the polymer chains was also quantified [9,12] and shown as a function of time in Figure 15. The swelling-deswelling behavior as a function of temperature is similar in the particles shown in Figure 14. The analysis as a function of time is necessary to ascertain the speed of response of the particles to the change in the temperature. As it can be seen, the particles achieve a great extent of swelling (up to 350%) within half an hour, and after that the swelling starts to level off. The deswelling was achieved within 20–30 minutes. The figure shows particles from Figures 5 and 6.

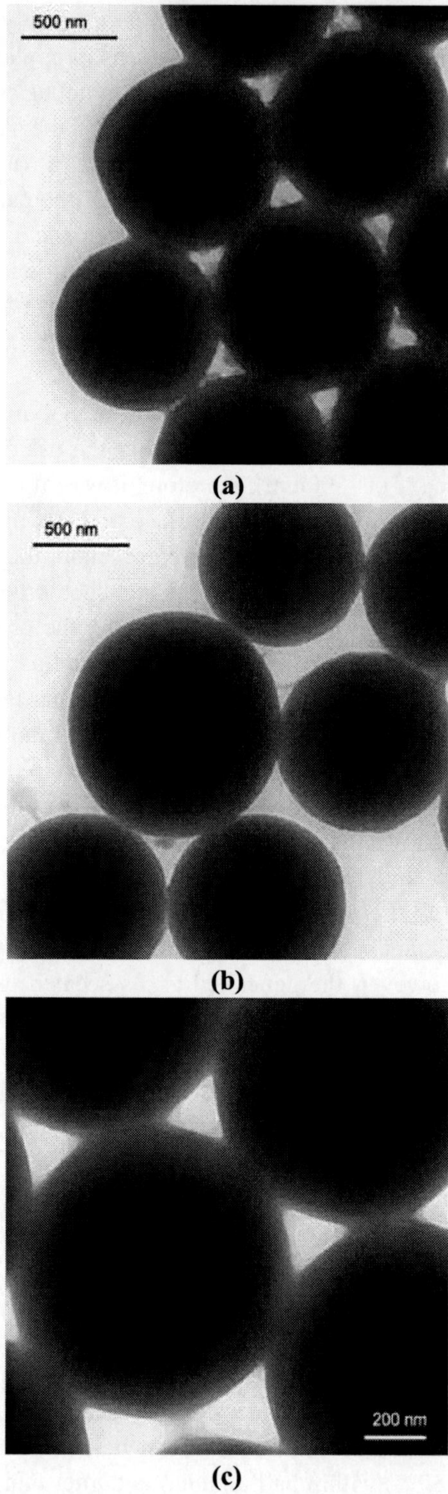

Figure 9. Effect of temperature on the brushes generated by increasing amount of N-isopropylacrylamide, as shown for the particles corresponding to Figures 8a, b and d.

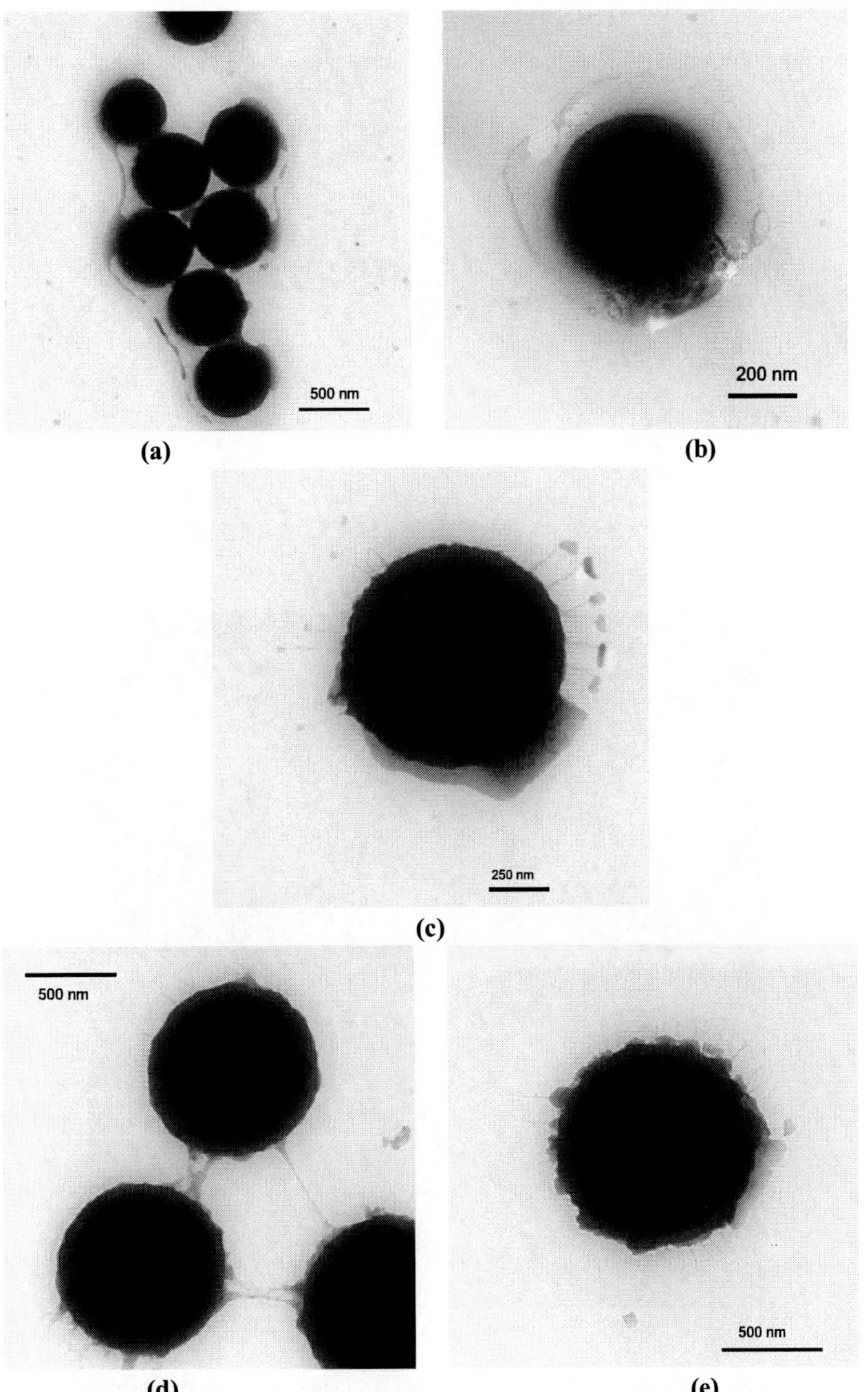

Figure 10. Effect of salt on the particles generated by increasing the amount of isopropylacrylamide in the system. Particles in (a) and (b) correspond to particles of Figure 8a, particles in (c) correspond to Figure 8b, and particles in (d) and (e) correspond to Figure 8d.

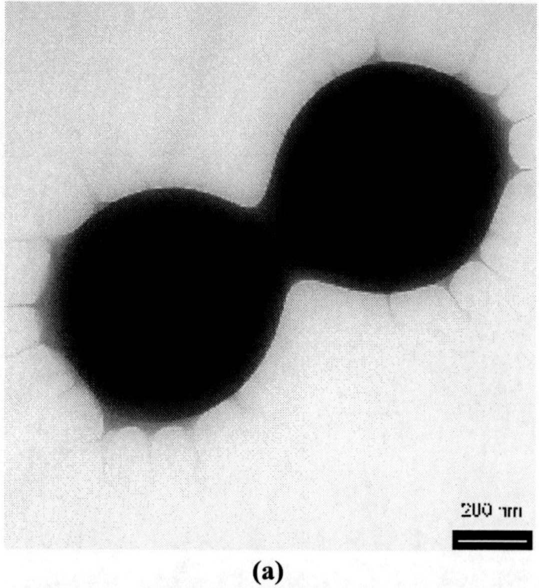

(a)

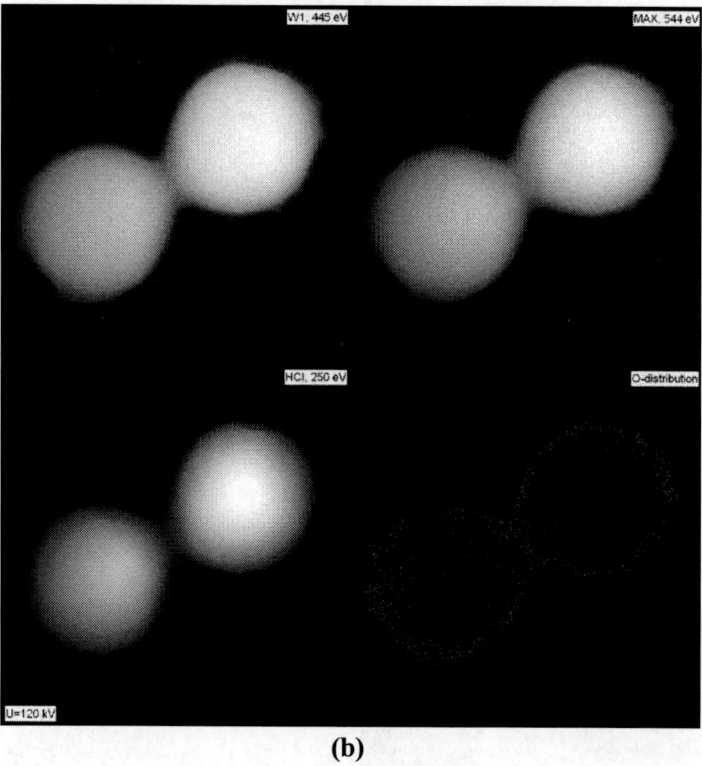

(b)

Figure 11. EELS analysis of the particles: (a) particles seen in the microscope and (b) the same particles analyzed for the presence of oxygen atoms corresponding to isopropylacrylamide (the image with dots on the bottom right side).

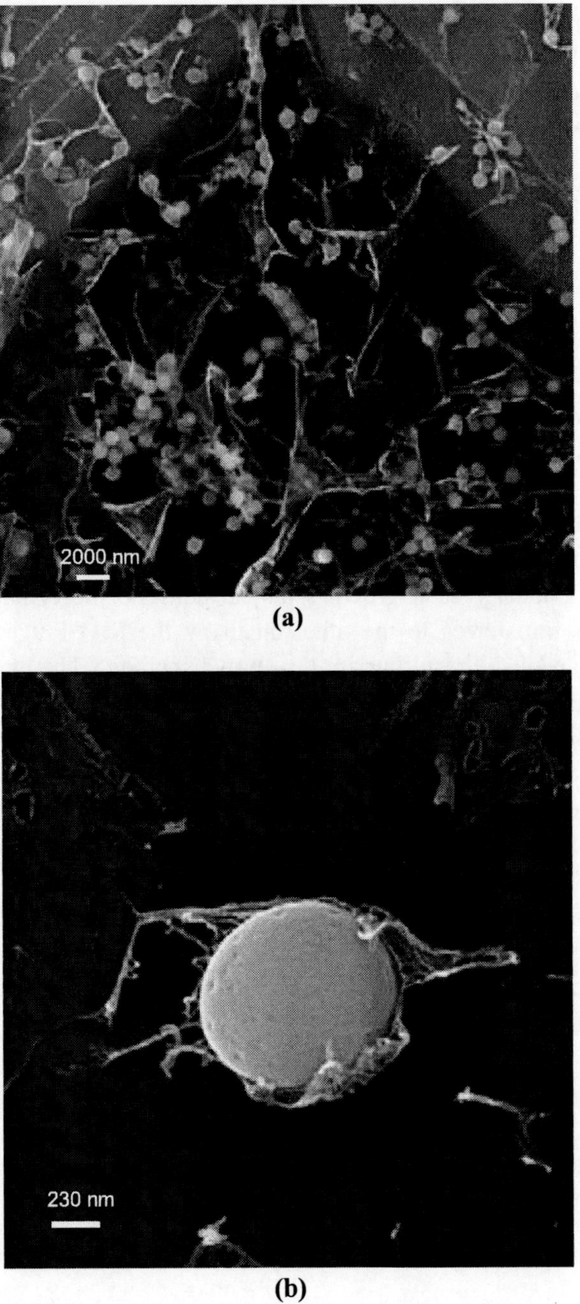

Figure 12. Cryo SEM analysis of the particles modified with brushes of poly(N-isopropylacrylamide).

B.3. Radical Addition Fragmentation Transfer

Radical addition fragmentation transfer (RAFT), which works on the method of controlled transfer of the growing chains in order to eliminate the irreversible termination and to maintain livingness of the system, has also been successfully used to generate polymers with special morphologies or architectures by using emulsion polymerization. Although similar initial problems were faced, as mentioned in the case of atom transfer radical polymerization, but by the development of special system specific RAFT agents, many more advancements have been achieved in recent years. In such a study, core shell functional particles were achieved with RAFT in emulsion using O-ethylxanthyl ethyl propionate as a RAFT agent [13]. The polymerization reactions were carried out in the presence of poly(methyl methacrylate) seed particles of predetermined number and size distribution, as the use of seed helps to control the polymerization reactions. The seed was added first with styrene monomer to form polystyrene block, to which butyl acrylate was then added. Styrene was added in batch conditions, whereas butyl acrylate was added slowly to the emulsion system in order to avoid the buildup of a high concentration of monomer in this system, as a high concentration may lead to a reduction in the control of the livingness of the polymerization reaction. Figure 16 is a representation of the generation of poly(methyl methacrylate)-b-polystyrene-poly(butyl acrylate) block copolymers by using xanthates as RAFT agent. It was observed that when the monomers were added in batch mode, the block polymerization is totally out of control using the xanthates RAFT agent and polydispersity values can reach up to 2. However, by slower semi batch mode of addition of butyl acrylate to the emulsion system, ideally each growing radical adds no more than one monomer unit before the radical is transferred to the other chain by the RAFT agent indicating that the chains grow simultaneously resulting in low polydispersity. The polystyrene block was obtained with a molecular weight of 7,000 and a polydispersity of nearly 2. However, after the formation of block of poly(butyl acrylate), the molecular weight increased to 20,000 and the polydispersity was much reduced to a value of 1.3. It is, thus, expected that in the case of semi batch addition of styrene to the system, the polydispersity can further be reduced. It was confirmed that the molecular weight of each block could be independently controlled by using the same RAFT agent. Furthermore, it was reported that final latex particle size can be controlled by changing the number as well as size of the seed particles, which also acts in the similar way as changing the amount of two monomers, which form the core and shell of the particles. It was observed that very high purity of the block copolymers was obtained by using this method, though very small amount of contamination from the other monomer was present. Therefore, one must be careful to ensure that the seed particles are further polymerized and there is no generation of other family of particles by secondary nucleation.

C. Considerations in the Generation of Structured Latexes with Living Miniemulsion Polymerization

C.1. Nitroxide Mediated Polymerization

Gradient copolymers of styrene with butyl acrylate were reported by using SG1-based alkoxyamine derived from methyl acrylate [14]. To achieve miniemulsion polymerization using nitroxide mediated route, monomers, alkoxyamine, hexadecane costabilizer, high molecular weight polystyrene and some free SG1 alkoxyamines were added to the aqueous phase which contained surfactants and the buffer. The surfactants used were sodium dodecyl sulphate, Forafac ($C_8F_{17}CH_2CH_2COO^-K^+$) and Dowfax 8390, which is a mixture of mono- and di-hexadecyl disulfonated diphenyloxide disodium salt. The miniemulsion was generated by the use of ultrasonification and the polymerization reaction was initiated by using high temperature of 125°C. It was remarked that, owing to the different reactivity ratios of the monomers, the composition drift affects the distribution of monomers in each chain but the composition distribution in the whole system is not affected, which is not the case when the polymerization is achieved by using conventional emulsion polymerization mode. This results in a gradient, but controlled composition of the monomers in the resulting polymer chains. It was also experimentally confirmed, as the compositional distribution was much narrower than that achieved in conventional emulsion polymerization, where there is practically no control in the compositional distribution. The extent of styrene was also observed to decrease as a function of time or monomer conversion, further confirming the generation of controlled gradient copolymers by using this nitroxide mediated method. As reactivity ratios of styrene and butyl acrylate were observed to be 0.81 and 0.23 at 120°C in bulk, it is obvious that the initial chains formed would be extremely rich in styrene, and as the polymerization proceeds, the amount of less reactive butyl acrylate keeps on increasing until the point is reached where the chains are very rich in butyl acrylate monomer molecules. Another confirmation regarding the synthesis of gradient copolymers was achieved from the glass transition temperature of the copolymer chains. A single glass transition temperature peak was observed, indicating that only one phase existed in the system, which confirms the generation of copolymer. The glass transition temperature was further reduced as a function of time or monomer conversion, indicating that the amount of butyl acrylate was increasing in the copolymer chains which led to the reduction in this temperature.

Block copolymers of polystyrene and poly(butyl acrylate) were also reported by using the nitroxide mediated polymerization route [15]. 2,2,6,6-tetramethylpiperidine-N-oxyl (TEMPO), as well as the more water soluble version 4-hydroxy-2,2,6,6-tetramethylpiperidine-N-oxyl (OH-TEMPO), were employed as nitroxides for the polymerization reactions. The TEMPO nitroxide has not been very successful in synthesizing living polymers with acrylates and methacrylates, because the trapping rate for butyl acrylate chains with the nitroxide is much faster, leading to the accumulation of the dormant species, thus leading to a reduction in the polymerization rate. However, the authors suggested that by using miniemulsion mode of polymerization, it is possible to control the partioning of the nitroxide into the organic and aqueous phases, which can help to solve the problem. In the miniemulsion polymerization reactions in the study, hexadecane was used as costabilizer and sodium dodecyl benzene-sulfonate was used as surfactant. To achieve the block copolymers, polystyrene block was first generated, followed by the addition of the second block of poly(butyl acrylate). Trials with both water and oil soluble initiators of potassium persulphate and benzoyl peroxide were performed along with both types of nitroxides, i.e., TEMPO and OH-TEMPO. Polystyrene block generated by using potassium persulphate initiator was observed to generate higher molecular weight compared to the benzoyl peroxide initiated polymerization. It was also observed that the polymer chains generated by using TEMPO as nitroxide had a lower molecular weight in comparison with the polymer chains end capped with OH-TEMPO. When the copolymerization was achieved, OH-TEMPO-mediated polymers had a much higher conversion of butyl acrylate than that in TEMPO-mediated runs. For the copolymers achieved by using benzoyl peroxide as initiator and OH-TEMPO as nitroxide, it was noted that a contamination of the copolymer with poly(butyl acrylate) was observed, owing to the greater partioning of OH-TEMPO into the aqueous phase. It was estimated from these results that only 38% of the total OH-TEMPO was expected to be present in the polymer particles, whereas the remaining 62% was present in the aqueous phase. On the other hand, owing to the more hydrophobic nature, 96% of TEMPO was expected to be present in the particles, leaving only 4% in the aqueous phase. As a result of higher water partitioning of OH-TEMPO, its amount in the particles may not be enough for the controlled polymerization, and it was thus observed that some of the poly(butyl acrylate) chains were generated and could grow to higher molecular weight owing to the fast rate of polymerization of butyl acrylate. Figure 17 shows the various permutations of the nitroxides and the initiators used in the study to generate miniemulsion particles. Similar studies to generate polystyrene-block-poly(butyl acrylate) copolymer particles have also been reported [16].

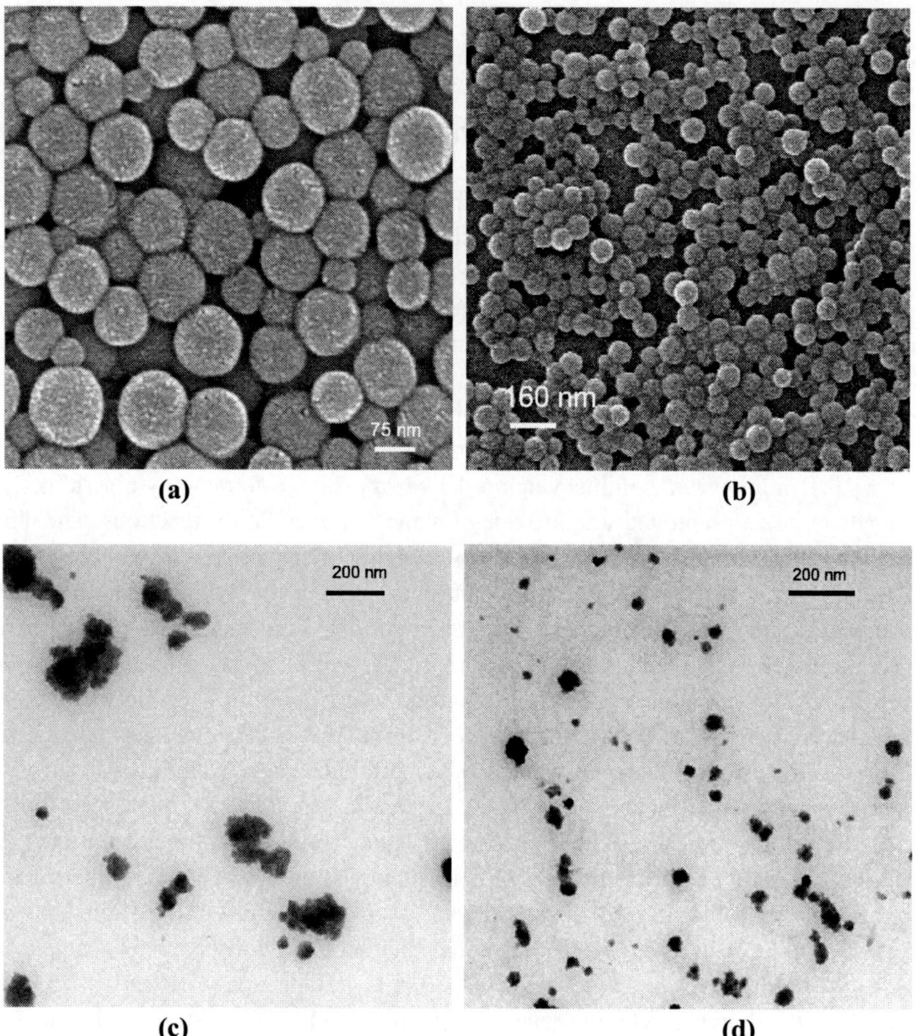

Figure 13. (a) Crosslinked polystyrene latex particles generated by the use of surfactant, (b) particles of Figure 13a surface modified with ATRP initiator, and (c) and (d) are the particles of Figure 13b after grafting with poly(N-isopropylacrylamide) brushes.

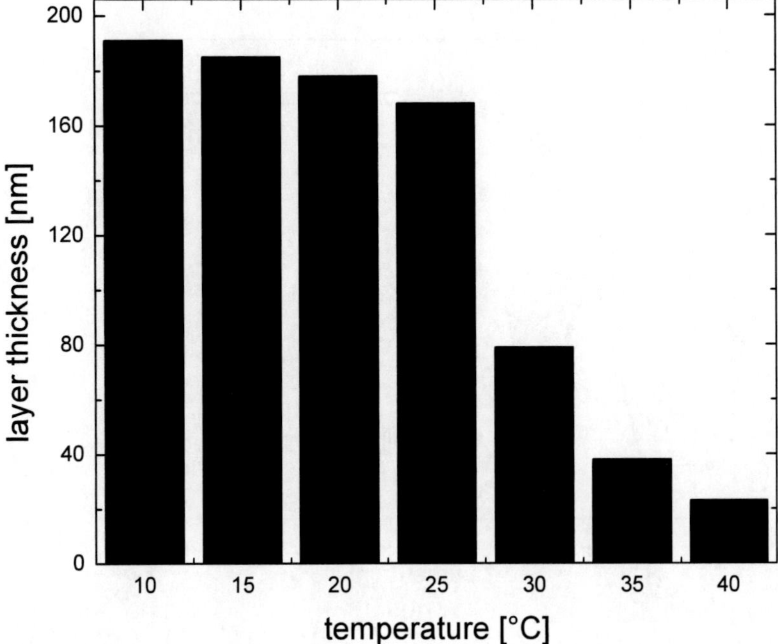

Figure 14. The thickness of the brushes of poly(N-isopropylacrylamide) around the particles and its behavior as a function of temperature.

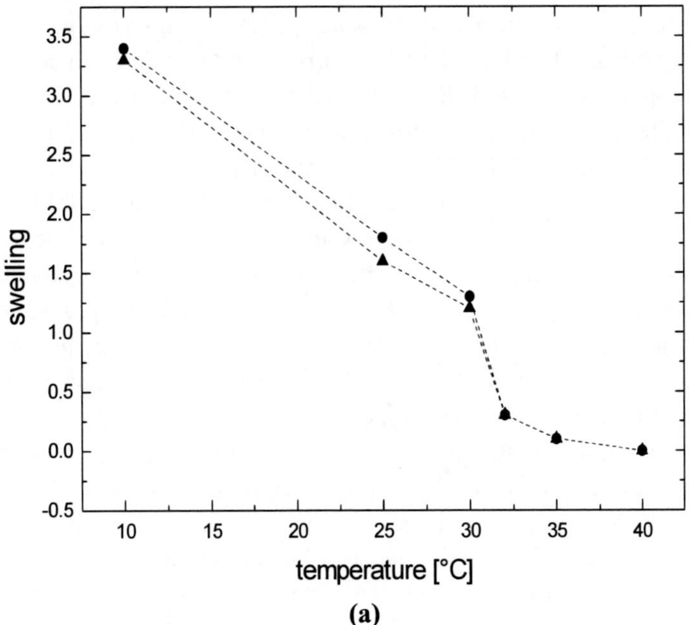

(a)

Figure 15. (Continues on next page)

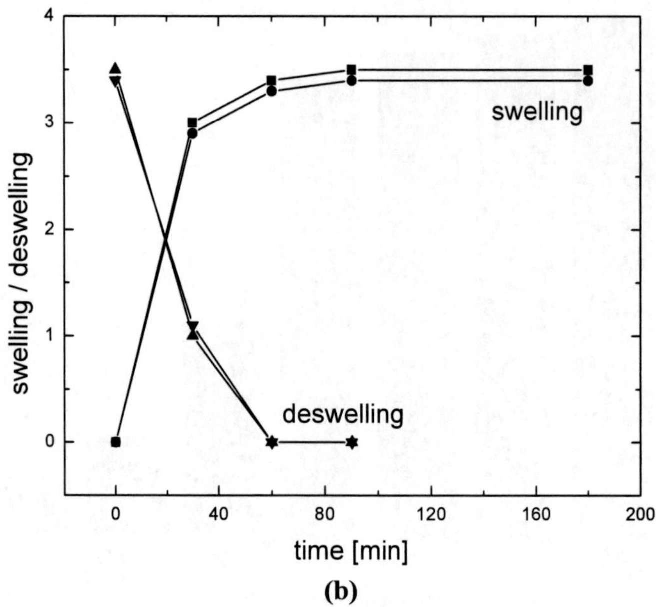

(b)

Figure 15. Swelling and deswelling of PNIPAAM brushes grafted from the surface of the particles as a function of (a) temperature and (b) time.

C.2. Atom Transfer Radical Polymerization

Simultaneous reverse and direct atom transfer radical polymerization of monomers like styrene, methyl acrylate and butyl acrylate was carried out to generate linear as well as three arm or star block copolymers [17]. AIBN was used as a conventional free radical initiator in this advanced simultaneous reverse and direct atom transfer radical polymerization process, whereby the radicals generated from the free radical initiator are used to initiate the polymerization, and the growing radicals then react with $CuBr_2$ to form dormant species. The generated CuBr then reacts again with these species to propagate the polymer chains by normal or direct atom transfer radical polymerization. The use of such techniques allows one to reduce the amount of catalyst by one tenth the amount generally employed in conventional atom transfer radical polymerization. First the macroinitiators with methyl acrylate, styrene and butyl acrylate were generated by bulk ATRP, which were later reacted further in miniemulsion to form block copolymers. Linear or star copolymers were formed depending on functionality of the initial ATRP initiator used in the generation of the macroinitiators. It was also reported that after the linear block copolymerization to generate PS-block-PMA, an average of 18% of the unreacted polystyrene macroinitiator was present in the final copolymer. This was suspected to occur because of some non-halogenated chains (or chains not capped with ATRP initiator) generated during the synthesis of macroinitiators. However, the amount of unreacted macroinitiator chains in the final copolymer was much lower when multi arm blocks were synthesized. Owing to the presence of free radical initiator AIBN present in the system, it was observed that some amount of homopolymer from the monomer forming the second block in the copolymer particles was generated.

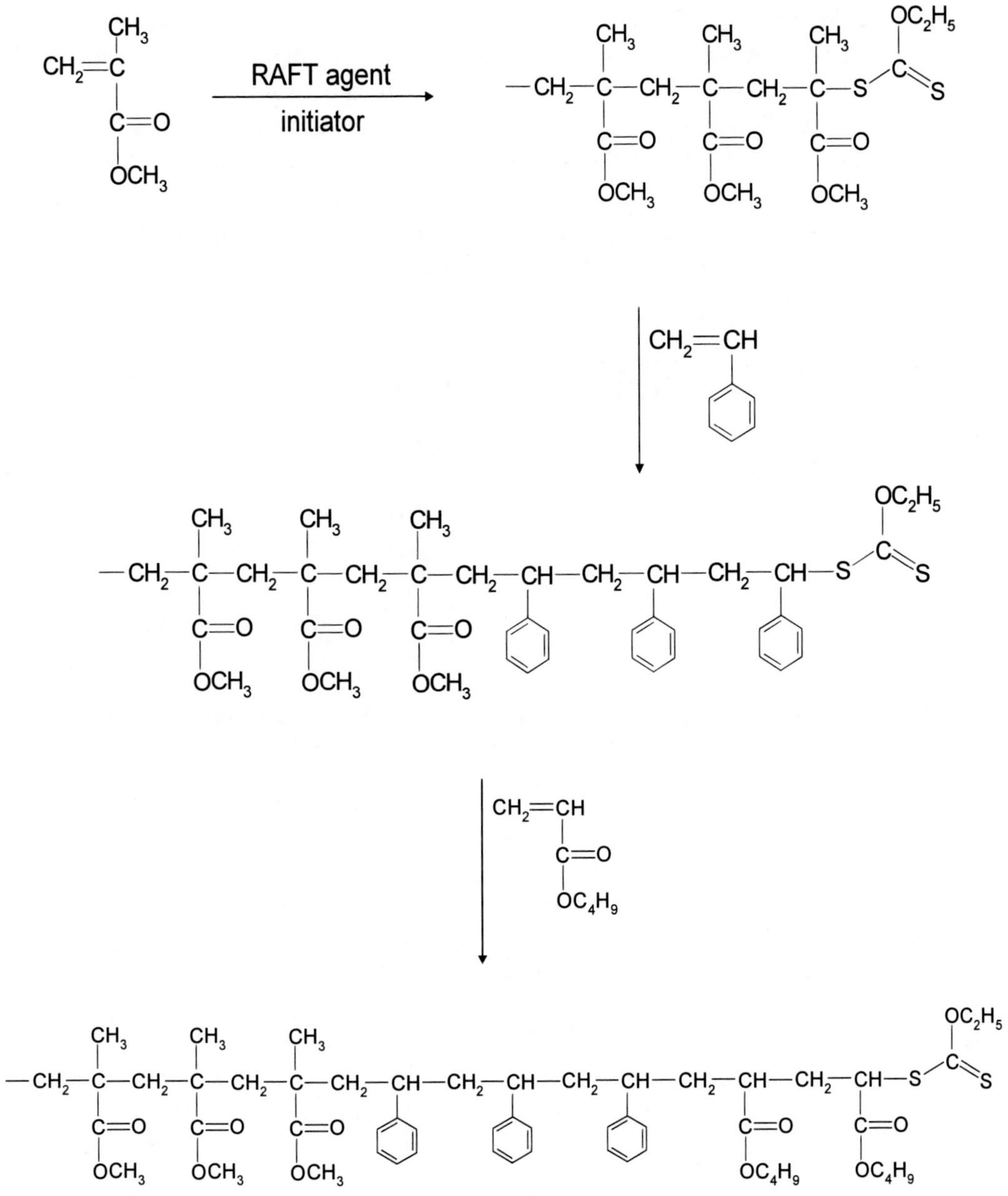

Figure 16. RAFT polymerization process for the generation of triblock copolymer of poly(methyl methacrylate), polystyrene and poly(butyl acrylate) [13].

Gradient copolymers from butyl acrylate and butyl methacrylate were also generated by using simultaneous reverse and direct atom transfer radical polymerization [18]. 4,4′,4″-Tris(5-nonyl)-2,2′:6′,2″-terpyridine (tNtpy) and bis(2-pyridylmethyl)octadecyl amine (BPMODA) were used as ligands. Hexadecane was used as cosurfactant and Brij 98 was used as surfactant. AIBN was used as conventional free radical initiator, as used above, and ethyl 2-bromoisobutyrate (EBiB) was used as ATRP initiator. It was confirmed in the study that the two monomers reacted at different rates. Butyl methacrylate was observed to be consumed faster in the beginning of the polymerization; however, owing to the reduction of its concentration in monomer mixture, its rate of consumption was also decreased, which resulted in the formation of chain ends richer in butyl acrylate. It was also found that the overall polydispersities in the chain lengths were low, which resulted because of fast exchange between the growing and the dormant chains. It was also remarked that, in the case of miniemulsion polymerization, a small amount of $CuBr_2$ complex may diffuse out of the monomer droplets, thus, enhancing termination reactions.

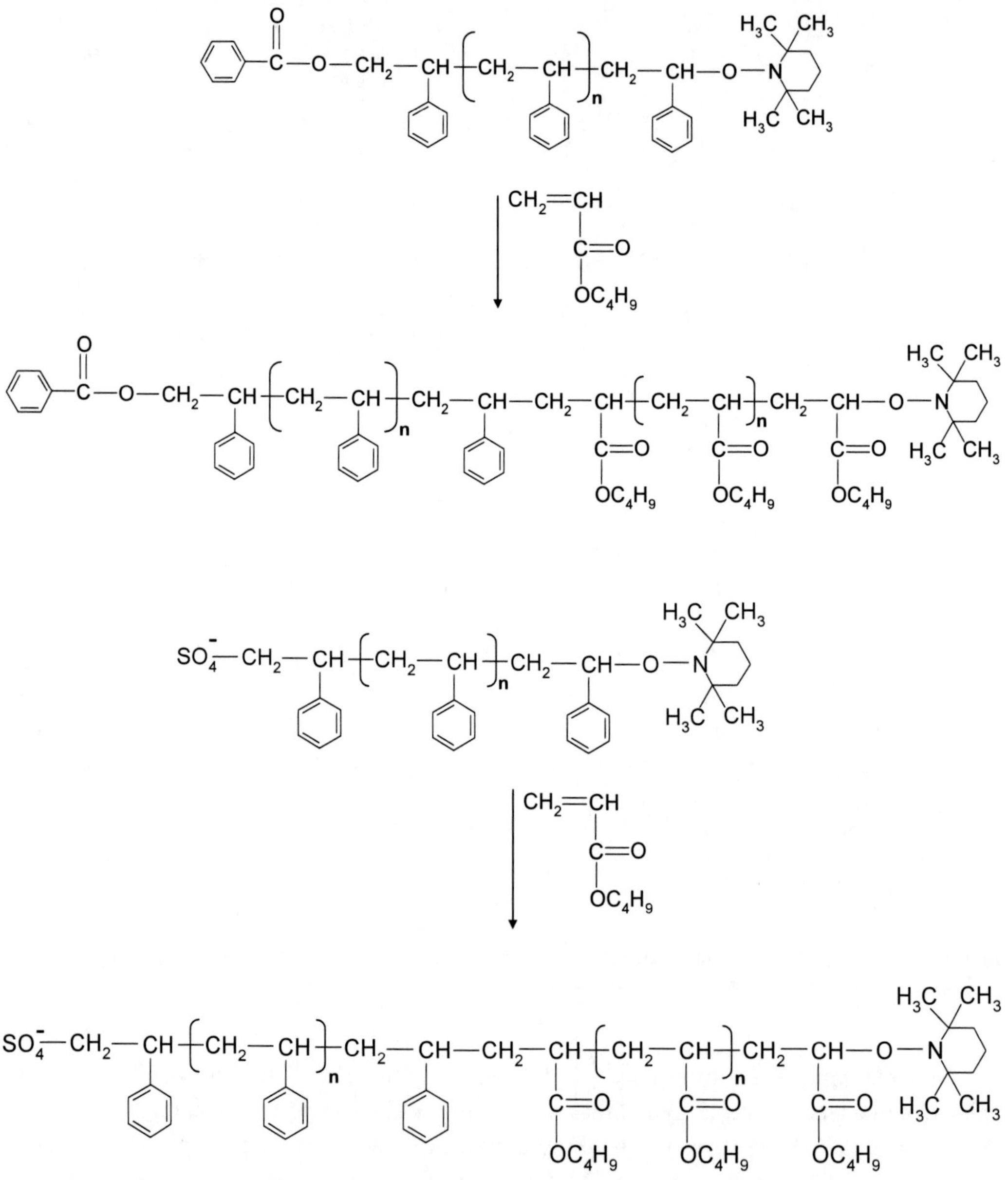

Figure 17. Block copolymers using TEMPO and different initiators [15].

Activator generated by electron transfer (AGET) atom transfer radical polymerization was also used to generate linear block as well as three arm block copolymers with polystyrene and poly(methyl acrylate) [19]. Ethyl 2-bromoisobutyrate (EBiB) and 1,1,1-tris(4-(2-bromoisobutyryloxy)-phenyl)ethane were used as ATRP initiators. Pentamethyldiethylenetriamine (PMDETA) and bis(2-pyridylmethyl)octadecylamine were used as amines. Ascorbic acid was used as a reducing agent, which reacts with $CuBr_2$ to generate CuBr, which can then be complexed with ligand to catalyze the ATRP reaction. To generate block copolymers, a macroinitiator containing a block of poly(methyl acrylate) was synthesized by using bulk atom transfer radical polymerization. Use of ethyl 2-bromoisobutyrate (EBiB) is required for the generation of linear block copolymers, whereas 1,1,1-tris(4-(2-bromoisobutyryloxy)-phenyl)ethane was used for the generation of three arm block copolymers. The second block was subsequently generated by using the macroinitiator along with monomer, ascorbic acid, hexadecane cosurfactant and Brij 98 surfactant in miniemulsion. It was suggested that the AGET ATRP method is much more beneficial than the earlier developed technique of simultaneous reverse and direct ATRP for the generation of block copolymers. In the case of simultaneous reverse and direct ATRP, the generation of a small amount of homopolymer is always observed, owing to the use of conventional free radical initiator. As there is no requirement of such conventional free radical initiator in the case of the AGET polymerization technique, the copolymers achieved by this method are, therefore, free from any homopolymer generated during the course of polymerization. Thus, the AGET technique does not require the handling of air sensitive CuBr and also generates more controlled block copolymers. The absence of any homopolymer formed in the polymerization reaction was experimentally confirmed. The macroinitiator as well as the linear block copolymer P(MA)-b-PS (poly(methyl acrylate)-block-polystyrene) had different molecular weights but similar elution volumes in 2D chromatograms, confirming that the polystyrene block was chromatographically invisible in the conditions critical for polystyrene. Three arm copolymer $[P(MA)\text{-}b\text{-}PS]_3$ could similarly be generated, and there was also no formation of the homopolymer in the polymerization system. Figures 18 and 19 show in detail the generation of linear and three arm block copolymers generated in the study mentioned above.

C.3. Reversible Addition Fragmentation Chain Transfer Polymerization

Miniemulsion polymerization with radical addition fragmentation chain transfer was reported to synthesize well-defined gradient copolymers from styrene and methyl methacrylate and poly[styrene-b-(styrene-*co*-methyl methacrylate)] block copolymer [20]. RAFT agent used in the study was 1-phenylethyl phenyldithioacetate (PEPDTA), whereas sodium dodecyl sulphate stabilizer and potassium persulphate initiator were used as hexadecane costabilizer. To synthesize the block copolymer, RAFT copolymerization of styrene and methyl methacrylate was carried out in miniemulsion. After the copolymerization was completed, a given amount of styrene was added to the latex. It was observed that as the reactivity ratios of methyl methacrylate and styrene are respectively 0.46 and 0.52, the copolymerization equation predicted an azeotropic point at a concentration of 0.471, where copolymerization was predicated to occur without a drift in monomer composition. When the mole ratio of methyl methacrylate in the feed was lower than 0.471, methyl methacrylate was observed to be consumed faster during the course of copolymerization. However, in the opposite case, styrene was observed to be consumed faster. It was also reported that much higher amounts of surfactant (SDS) and costabilizer (HD) levels were required to achieve colloidal stability in miniemulsion polymerization by RAFT compared to conventional miniemulsion polymerization. It was observed that the evolution of the copolymer composition could be confirmed to fit well with the theoretical predictions. As the copolymer composition changes during the polymerization, it was therefore opined that gradient copolymer instead of random copolymer would actually be generated during the course of the copolymerization. Overall, the behavior of the copolymerization process was observed to be dependent on the monomer feed ratios. RAFT miniemulsion copolymerization was very stable through out the polymerization process, when styrene was present in more amount than methyl methacrylate. However, when the opposite concentration was used—i.e., more methyl methacrylate in the feed—the RAFT miniemulsion polymerization was observed to be marginally out of control in the later stages of the copolymerization.

AB and ABA type block copolymers from styrene and butyl acrylate were produced by using RAFT in miniemulsion processes, and for this purpose a monofunctional RAFT agent, benzyl butyl trithiocarbonate, as well as a difunctional RAFT agent, butyl 4-({[(butylthio)carbonothioyl]thio} methyl) benzyl trithiocarbonate, were used [21]. To generate the miniemulsion, RAFT agent, AIBN initiator and hexadecane costabilizer were dissolved in the monomer, and this solution was added to an aqueous solution of surfactant followed by sonication. To generate the block copolymer, a second batch of monomer was

$$CuBr_2/L + \text{Reducing Agent} \longrightarrow CuBr/L$$

$$CuBr/L + \underset{\text{(ATRP initiator)}}{R-Br} \longrightarrow CuBr_2/L + R^{\bullet}$$

$$R^{\bullet} + nM_1 + CuBr_2/L \longrightarrow R-P(M_1)-Br + CuBr/L$$

or

$$CuBr/L + \underset{\text{(macroinitiator)}}{P(M_1)-Br} + M_2 \longrightarrow P(M_1)-P(M_2)^{\bullet} + CuBr_2/L$$

$$\downarrow$$

$$P(M_1)-(M_2)-Br + CuBr/L$$

Figure 18. AGET ATRP process for the generation of diblock copolymer particles.

added to the latex once the first batch of monomer was polymerized. The monofunctional RAFT agent would generate a block of first monomer, to which a batch of second monomer would be subsequently added to form a diblock copolymer. However, by the use of difunctional RAFT agent, two blocks could be added to the first block on each side owing to the two functional groups, thus leading to the synthesis of triblock copolymer. It was also confirmed that the particles containing the first block chains, which also acted as seed for the block copolymer production, were living and were fully extended to form the block of the second polymer. It was observed that block copolymerization reactions yielded different results depending on which block was synthesized first, i.e., if polystyrene was the first block or poly(butyl acrylate) formed the first block. When poly(butyl acrylate) was formed first and was used as seed to chain extend with styrene, a small amount of styrene homopolymer was observed to form during the course of polymerization. On the other hand, when polystyrene formed the first block and was extended with butyl acrylate, the broader distribution made it difficult to detect the formation of homopolymer. It was stressed that the order of block formation is important when the relative leaving group abilities of the monomers are different. It was opined that polystyrene should form the first block, owing to the better leaving ability of the polystyryl propagating radical, leading to preferential fragmentation in favor of formation of a block copolymer, thus, eliminating the possibility of generation of homopolymer of butyl acrylate. However, polystyrene would form homopolymer preferentially in the case of butyl acrylate forming the first block. It was also remarked, however, that although polystyrene forming the first block is beneficial for the polymerization process, better colloidal stability is achieved when butyl acrylate forms the seed or first block followed by chain extension with styrene. The effect of different surfactants was also indicated in the study. By using Brij 98 as surfactant in the miniemulsion polymerization of butyl acrylate, stable latexes were obtained; however, when Igepal CO-990 was used a surfactant in the RAFT mediated miniemulsion polymerization of styrene, the colloidal stability was a problem, and the block copolymers subsequently formed by this system were not optimum.

Multi functional RAFT agents were used to synthesize multiblock copolymers from butyl acrylate and iso-octyl acrylate monomers [22,23]. The various RAFT agents used were S-(1,4-phenylenebis(propane-2,2-diyl)bis(N,N-butoxycarbonylmethyldithiocarbamate)), S-(1,4-phenylenebis(propane-2,2-diyl)bis(N,N-'Kraton'carbonyl-methyldithiocarbamate)) where Kraton is a mono hydroxyl end capped poly(ethylene-co-butylene), and poly(S-(1,4-phenylenebis(propane-2,2-diyl)) bis(N,N-decoxycarbonylmethyldithiocarbamate)). The last RAFT agent was poly(RAFT) agent with roughly 12 RAFT moieties per molecule. Using the first functional RAFT agent, it was shown to generate triblock copolymers in two polymerization steps. The second RAFT agent formed triblock copolymer in merely one polymerization step, whereas the poly(RAFT) agent was observed to form multi-block copolymers in two sequential polymerization steps. In the case of the use of poly(RAFT) agent for the miniemulsion copolymerization, it was reported that the end product had, on average, the structure of poly(iso-octyl acrylate)-block-poly(butyl acrylate)(-block-poly(iso-octyl acrylate)) di and/or triblock copolymers. Although the number of RAFT moieties was much larger per molecule, the number of blocks generated as result was much lower owing to irreversible termination reactions.

In another interesting study, block copolymer poly(fluoroalkyl mathacrylate)-block-poly(butyl methacrylate) particles were synthesized by a similar two-step process described above [24]. In the first step, a RAFT polymerization in bulk was carried out to form homopolymer of poly(fluoroalkyl mathacrylate) (PFAMA) or poly(butyl methacrylate) (PBMA) which was end-capped with a dithiobenzoyl group. Subsequently, this macro chain transfer agent was swollen with second monomer, and this monomer was polymerized in miniemulsion. AIBN was used as initiator, whereas 2-cyanoprop-2-yl dithiobenzoate (CPDB) was used as RAFT agent. It was also tried to synthesize the similar copolymer particles in emulsion; however, hindrance of the diffusion of flouro monomer through the aqueous phase does not allow the controlled polymerization. This was not the case in miniemulsion, where the diffusion process is not required and the nucleation can be achieved in monomer droplets.

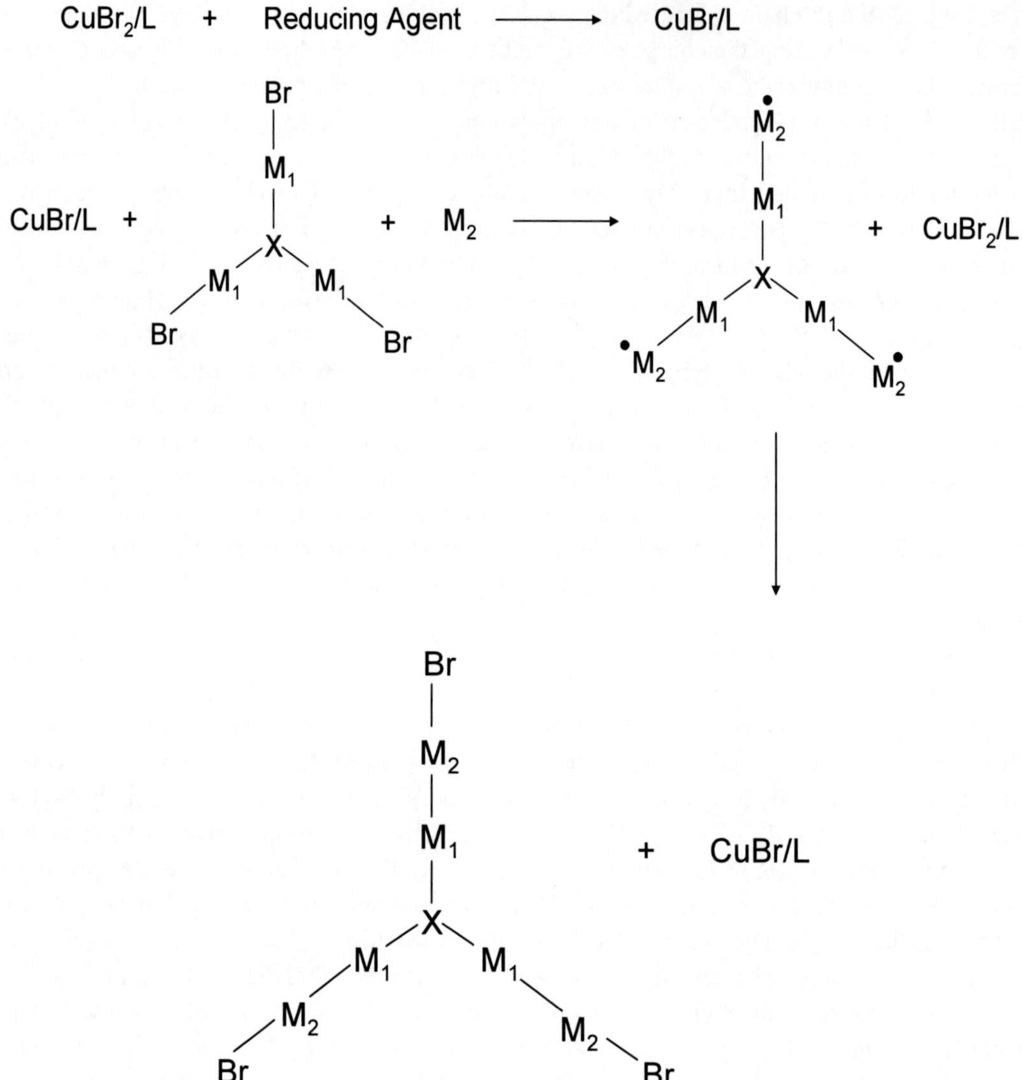

Figure 19. AGET ATRP process for the synthesis of branched block copolymers.

REFERENCES

[1] M. F. Cunningham (2008). *Progress in Polymer Science*, 33, 365-398.

[2] J. Nicolas, B. Charleux, O. Guerret, and S. Magnet (2005). *Macromolecules*, 38, 9963-9973.

[3] K. Matyjaszewski and T. P. Davis (2002). Editors, *Handbook of Radical Polymerization*, John Wiley & Sons, Inc., New Jersey.

[4] F. J. Schork, Y. Luo, W. Smulders, J. P. Russum, A. Butté, and K. Fontenot (2005). *Advances in Polymer Science*, 175, 129-255.

[5] H. Eslami and S. Zhu (2006). *Journal of Polymer Science, Part A: Polymer Chemistry*, 44, 1914-1925.

[6] M. Okubo, H. Minami and J. Zhou (2004). *Colloid and Polymer Science*, 282, 747-752.

[7] V. Mittal, N. B. Matsko, A. Butté, and M. Morbidelli (2007). *Polymer*, 48, 2806-2817.

[8] V. Mittal, N. B. Matsko, A. Butté, and M. Morbidelli (2007). *European Polymer Journal*, 43, 4868-4881.

[9] V. Mittal, N. B. Matsko, A. Butté, and M. Morbidelli (2008). *Macromolecular Materials and Engineering*, 293, 491-502.

[10] 10 J. N. Kizhakkedathu, R. Norris-Jones, and D. E. Brooks (2004). *Macromolecules*, 37, 734-743.

[11] J. N. Kizhakkedathu, A. Takacs-Cox, and D. E. Brooks (2002). *Macromolecules*, 35, 4247-4257.

[12] V. Mittal, N. B. Matsko, A. Butté, and M. Morbidelli (2008). *Macromolecular Reaction Engineering*, 2, 215-221.

[13] W. Smulders and M. J. Monteiro (2004). *Macromolecules*, 37, 4474-4483.

[14] C. Farcet and B. Charleux (2002). *Macromolecular Symposia*, 182, 249-260.

[15] K. Tortosa, J.-A. Smith and M. F. Cunningham (2001). *Macromolecular Rapid Communications*, 22, 957-961.

[16] B. Keoshkerian, P. J. MacLeod and M. K. Georges (2001). *Macromolecules*, 34, 3594-3599.

[17] M. Li, N. M. Jahed, K. Min, and K. Matyjaszewski (2004). *Macromolecules*, 37, 2434-2441.

[18] K. Min, M. Li and K. Matyjaszewski (2005). *Journal of Polymer Science, Part A: Polymer Chemistry*, 43, 3616-3622.

[19] K. Min, H. Gao and K. Matyjaszewski (2005). *Journal of the American Chemical Society*, 127, 3825-3830.

[20] Y. Luo and X. Liu (2004). *Journal of Polymer Science, Part A: Polymer Chemistry*, 42, 6248-6258.

[21] A. Bowes, J. B. McLeary and R. D. Sanderson (2007). *Journal of Polymer Science, Part A: Polymer Chemistry*, 45, 588-604.

[22] R. Bussels, C. Bergman-Göttgens, J. Meuldijk, and C. Koning (2004). *Macromolecules*, 37, 9299-9301.

[23] R. Bussels, C. Bergman-Göttgens, J. Meuldijk, and C. Koning (2005). *Polymer*, 46, 8546-8554.

[24] X. Zhou, P. Ni, Z. Yu, and F. Zhang (2007). *Journal of Polymer Science, Part A: Polymer Chemistry*, 45, 471-484.

Chapter 8

Characterization of Latexes by Microscopy

A. Introduction

Characterization of surface morphology and size of the latex particles is of extreme importance in establishing the success of the polymerization processes. When, for example, core shell particles are generated, the presence of a shell on the core can be established by the increase in the diameter of the particles and by change in the surface morphology. The size of the particles can also be used to ascertain how much of the amount of originally added monomers has been polymerized on the surface. Similarly, the secondary nucleation processes or the non-polymerization of a part of the monomer can be ascertained by using these methods of characterization. Thus, the combination of the qualitative and quantitative means is essential in understanding the characteristics of the system. This information on size as well as surface morphology is very important, as it defines the application of the latex particles. For example, for many applications only small sized particles can be used, and in many applications a combination of two different sized latexes is required. Similarly, the morphology can also provide interesting information on the mechanism of polymerization. For example, depending on the generated morphologies (i.e., smooth surface, hemispherical, inverse morphology, strawberry, orange-peel, moon crater morphology, etc.), one can obtain information about the probable routes of the reaction. Apart from that, samples can be taken at different intervals during the course of a polymerization reaction, and the analysis of these samples for size and morphology can help in understanding the evolution of the polymer particles. There are a number of methods that can be used to characterize the quantitative as well as qualitative aspects of polymer latex particles, and generally these methods are different from each other owing to the different process requirements in sample preparation or analysis. Therefore, it is quite possible that one method yields a different result from other methods, especially in the case of size analysis. For example, the size distribution achieved from the light scattering methods is, on average, a little higher than the sizes determined through scanning or transmission electron micrographs. The particles in microscopy methods tend to be a bit shrunken in size, possibly because of the use of a high energy beam for the analysis. Also, as in the case of light scattering methods, a large number of particles constitute the sample and, thus, the result is an average signal from these particles; but, on the other hand, particle size analysis by microscopy cannot take into account a large number of particles. Similarly, the morphology achieved in the microscope can be different from the morphology predicted by hydrodynamic methods of characterization such as field flow fractionation. In the microscope, the particles are dry and under high energy beam, which can generate different structures on the surface or can disturb the original structure, whereas in the hydrodynamic methods of characterization, the particles are in the original surroundings. However, the use of such hydrodynamic methods also requires the addition of salts to the sample to provide stability of the system, and this can also affect the morphology of the system. Unfortunately, there is no one universal characterization method that can be applied to all of the systems. In real cases, a combination of two or more methods is required to achieve a better understanding. Figure 1 shows the wide variety of various qualitative and quantitative methods of characterization of polymer latex particles. The number of different techniques developed itself confirms the importance of the generation of information regarding the particles.

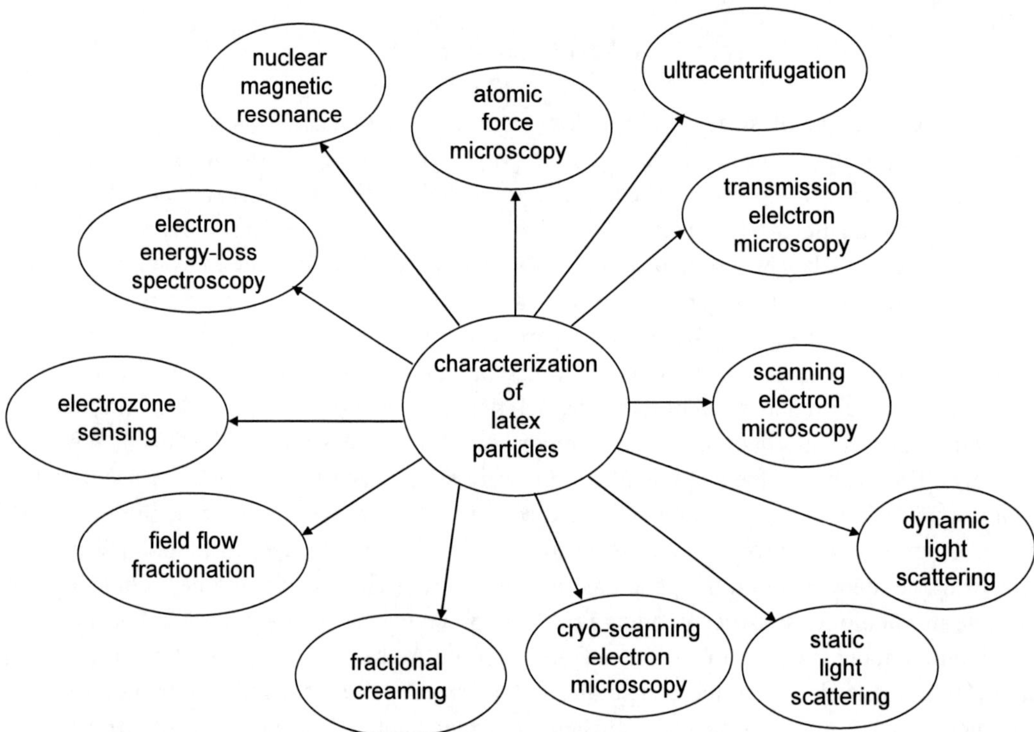

Figure 1. Various methods for the characterization of polymer latex particles.

Even though polymers may not be stable for long scanning times under the high energy electron beams and the beams may cause permanent damage to the structure of the surface, still the microscopy means of characterization of the morphology of the particle are common. And in many instances, microscopy is the only means of characterization of the particles surface morphology, e.g., when there is present a small extent of aggregation of the particles, the size determination may be faulty as the combined signal from a multitude of particles would be presented by the scattering equipment and it can be wrongly taken as the size of the primary particles. Also, when the core shell structures are generated, it is difficult to characterize how much shell has been generated on the core by the use of other methods and in

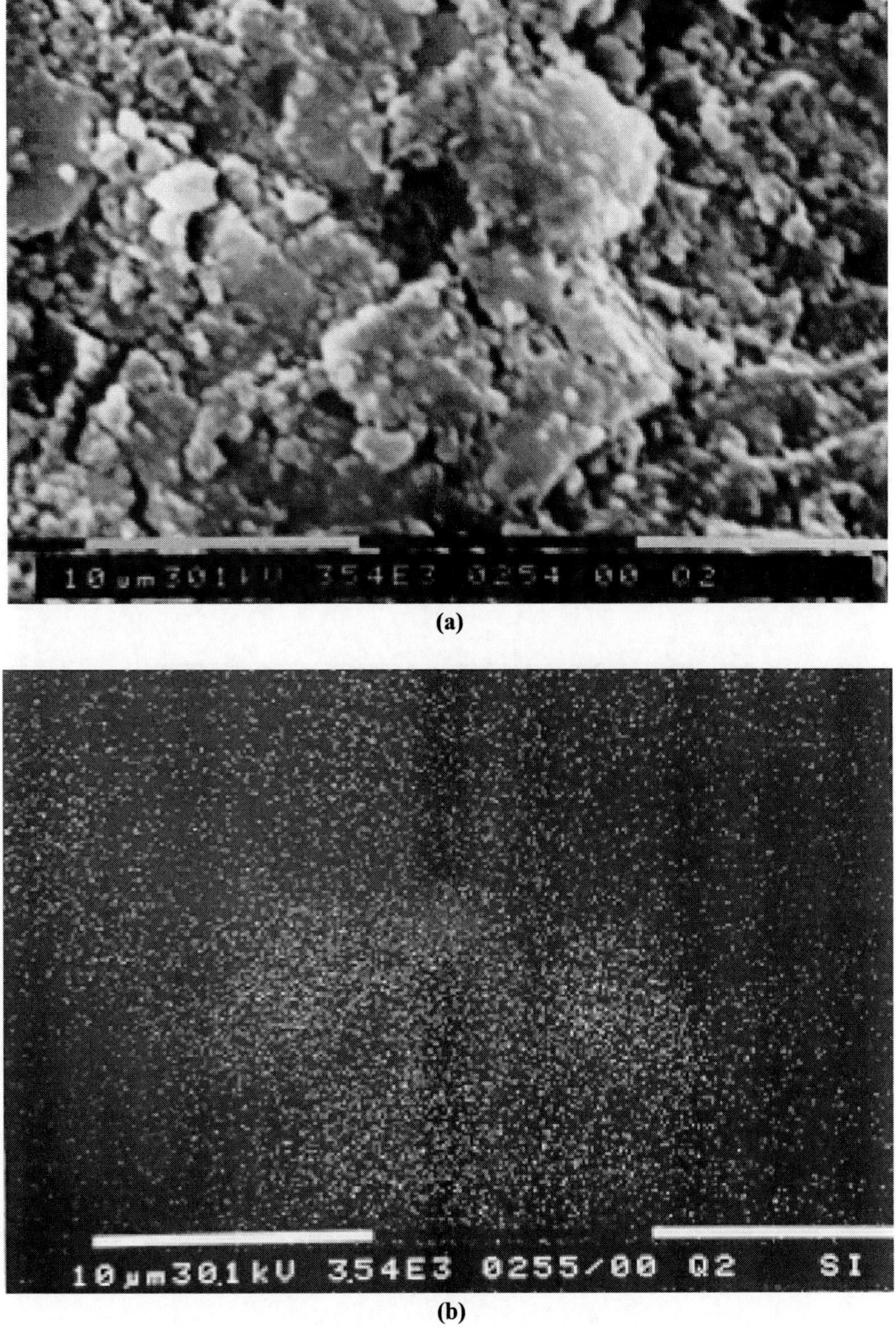

Figure 2. (a) SEM micrograph of a polymer/inorganic composite material and (b) energy dispersive X-ray analysis of the micrograph in 2a for the presence of silicon.

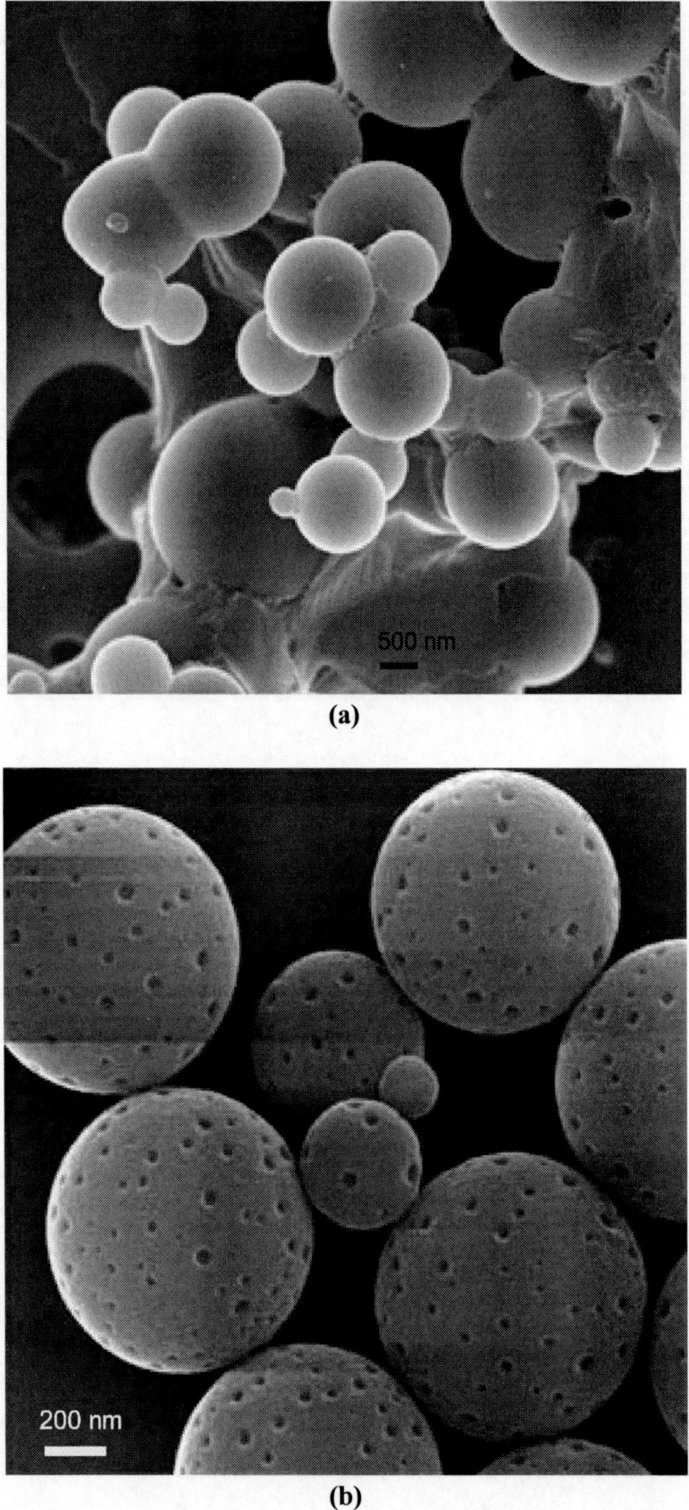

Figure 3. SEM micrographs of various polymer latex systems illustrating various properties or morphologies, like (a) aggregated and secondary nucleated particles and (b) peculiar orange-peel morphology.

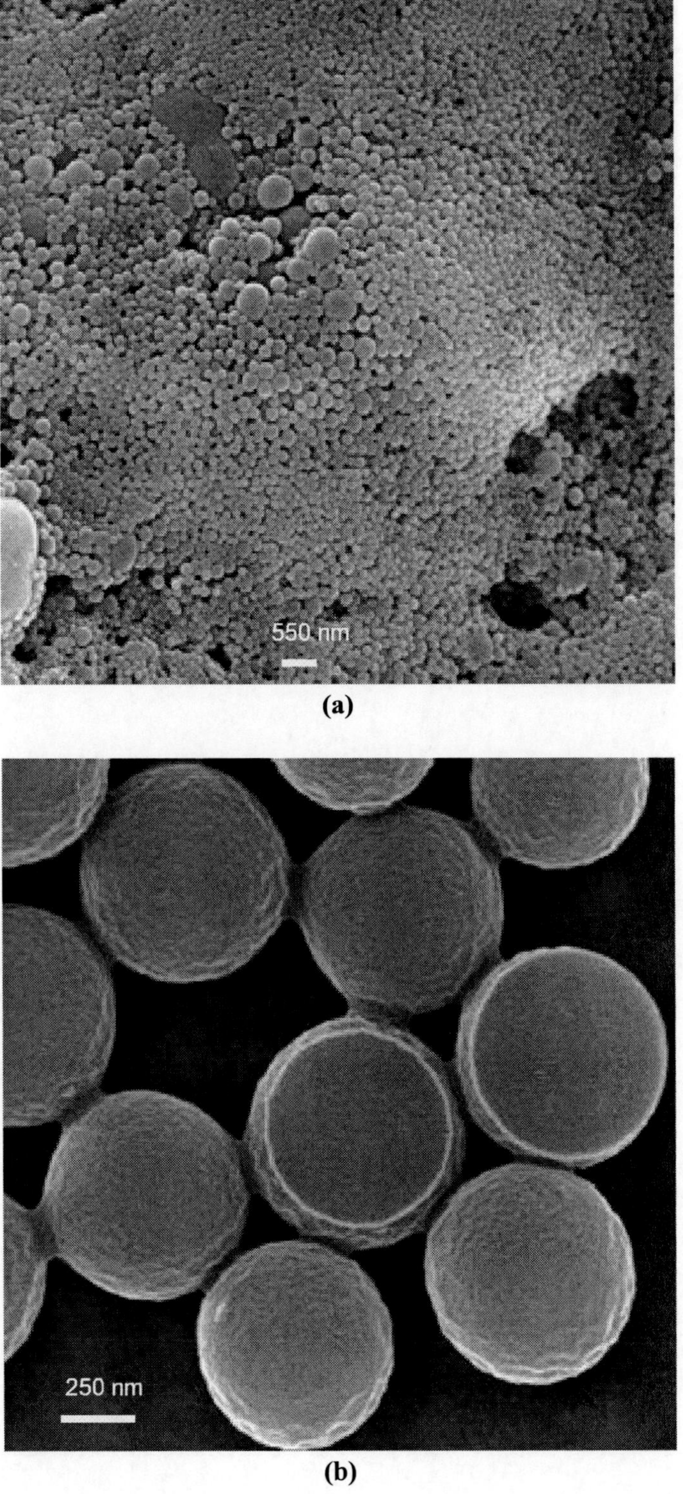

Figure 4. SEM micrographs of various polymer latex systems illustrating various properties or morphologies, like (a) polydispersity and (b) flat particles.

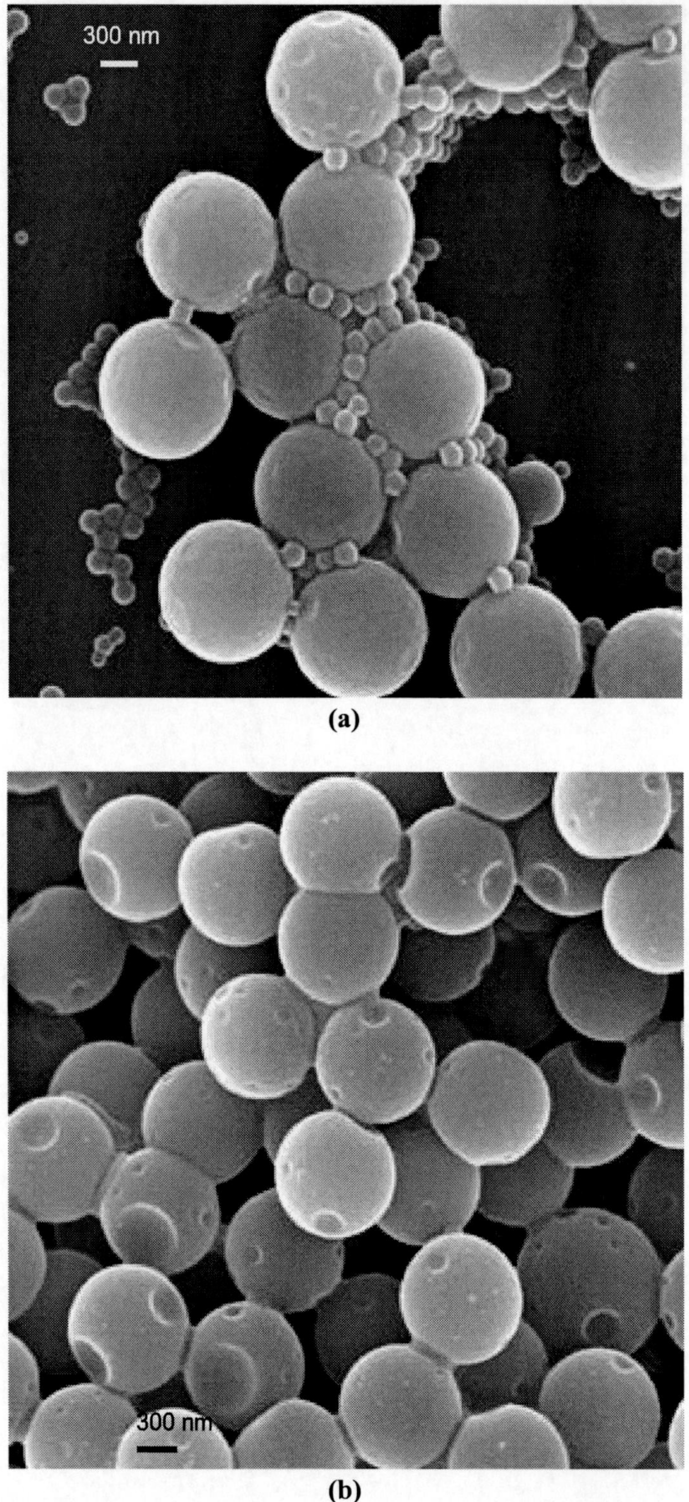

Figure 5. SEM micrographs of various polymer latex systems illustrating various properties or morphologies, like (a) secondary nucleation and (b) moon-crater morphology.

microscopy, it can be possible to ascertain it by using spectroscopy techniques coupled with microscopes, in which the thin section of the particles embedded in epoxy matrix can be analyzed for the presence of different atoms at different locations, thus, locating the monomers present in the core as well as shell. Also, the use of different microscopy techniques for the particle characterization makes this method very powerful, e.g., one can have the analysis of the particle in the transmission mode (when the sample is thin enough), or in the back scattered electron analysis mode. Similarly, energy dispersive X-ray analysis can also be achieved from the characteristic X-rays emitted by the elements present on the surface of the particles. The particles can also be prepared for analysis under cryo conditions which helps to retain the original surrounding of the particle. Not only scanning electron microscopy and transmission electron microscopy, atomic force microscopy can also be used for the polymer latex systems and is a useful tool in characterizing the functionalized systems. The following paragraphs explain the application of various microscopy techniques of the analysis of the latex particles.

B. Scanning Electron Microscopy

Scanning electron microscopy is the technique of choice for the general characterization of latex particles for their size as well as morphology evaluation. The sample preparation is not cumbersome, and the particles can be easily adsorbed on the freshly etched copper grids coated with carbon or collodium, which can subsequently be sputter coated with a thin layer of platinum. The particles can then be analyzed under the microscope at voltages of 10–20 kV on average, although a higher voltage can also be used. One must be careful that the coating of platinum is not thick enough to mask the surface features of the sample, and should be uniform enough not to cause excessive charging of the sample. Different possibilities of characterization are possible, depending upon the various interactions of the electrons with the particles [1-4]. For example, compositional contrast can be evaluated by the imaging of backscattered electrons. One can also have a transmitted electron analysis facility coupled with scanning electron microscope, and then the microstructures inside the particles can also be revealed, e.g., in core shell particles. Secondary electron imaging provides the information on the topography and particle sizes, etc., whereas energy dispersive X-ray analysis can be carried out to quantify the elemental composition in the sample. Figure 2 is an example of the use of energy dispersive X-ray analysis to find out the various elements in a composite material.

Figures 3–5 show the various latex systems which can be characterized by the use of scanning electron microscopy. Figure 3a is a representation of the system in which the polystyrene latex particles were generated by the surfactant free polymerization and were subsequently swollen with a second batch of styrene monomer and divinylbenzene crosslinker. These swollen particles were then gelated by the addition of salt, and subsequently the gelated system was crosslinked by polymerizing the monomers used to swell the seed polymers. The ideal morphology would have been the seed particles crosslinked with each other to form a network. However, in this case, although the network is formed (as seen by the particles joined to each other), a lot of smaller particles can also be seen, which indicates that the monomers used to swell the seed may also have generated new particles. Thus, SEM analysis allows understanding of the system morphology. Similarly, Figure 3b shows SEM micrographs of the polystyrene seed latex particles which were functionalized on the surface with another polymer shell generated from a hydrophobic monomer that does not form its own particles owing to its colloidal instability. In the system, the seed particles were roughly 800 nm in diameter, depicted by the large particles in the figure. Although the small particles are also visible in the micrograph, these particles are not expected to be a result of secondary nucleation similar to that seen in Figure 3a. Owing to the colloidal instability of the polymer chains formed from this hydrophobic monomer, these polymer chains cannot form secondary particles, therefore, collapse on the surface of the seed particles and, thus, subsequently polymerize on the surface. Therefore, the small particles in this case would most probably be present from the seed itself.

Figure 4a is an example of latex that is polydispersed in nature. Particles of numerous different sizes are present. The other morphologies can also be evaluated in the case of microscopy and cannot be analyzed when using other quantitative methods of analysis such as light scattering. As an example, as shown in Figure 4b, a sample of polystyrene particles are flat from one side, thus forming hemispheres. These particles, when analyzed by light scattering, lead to a distribution curve wherein the diameter of these non-spherical particles has been estimated by the software to be spherical and, as a result, the diameters from the light scattering experiment are different from the actual diameter of the particles.

Figure 5 is a similar example wherein the particles are generated with very peculiar surface morphologies. Though the particles are spherical in nature, there are very subtle changes on the surface of the particles which can only be analyzed by the use of a microscope.

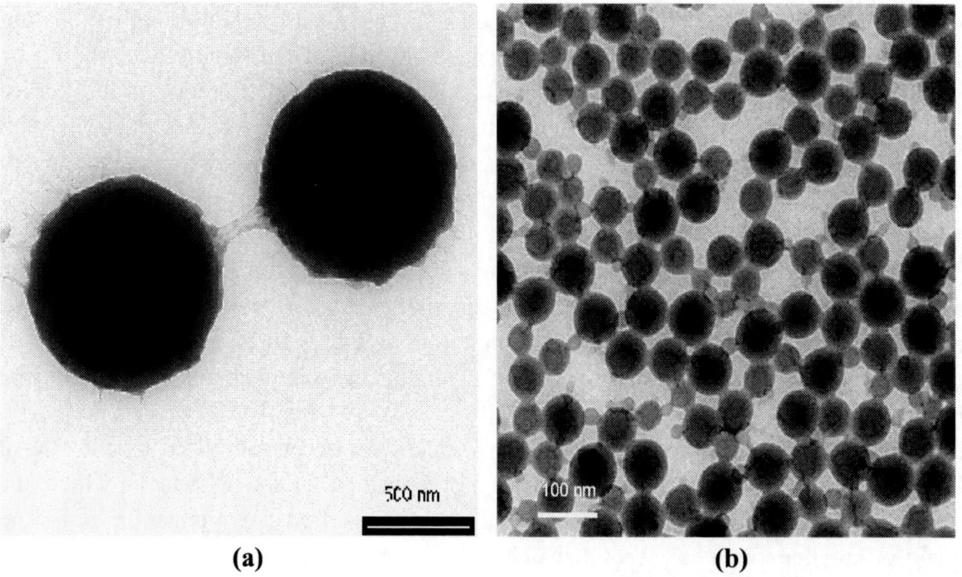

Figure 6. TEM micrographs of various polymer latex systems illustrating various properties or morphologies, like (a) surface functionalized particles and (b) polydispersity in the particle sizes.

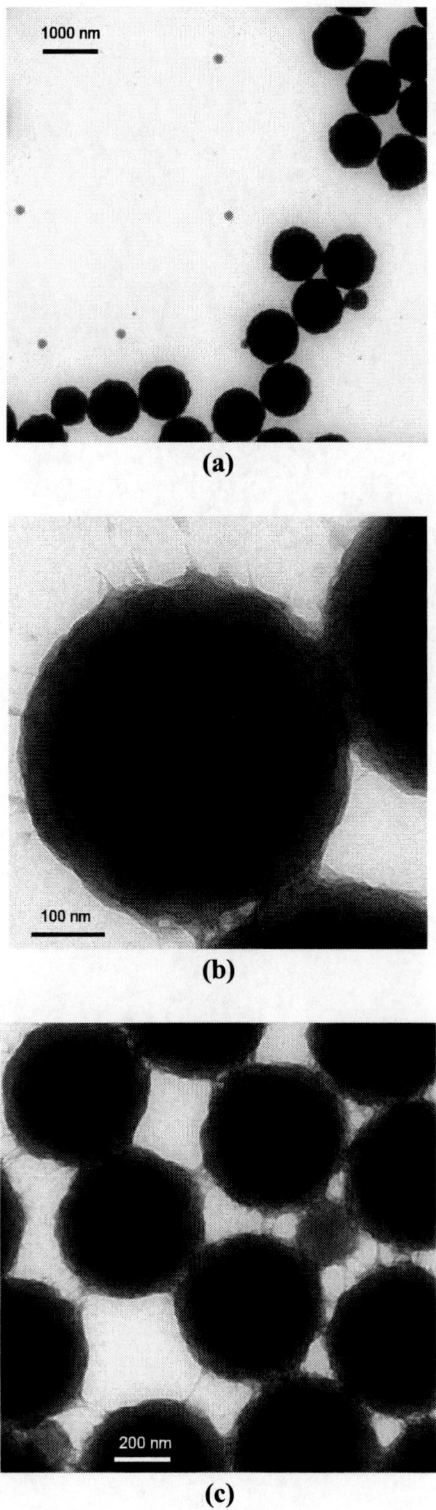

Figure 7. TEM micrographs of various polymer latex systems illustrating various properties or morphologies of surface functionalized particles.

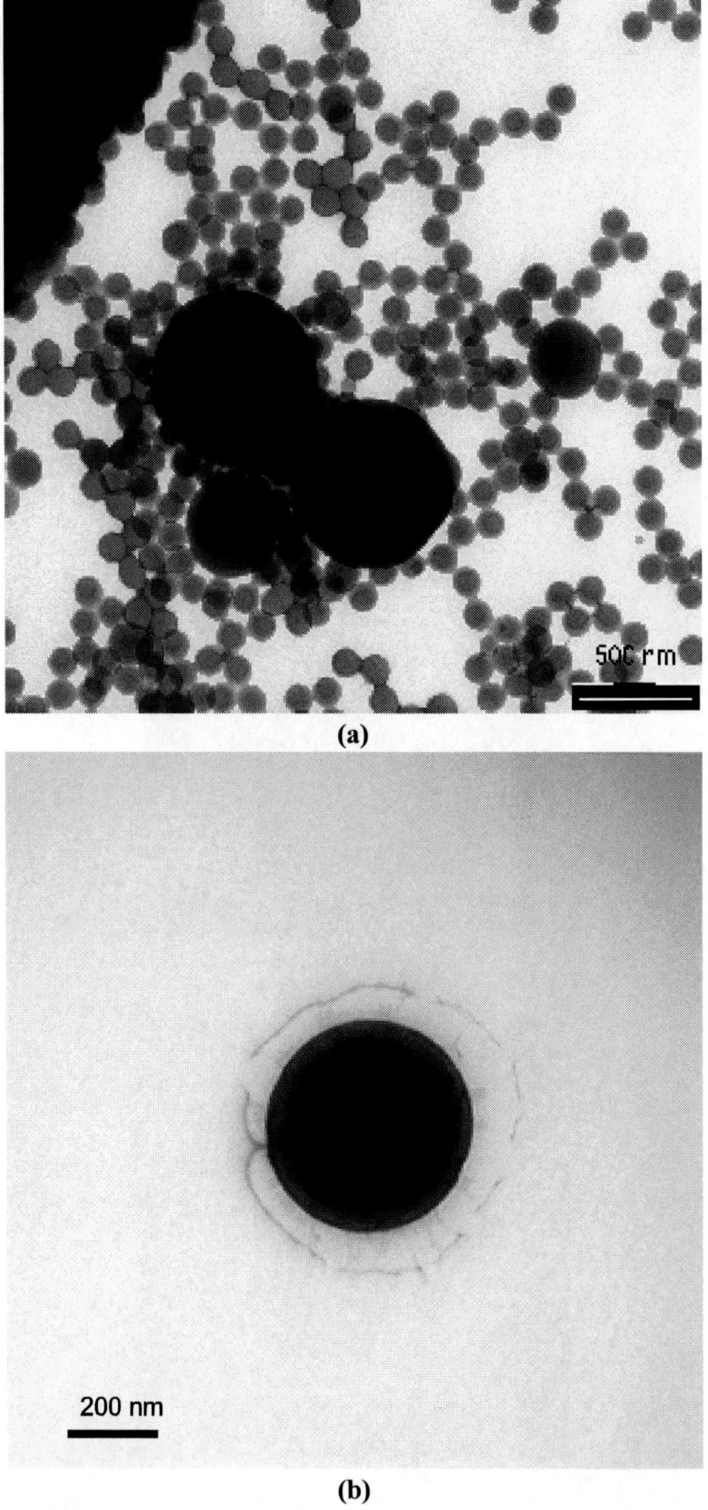

Figure 8. TEM micrographs of various polymer latex systems illustrating various properties or morphologies, like (a) secondary nucleated particles and (b) polymer particles with grafted brushes from the surface.

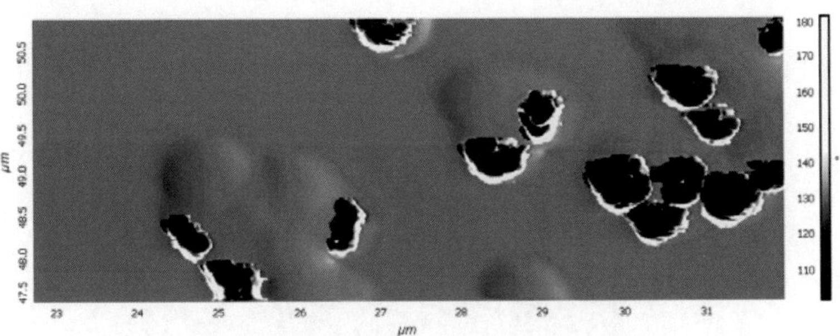

(a)

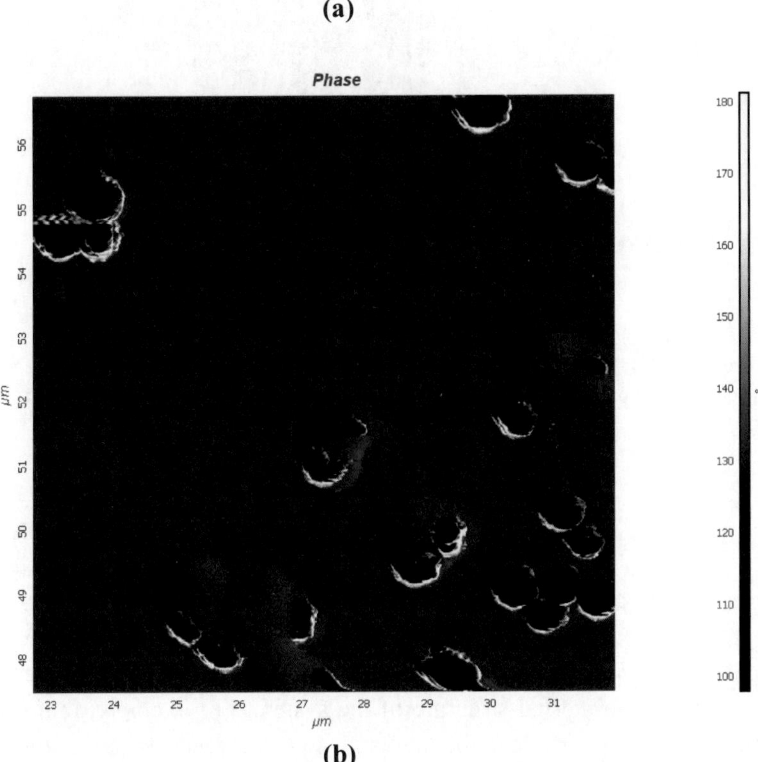

(b)

Figure 9. AFM micrographs of the surface functionalized polymer particles corresponding to the TEM micrograph of Figure 8b.

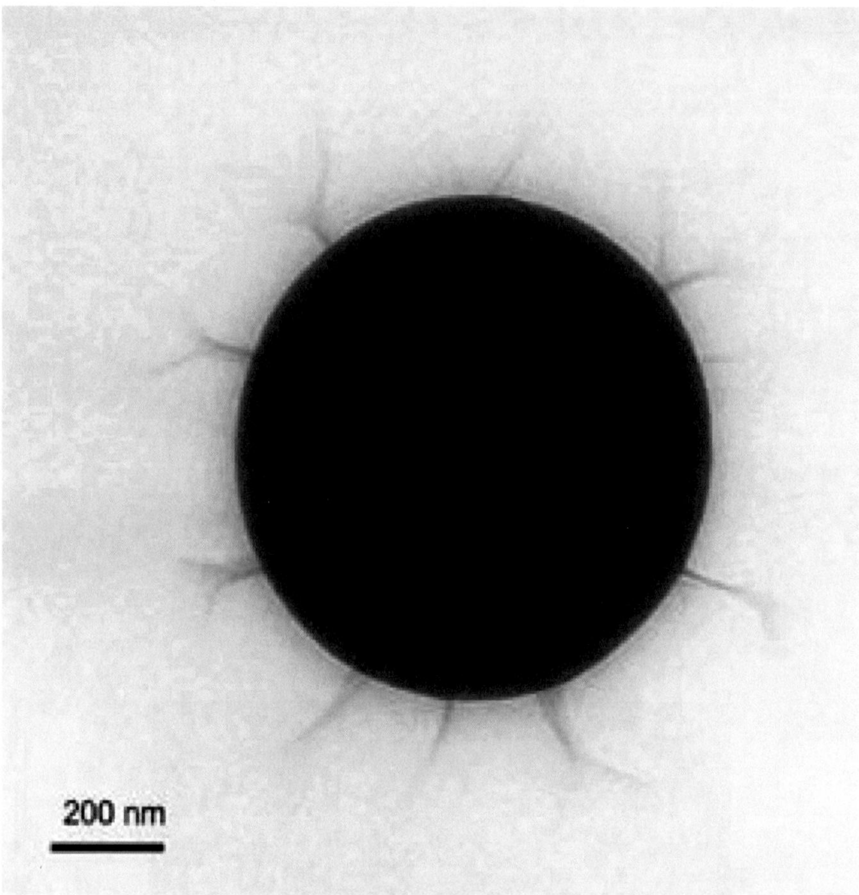

Figure 10. TEM micrograph of a single particle that was surface functionalized by grafting poly(N-isopropylacrylamide) brushes from the surface. The particle was then analyzed by EELS to ascertain the presence of carbon in the periphery of the particle as shown in Figure 11.

C. Transmission Electron Microscopy

As mentioned above, transmission electron microscopy can also be coupled with the scanning electron microscope, although only transmission electron microscopes are very common. The transmission electron microscope is a very useful technique, as it provides the information about the internal structure of the system. For example, thin sections of particle embedded in epoxy matrix can be cut and can analyzed under a transmission electron microscope to achieve valuable information about the architecture of the particle, e.g., in core shell particles. So, in comparison, the scanning electron microscope can define the surface characteristics of the particles, but the transmission electron microscope is more useful in providing internal microstructure information. One can also carry out transmission electron microscopy in the dark field or bright field imaging mode. This helps to achieve contrast for analyzing the microstructure or identifying possible defects in the system. Energy dispersive X-ray analysis can also be carried out in this case, and the elemental composition in the material can be established. Apart from that, as described in the next sections, the transmission electron microscope can also be coupled with a spectroscope, and the elements present in the materials can be identified by using the electron energy loss values for different elements. The method of preparation of the grids for routine evaluation of the particles is very straightforward. Carbon coated copper grids that are freshly glow discharged are placed on the droplets of particle suspension for roughly 20–30 seconds and are then washed with distilled water. The staining agents can also be used to improve the contrast in the images. Negative staining agents such as uranyl acetate are more preferred as they do not interact with the particle and only affect the surroundings. The positive staining agents, on the other hand, interact with the particles to improve their contrast, but in this process the fine microstructure of the particles surface may be lost or disturbed.

Figure 6–8 are examples of various types of different morphologies of the particle surfaces which can be characterized by transmission electron microscopy [1-4]. Although in many cases it complements the results of the scanning electron microscope—e.g., to define the particle size or polydispersity—there are, however, some instances in which only the use of a transmission electron microscope can provide insight into the system, as the system's characteristics are not visible in a scanning electron microscope. An example is shown in Figure 6a, where the polystyrene seed particles were grafted with the brushes of poly(N-isopropylacrylamide) polymer from the surface using living or controlled polymerization reactions (using atom transfer radical polymerization). The polystyrene seed particles were first covered with a thin shell of a copolymer of styrene and another acrylic monomer which, on the other end of the molecule, had an atom transfer radical polymerization initiation site. These sites can then be used to grow the polymer brushes from the surface. These brushes are very fine and less compact by nature and, thus, are totally invisible in the scanning electron microscope. However, as seen in the transmission electron micrograph of Figure 6a, the periphery of the particles clearly indicates the polymer brushes grafted from the surface, thus, making transmission electron microscopy a very important technique for the analysis of such functional systems. Figure 6b is an example of polymer latex formed by conventional emulsion polymerization and where the size of polymer particles was not controlled, thus, leading to polydispersity. It is interesting to note that the measurements of particle size in the dynamic light scattering for these samples did not indicate large polydispersity. Owing to smaller particle sizes, the particles are also visible in the micrograph as more transparent in comparison with the larger particle shown in Figure 6a. Also, one would notice that there is a presence of very small-sized particles that seem to be sticking to the periphery of the bigger particles.

Figure 7 similarly shows the transmission electron micrographs of particles with different surface morphologies. These morphologies with subtle changes on surface can be easily detected either by the use of scanning electron microscope or transmission electron microscope. The micrograph in Figure 7b consists of copolymer particles of polystyrene and poly(N-isopropylacrylamide), and poly(N-isopropylacrylamide) being hydrophilic in nature, diffuses to the surface of the particles during the course of polymerization. The surface morphology, therefore, completely changes, and as the amount of poly(N-isopropylacrylamide) is increased in the system, the particles tend to become sticky and start aggregating with each other, as shown in the micrograph (c). There is also a possible presence of particles only consisting of poly(N-isopropylacrylamide). It is also important to note that although the scanning electron microscope could be used to analyze this system, the true contrast and interactions of poly(N-isopropylacrylamide) on the surface with particles was possible only in the transmission electron microscope.

Figure 8a is a representation of the system wherein a large amount of secondary nucleation can be seen. The polymerization of the seed particles was carried out first, and these seed particles were then subjected to another round of polymerization to form a thin shell on the particles to generate core shell particles. The secondary batch added to form the shell on the surface of the particles is required to polymerize on the surface on the seed particles to achieve the required morphology. However, as can be seen in the figure, there is a presence of a large amount of small particles, indicating that the shell-forming monomers may have partially or fully formed the secondary particles. Indeed, no change in the size of the core particles could be detected before and after the shell-forming step, indicating that the whole batch of shell-forming monomers formed the small particles. This information is very important in gaining better insight into the interaction or compatibility of the seed particles with the monomers forming the shell. Figure 8b is another interesting example in which the brushes of a polymer were grafted from the surface of the particles. The particles were initially modified with an initiator, and in the subsequent step the brushes were grafted. It was only through a transmission electron microscope that these brushes could be recognized. The brushes were more prominent when a negative staining agent was used, which helped to improve the contrast of the background. The brushes were completely invisible in the scanning electron microscope, although when cryo scanning mode was used, the full length of the brushes around the particles could be observed, confirming that although the transmission electron microscope was able to detect the brushes, the brushes were somewhat shrunken due to dehydration of the brushes on the grids or possibly due to the effect of the staining agent.

D. Atomic Force Microscopy

Atomic force microscopy has become, in recent years, a very important tool for the characterization of polymer latex particles. It offers the possibility of collection of qualitative and quantitative information on the various aspects of the polymer particles, such as size, topography, morphology, roughness and the confirmation of the surface functionalization. Atomic force microscopy can also be used to characterize particles in air and even in liquid dispersions. The advantage of atomic force microscopy over the other techniques is the visualization of the particles in three dimensions, so it is possible to measure the height of the nanoparticles quantitatively.

Figure 9 is a representation of the system whereby the latex particles are functionalized on the surface by grafting polymer chains from the surface. These polymer chains have totally different morphology compared to the seed or core particles, and also have a less compact form on the surface of the particles. For these reasons, the functionalization on the particles' surface could be recognized in atomic force microscopy. The black particles in these micrographs are the polymer core particles, and the totally transparent bulge surrounding the particles corresponds to the grafted polymer brushes on the surface.

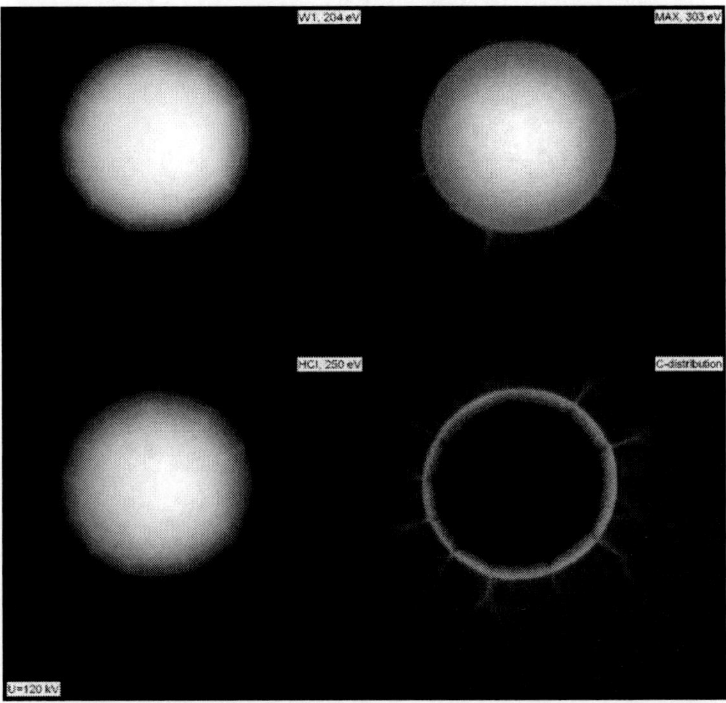

Figure 11. EELS micrograph of particle shown in Figure 10 for the distribution of carbon. The TEM grid is also coated with carbon, therefore the background also provides the signal of the carbon atom. The particles are larger than 500 nm diameter; therefore, the core of the particle does not generate any carbon signal because there is no electron transmission.

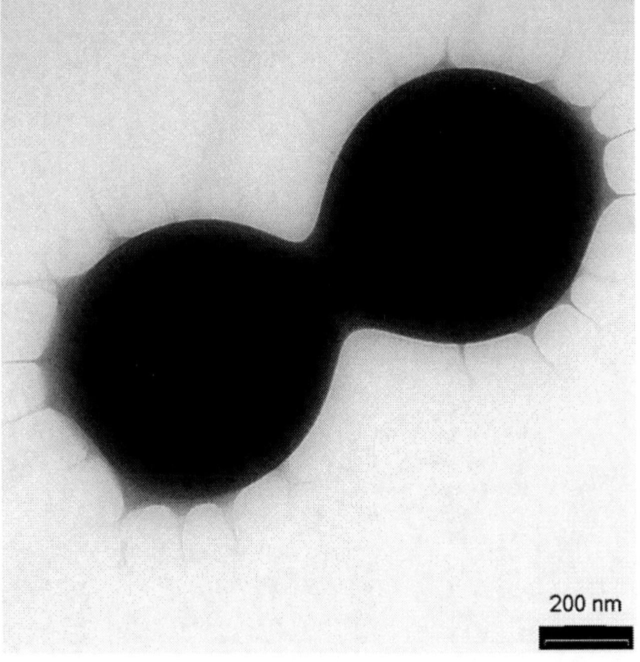

Figure 12. The TEM micrograph of particles which were surface functionalized by grafting of poly(N-isopropylacrylamide) brushes from the surface. The particles were then analyzed by EELS to ascertain the presence of oxygen in the periphery of the particle as shown in Figure 13.

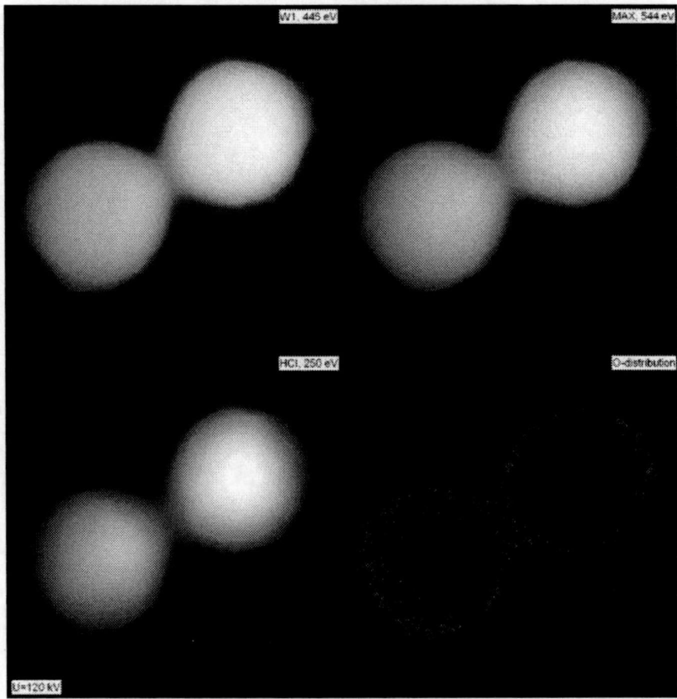

Figure 13. EELS micrograph of particles shown in Figure 12 for the distribution of oxygen. As the oxygen atoms are present only in poly(N-isopropylacrylamide) and not in the polystyrene core, therefore, the signal of oxygen around the particles would confirm the successful grafting of poly(N-isopropylacrylamide) on the particles.

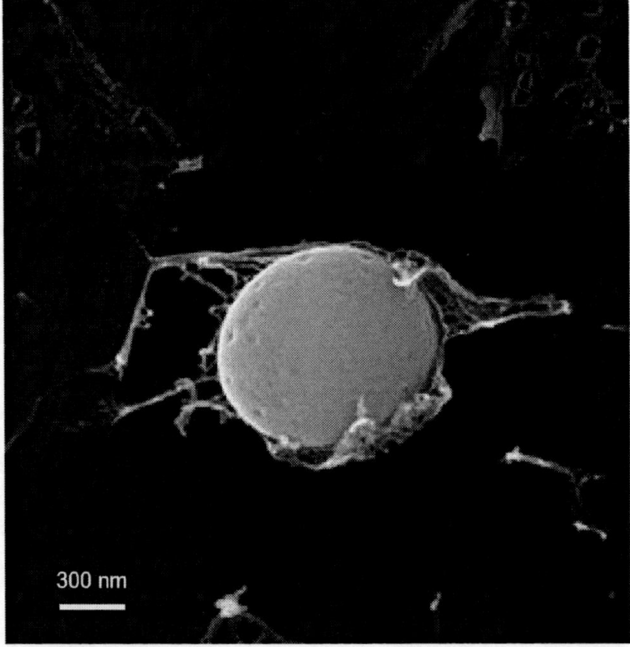

Figure 14. Cryo scanning electron micrographs of polystyrene core particles surface grafted with brushes of poly(N-isopropylacrylamide). The use of cryo SEM facilitates the analysis of the full length of polymer brushes, which may not be possible to observe in non-cryo SEM and TEM operations.

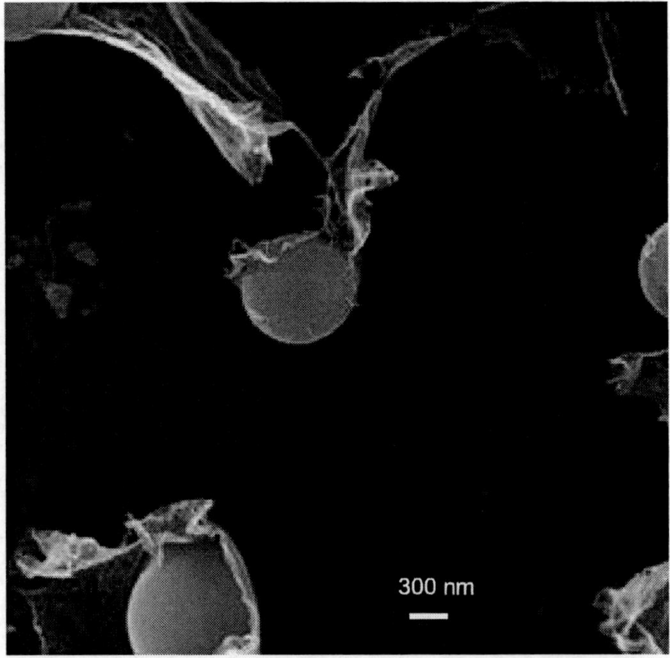

Figure 15. Cryo scanning electron micrographs of polystyrene core particles' surface grafted with brushes of poly(N-isopropylacrylamide). The use of cryo SEM facilitates the analysis of the full length of polymer brushes, which may not be possible to observe in non-cryo SEM and TEM operations.

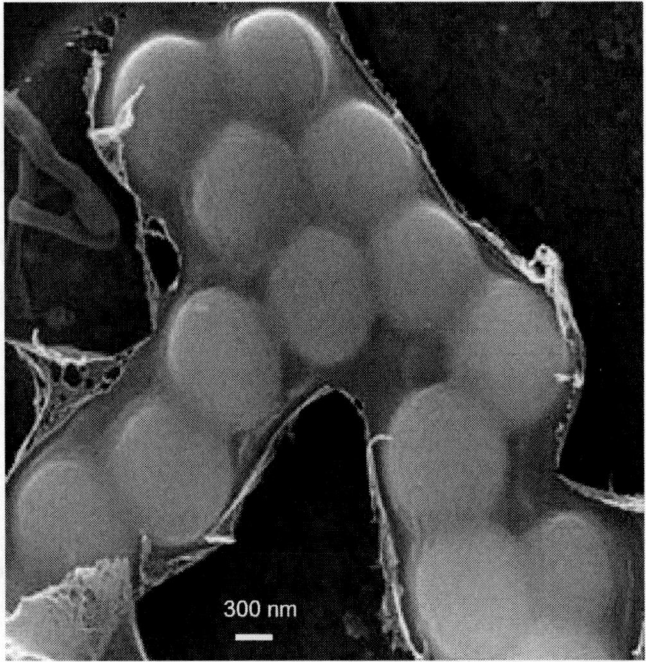

Figure 16. Cryo scanning electron micrographs of polystyrene core particles surface grafted with brushes of poly(N-isopropylacrylamide). Special features could be observed: e.g., in this case, it is clear that the particles aggregated before grafting and the whole aggregates were subsequently grafted with a poly(N-isopropylacrylamide) layer.

E. Electron Energy Loss Spectroscopy

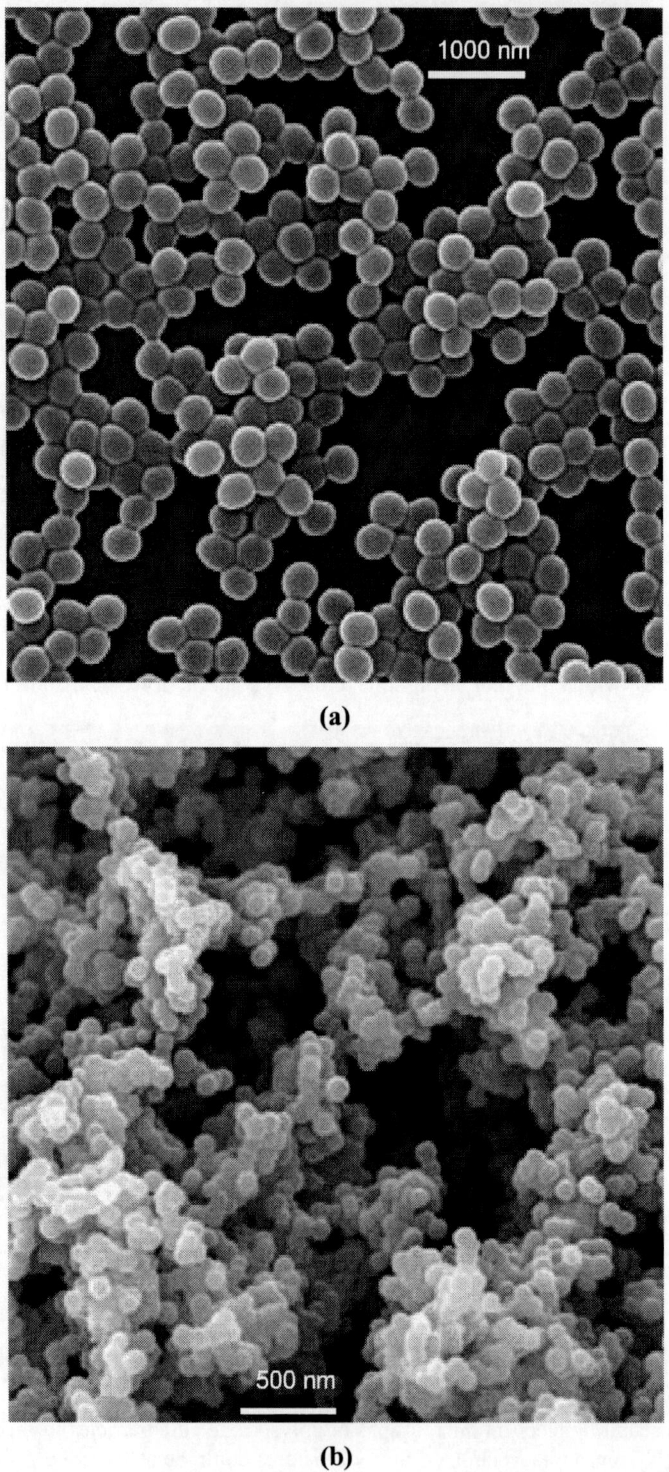

Figure 17. SEM micrographs demonstrating various polymer particle systems: (a) physical aggregates and (b) particle monoliths, where particles are chemically bound to each other.

Figure 18. SEM micrographs demonstrating various polymer particle systems: (a) film of polymer particles on a flat substrate after heat treatment and (b) physical aggregates or concentrated latexes.

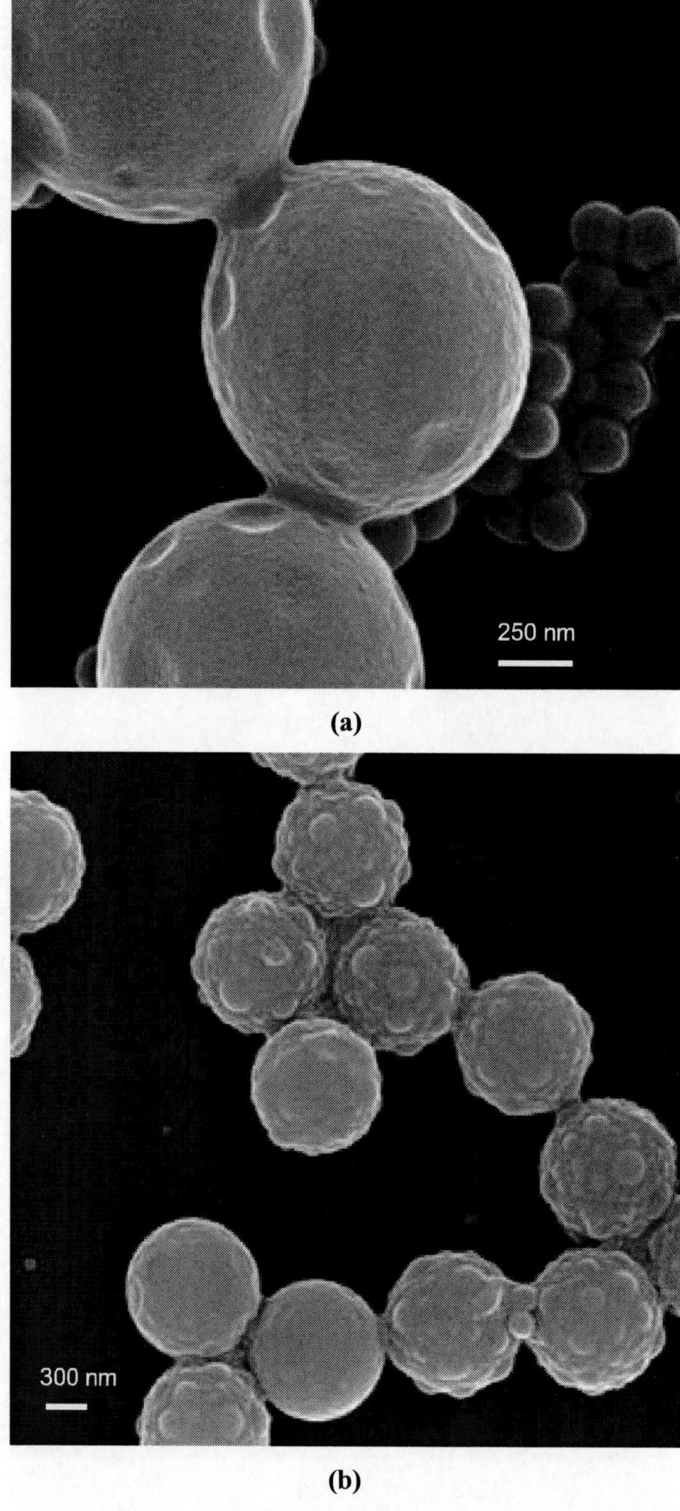

(a)

(b)

Figure 19. SEM micrographs demonstrating various polymer particle systems which have different morphologies depending on the reaction conditions indicating that the evolution of morphology could be microscopically followed.

Electron energy loss spectroscopy coupled with a transmission electron microscope is a very important tool for characterizing nano-particles for the elemental composition on the surface. Electron energy loss spectroscopy images can be recorded through TEM operated, for example, at 120kV and 10,000 magnifications. A 90 μm objective diaphragm is used, corresponding to a collection of semi-angles of 12.4 mrad. A series of energy-filtered images are recorded using a slow-scan CCD camera (1k x 1k pixels and 19 μm pixel size). Generally, an energy slit width of 15 eV is used. The series cover the following energy ranges: 258–303 eV(6C), 454–544 eV(8O), 317–410 eV(7N) and 1742–1970 eV(74W). Recording of the series is automated by the vision software packages hosted on a PC. During image series acquisition, the program controls the microscope and the CCD camera. Before and after the acquisition of an image series, both a black level image and HCl at 250 eV are recorded. A gain image that has been recorded with uniform illumination of the CCD camera and previously stored is used to correct for the pixel-to-pixel gain variations of the CCD camera.

Figures 10 and 11 correspond to a system of surface grafted polymer latex particles that have been analyzed by the use of electron energy loss spectroscopy. It is indeed the same system as characterized by the transmission electron microscopy. In this system, the polystyrene latex seed particles were grafted on the surface with the polymer brushes of poly(N-isopropylacrylamide) by using the controlled radical polymerization technique of atom transfer radical polymerization. Though it was possible to detect the brushes with transmission electron microscopy, it was of interest, however, to confirm the presence of poly(N-isopropylacrylamide) on the surface. Therefore, electron energy loss spectroscopy was carried out on the particle shown in Figure 10, and the distribution of carbon was analyzed on and around the particles, as shown in Figure 11. The carbon signal is distributed uniformly in the background of the particles owing to the deposition of a thin carbon layer on the grid. However, the carbon signal was very significant around the particles, indicating that the brushes were indeed not an artifact and contained carbon, which could have come only from the polymer brushes grafted onto the surface. Moreover, the presence of nitrogen and oxygen atoms specific to poly(N-isopropylacrylamide) structure could also be confirmed around the particles, further proving that the brushes indeed were of poly(N-isopropylacrylamide). Similarly, another system of particles is shown in Figures 12 and 13. In this case, the distribution of oxygen atoms around the particles was confirmed. Thus, in addition to the visual characterization of the surface functionalization, one can also confirm the functionalization through spectroscopic means.

F. Cryo-Scanning Electron Microscopy

Cryo-scanning electron microscopy is also a very specific technique used to gain further insights into latex systems in addition to the information gained through scanning electron microscopy. Cryo-scanning electron microscopy, however, is difficult to perform. For cryo SEM studies, the latex particles are first adsorbed on the grids, and these grids are immediately frozen by being plunged into liquid nitrogen. The grid is then introduced under liquid nitrogen to a pre-cooled freeze-drying device. After freeze drying, the samples are sputter-coated with 5 nm platinum, and grids are analyzed in the field emission in-lens microscope at -85°C.

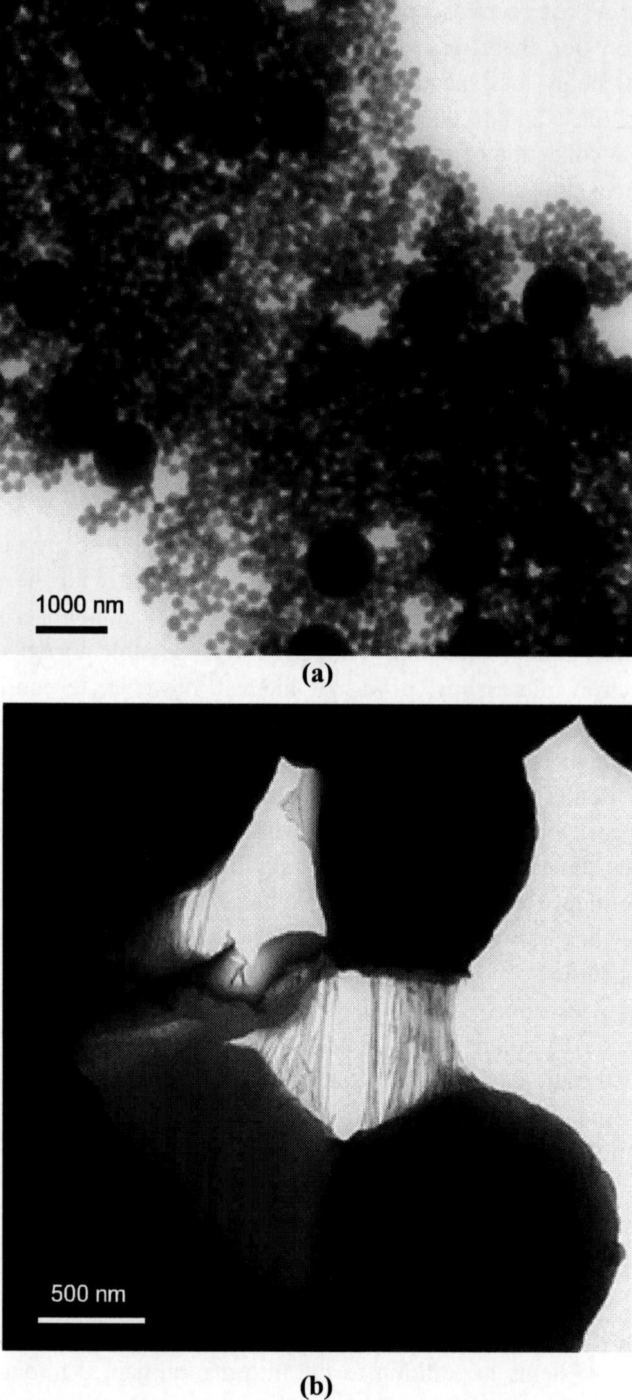

Figure 20. TEM micrographs demonstrating various polymer particle systems: (a) secondary nucleation generated during surface functionalization process and (b) debonding of surface functionalized polymer particles stuck to each other.

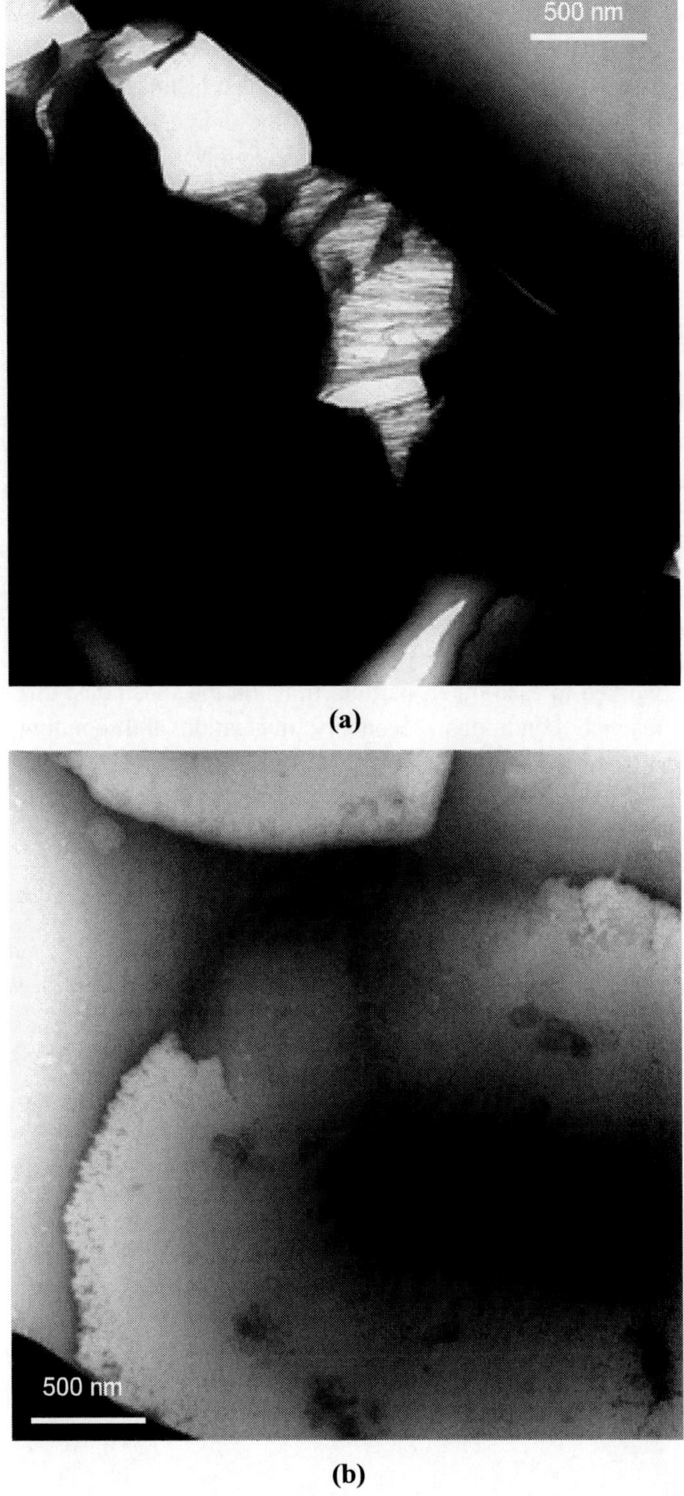

Figure 21. TEM micrographs demonstrating various polymer particle systems: (a) debonding of surface functionalized polymer particles stuck to each other and (b) analysis of the growth of polymer brushes from the polymer particles (black mass).

To compare cryo-scanning electron microscopy with transmission electron microscopy, similar surface grated particles, as analyzed with transmission electron microscopy, were analyzed by cryo-scanning electron microscopy. In the transmission electron micrographs, the grafted brushes in the surface were affected by the dehydration of the brushes as well as possibly by the staining agent. In cryo-scanning electron microscopy, as the particles retain exactly the original morphology as they were present in the aqueous phase, one can therefore have a true idea of the thickness of the brushes around the particles. As seen in Figures 14 and 15, the length of the brushes around the particles were much longer in length compared to the one seen in transmission electron microscopy. The brushes were very delicate in nature, as they collapsed or were permanently damaged as soon as they came under the high energy electron beam. Most of the particles were functionalized with thick layers of brushes around them, though some free polymer was also present in the system. Similarly, as seen in Figure 16, some special features could also be recognized. As shown in the micrograph, an aggregate of particles was formed first, and was collectively functionalized with a thick layer of brushes thereafter.

The chapter throughout has provided numerous examples to demonstrate the use and importance of electron microscopy methods for the characterization of latex particles. There are many more examples of particle systems that can be characterized by microscopy, as shown in Figures 17 to 21. These systems include physical aggregates, monoliths where particles are networked to each other, particle films on the substrates, different morphologies depending on reaction conditions, secondary nucleation and bonding-debonding of the surface functionalized particles, etc.

REFERENCES

[1] V. Mittal, N. B. Matsko, A. Butté, and M. Morbidelli (2007). *Polymer*, 48, 2806-2817.

[2] V. Mittal, N. B. Matsko, A. Butté, and M. Morbidelli (2007). *European Polymer Journal*, 43, 4868-4881.

[3] V. Mittal, N. B. Matsko, A. Butté, and M. Morbidelli (2008). *Macromolecular Materials and Engineering*, 293, 491-502.

[4] V. Mittal, N. B. Matsko, A. Butté, and M. Morbidelli (2008). *Macromolecular Reaction Engineering*, 2, 215-221.

Chapter 9

LATEX STABILIZATION, HIGH SOLIDS AND SCALE UP

A. INTRODUCTION

Latex stabilization is perhaps one of the most important properties that are required in any of the generated latexes; unless the latexes are destabilized deliberately, e.g., by the addition of salts or by shearing them by very high shear forces. Irrespective of the way the latexes are generated, i.e., either by using conventional emulsion polymerization or by using the controlled living polymerization, the stability concerns are of prime importance. In fact, the emulsions generated by specialized modes of latex generation, i.e., surfactant-free emulsion polymerization or controlled living polymerizations, are more sensitive to the destabilization. Stabilization is a property that is required during the polymerization as well as after the polymerization reactions. Better stability means a longer shelf life for the generated latexes, i.e., the ability of the latexes to stay stable for longer periods of time. For this reason, the latexes are also added with an additional amount of surfactant after the complete polymerization, also called post stabilization.

Achievement of high solids in the latexes is similarly of immense importance, as the greater extent of solids in the generated latex would allow the transportation of more and more materials easily from one place to the other. It would also allow the generation of more material by using the same amount of dispersed phase material, thus allowing one to save materials and energy. High solids are sometimes generated by synthesizing particles with two or three average particle sizes in order to help control the rheology of the dispersions used in various commercial applications. However, during this process one again must consider the stability of the latexes, as the greater extent of solids may cause the destabilization in the system to occur.

Another very important point of commercial significance is the scale up of the latex technology from the lab scale to the higher pilot plant or production scale. It is quite common that the reaction and processing condition which lead to highly controlled generation or functionalization of latex particles at the lab scale may be totally out of control when these conditions are used on a larger scale. One, therefore, must be careful regarding these processing conditions, especially in stirring, heating or cooling systems when latex is generated at higher scales. The following paragraphs deal with the important considerations of latex stabilization, high solid generation as well as the scale up of latexes.

B. Latex Stabilization

The latex particles need to be stabilized to prevent their aggregation. There are various mechanisms by which the stability of the latex particles can be achieved. The most common is the electrostatic stabilization of the particles, whereby the presence of charges on the surface of the particles keeps them away from each other owing to repulsive forces between the like charges. The second most important way to stabilize latexes is by steric stabilization, which can be achieved by the adsorption of non-ionic polymer chains on the surface of latex particles, which also acts to repel these particles from each other. There is also a possibility to combine the two processes of electrostatic stabilization and steric stabilization of polymer particles, whereby the particles are then said to be stabilized by electrosteric means. This mode of stabilization is generally the most effective way to control the stability and de-stability of the polymer particles.

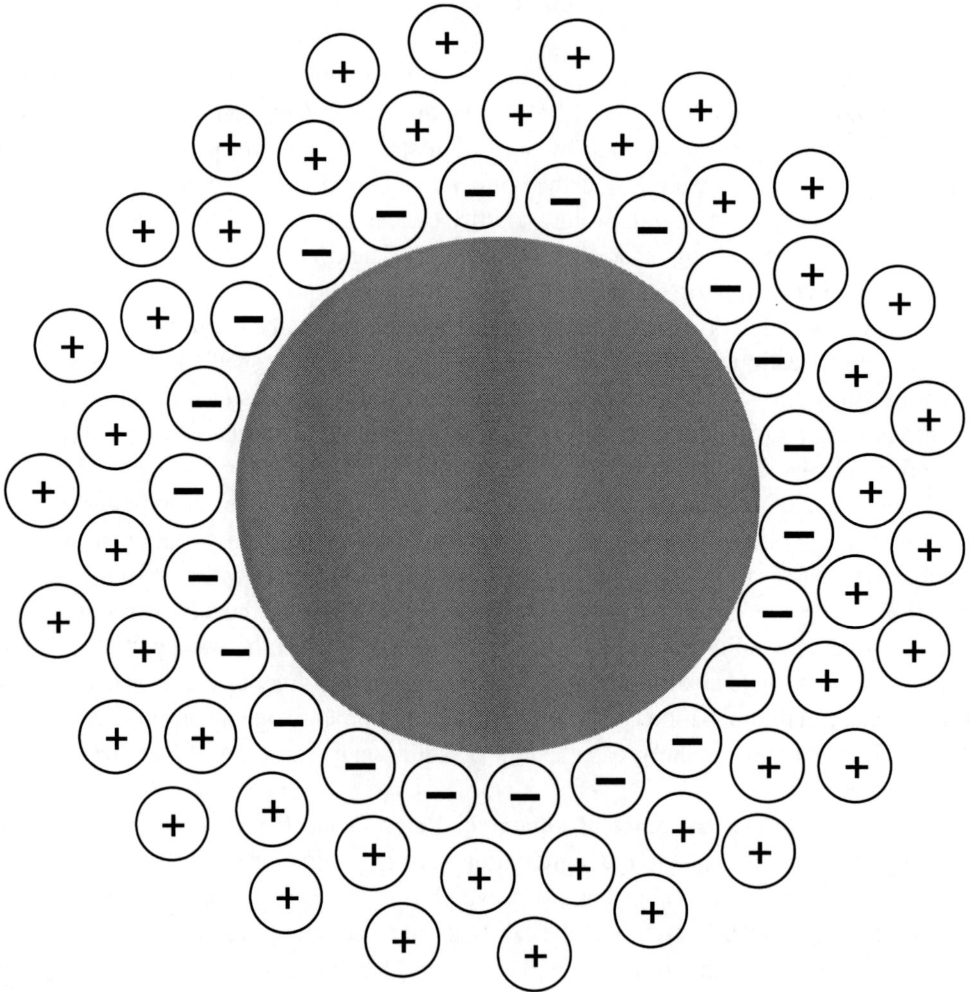

Figure 1. Electrostatic stabilization of colloidal particles.

Figure 1 represents such a system of electrostatic stabilization, where in the particles are stabilized by the formation of an electric double layer. The first charge comes on the surface of the particles during the polymerization or specific adsorption of these charges on the surface. These charges (negative in this case) are then balanced by the opposite charges, e.g., hydrated counter ions in this case. During polymerization, the charges on the surface of the particle can be achieved in a number of ways. The first proposition for achieving the electrostatic stabilization of the particles is the use of ionic initiators such as potassium persulphate. Potassium persulphate, when heated during the initiation of polymerization reaction, decomposes into two ionic radicals, as shown in equation 1. These radicals then initiate the polymerization reaction and, as a result, one end of the chains is always formed by this ionic pair. This ionic pair, owing to its water solubility, resides on the surface of the particles at the interface with the aqueous phase and provides stability to the latex particles against aggregation with other particles, due to the generation of repulsion between the same charges on all of the particles. Figure 2a is a representation of such a system. In fact, the case of emulsion polymerization achieved without the use of surfactant has only this mode of stabilization mechanism for the particles. The second proposition for achieving such an electrostatic stabilization is the use of ionic monomers for polymerization, as shown in Figure 2b. The monomers with ionic groups also lead to the stabilization of the particles, as many of the ionic pairs on the monomers reside on the surface of the resulting polymer particles, thus, providing electrostatic stability. As shown in Figure 2b, when a copolymerization of such monomers with more hydrophobic monomers is carried out, most of these ionic groups—owing to the higher hydrophilicity—tend to concentrate near the surface of the particles, thus, generating different morphology which may be useful in a number of applications of hydrophobic hydrophilic core shell particles. The methods described above for generating electrostatic stability in the latex particles chemically bind the charges on the surface of the particles. However, the physical adsorption of the charges on the surface of the particles is also one of the most important methods for achieving latex stabilization. The physical adsorption of the charges is achieved by the use of surfactants, used commonly during the emulsion polymerization. The surfactants are generally ionic in nature; the most commonly used ionic surfactant is sodium dodecyl sulfate. The ionic surfactants form micelles in the water phase at the beginning of the polymerization reaction, when their concentration in the aqueous phase is reached above the critical micelle concentration. The micelles are formed due to the alignment of the hydrophobic parts of the surfactant molecule away from the aqueous phase, and the hydrophilic parts in contact with the aqueous phase. This leads to a completely hydrophobic space in the interior of the micelles, which is a very suitable place for the monomer to diffuse during the emulsion polymerization. The polymer is formed in these micelles which are, therefore, subsequently called polymer particles. The surfactant molecules, which initially formed the boundary of the micelles, now become physically adsorbed on the surface of polymer particles, and provide electrostatic stability to the polymer particles. Figure 2c is an example of such a system in which the polymers stabilized by sodium dodecyl sulphate are shown with the surfactant physically adsorbed on the surface.

$$^{+}K\ O_3S{-}O{-}O{-}SO_3\ K^{+} \longrightarrow 2\ SO_4^{\bullet -}\ K^{+}$$

$$K^{+}\ SO_4^{\bullet -} + M \longrightarrow K^{+}\ SO_4^{-} - M^{\bullet} \quad (1)$$

K+ SO4-
K+ SO4-
SO4- K+
K+ SO4-
SO4- K+
K+ SO4-

K+ SO4-
K+ SO4-
SO4- K+
K+ SO4-
SO4- K+
K+ SO4-

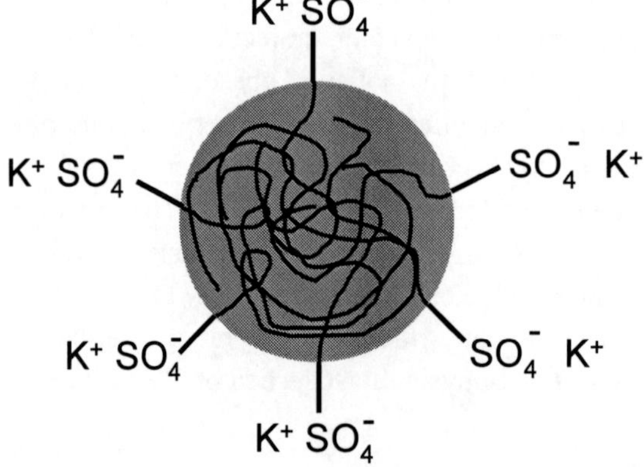

Figure 2a. Mode of initiation using potassium persulphate and the particles stabilized by the sulphate surface groups.

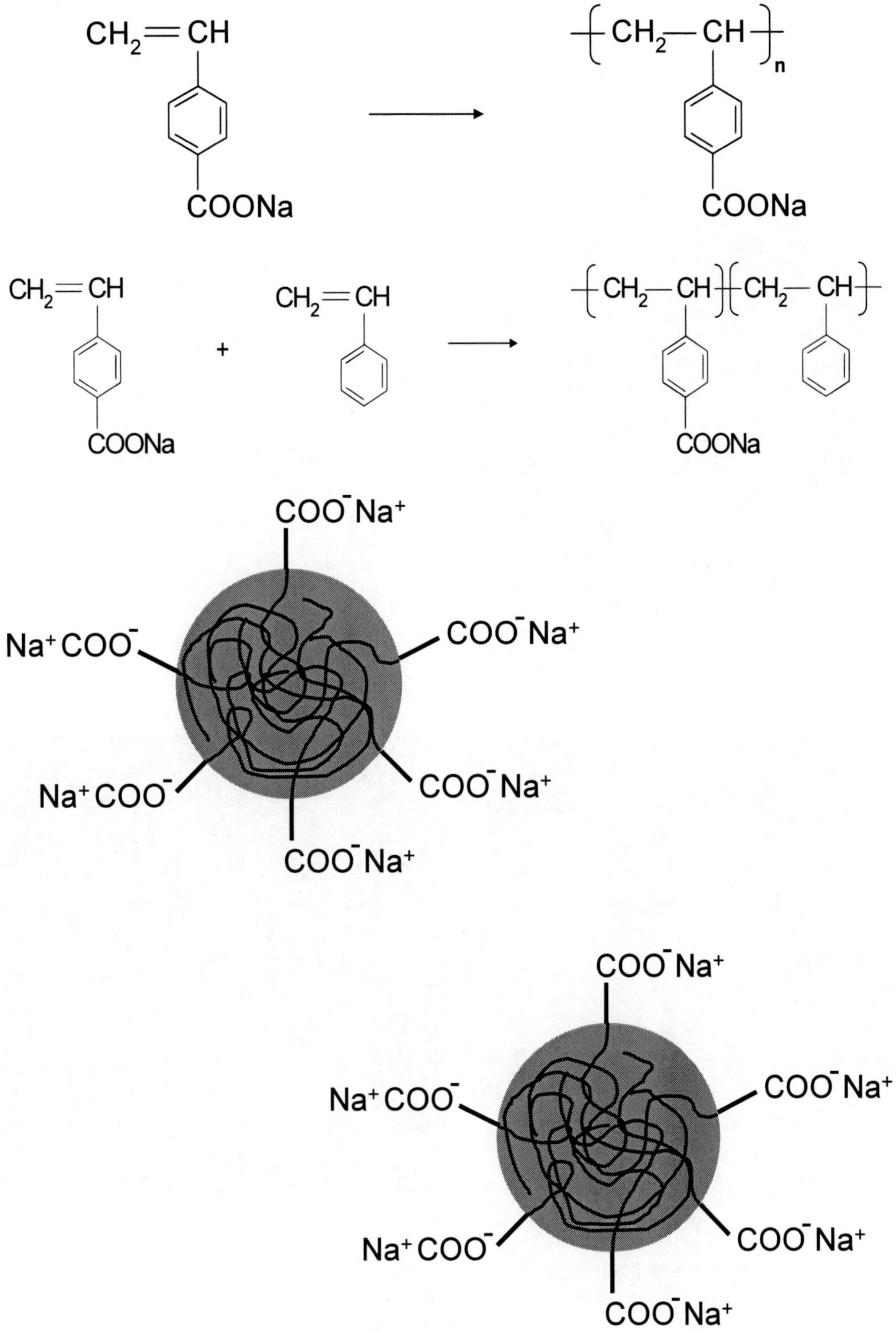

Figure 2b. Homopolymer or copolymer synthesized with hydrophilic monomer and the particles stabilized by these hydrophilic groups on the surface.

$$CH_3 - CH_2 - CH_2 \text{-----} CH_2 - SO_4^- \, ^+Na$$

$$CH_3 - CH_2 - CH_2 \text{-----} CH_2 - SO_4^- \, ^+Na$$
$$CH_3 - CH_2 - CH_2 \text{-----} CH_2 - SO_4^- \, ^+Na$$
$$CH_3 - CH_2 - CH_2 \text{-----} CH_2 - SO_4^- \, ^+Na$$
$$CH_3 - CH_2 - CH_2 \text{-----} CH_2 - SO_4^- \, ^+Na$$
$$CH_3 - CH_2 - CH_2 \text{-----} CH_2 - SO_4^- \, ^+Na$$

Figure 2c. Particles stabilized by surfactant.

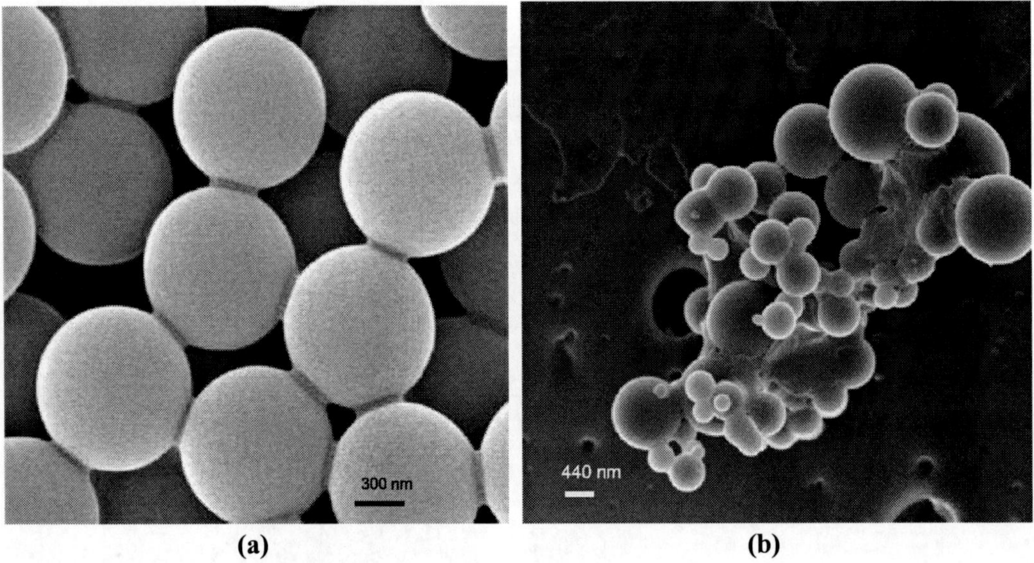

Figure 3. (a) Pure polystyrene particles and (b) same particles after the swelling with additional monomer and salt destabilization followed by polymerization.

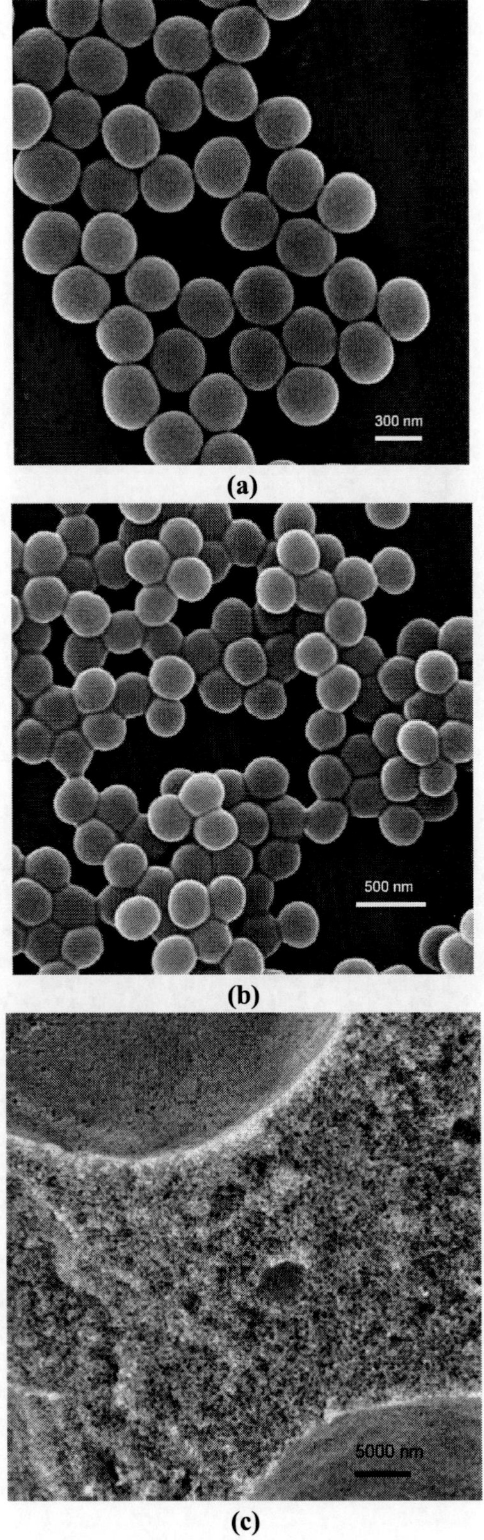

Figure 4. (a) Polystyrene particles and (b), (c) same particles destabilized by using high shear.

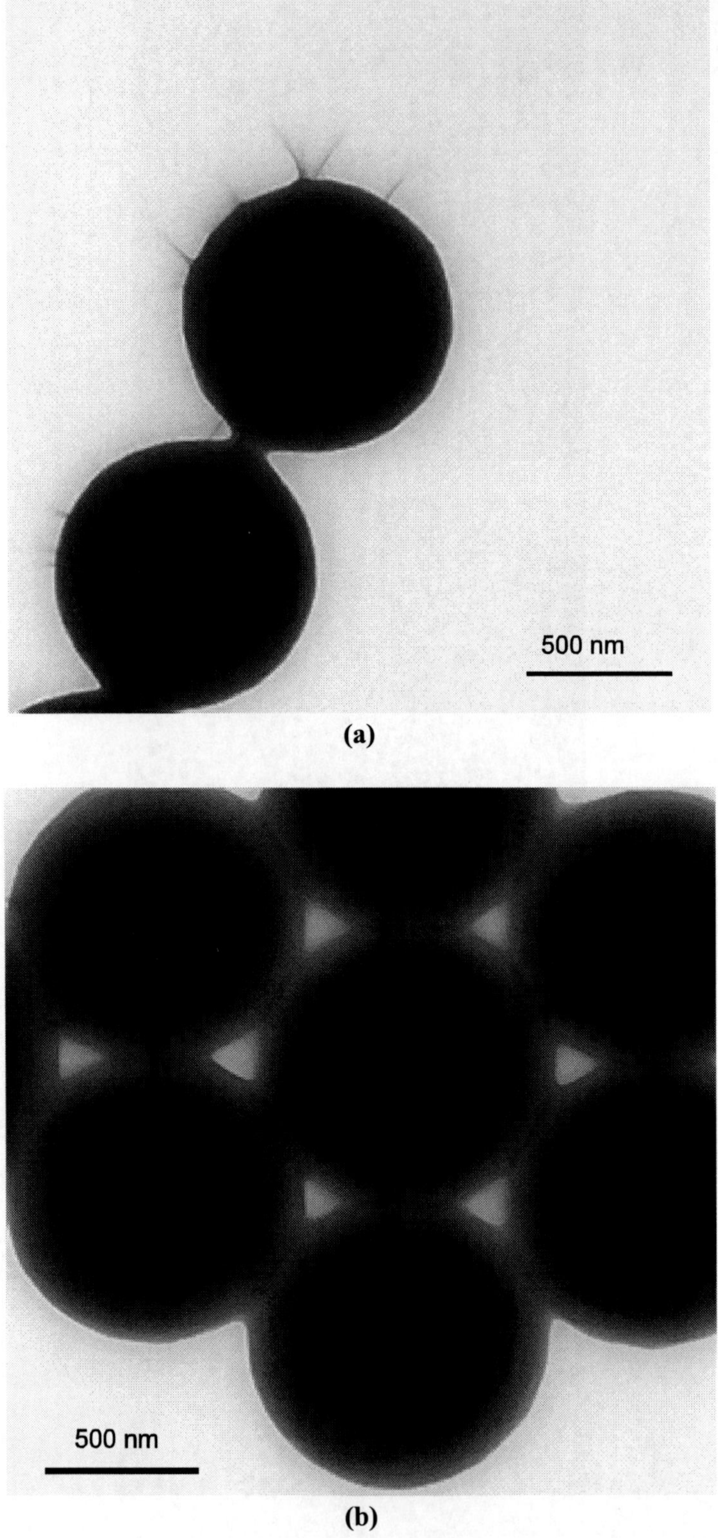

Figure 5. Surface functionalized polystyrene particles (a) stabilized at temperature below 32°C and (b) destabilized above 32°C.

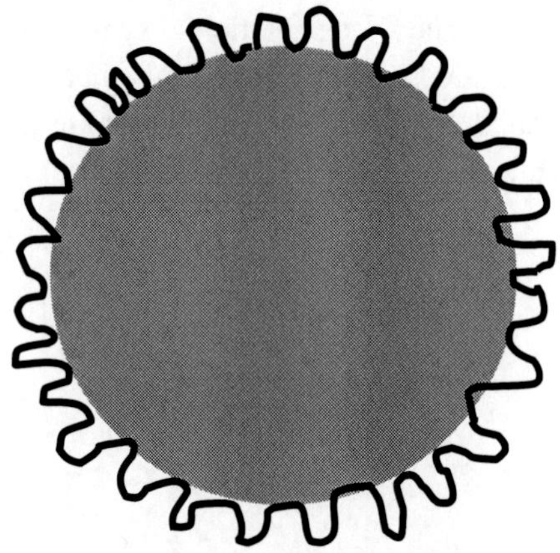

Figure 6. Steric stabilization of latex particle using polymer chains.

Although the immobilizations of charges on the surface of polymer particles act toward electrostatic stabilization of polymer particles, one should also be mindful that the addition of a small amount of ionic salts to the system can completely destabilize the system due to bridging of the electrical double layers around the polymer particles by the added ions. The additional amount of ions starts forming layers around the polymer particles, and in this process lead to aggregation of the polymer latex. The polymer latex is initially gelated by the addition of salt, i.e., the particles do not form extremely large aggregates and settle down. Instead, they lock the whole latex into a thick mass in which the particles form physical networks. However, the increased addition of salt can also bring about the total collapse or sedimentation of the particles. Therefore, to avoid unwanted generation of latex destabilization, one must be careful that no salt impurities are added along with other ingredients to the latex system. Due to its nature of destabilizing the particles, salt is indeed used in the case of surfactant free emulsion polymerization, as it destabilizes the nucleated particles and helps them to aggregate to a size where they become colloidally stable. In order to ensure better stability, an additional amount of surfactant is also added to the emulsion systems after the complete polymerization of the monomer, a process termed post stabilization, as it covers the surface of the particles that may have not homogenous coverage of the adsorbed surfactant. Figure 3 is a representation of the system that was coagulated by the addition of a small amount of salt to the system. The complete change of the state of the system before and after the addition of salt is completely visible.

Another important factor with a significant influence on the latex stabilization is the shearing of the latex by using high shears or sonication. Generally, the latexes are sheared or sonicated in order to remove the physical aggregates or lumps formed in the colloid system, thus, to achieve better dispersion. However, some special systems such as latexes generated without the use of surfactant can be totally destabilized by the use of shear or sonication, thus, requiring one to treat them with caution. As the latex particles generated by the surfactant free emulsion polymerization are only stabilized by the limited number of negative charges on the surface of the particles originating from the initiator fractions, they are therefore prone to destabilization as the higher extents of agitation leads to a higher number of collisions of these particles with each other and they are more susceptible to stick to each other in the absence of strong repelling forces on the surface. Figure 4 shows such a system of polystyrene latex particles synthesized by emulsifier free polymerization, which were aggregated by the use of high shear. One point to note here is that although the aggregates formed as a result of high shear are purely physical in nature, they are, however, also quite stable and do not break into primary particles.

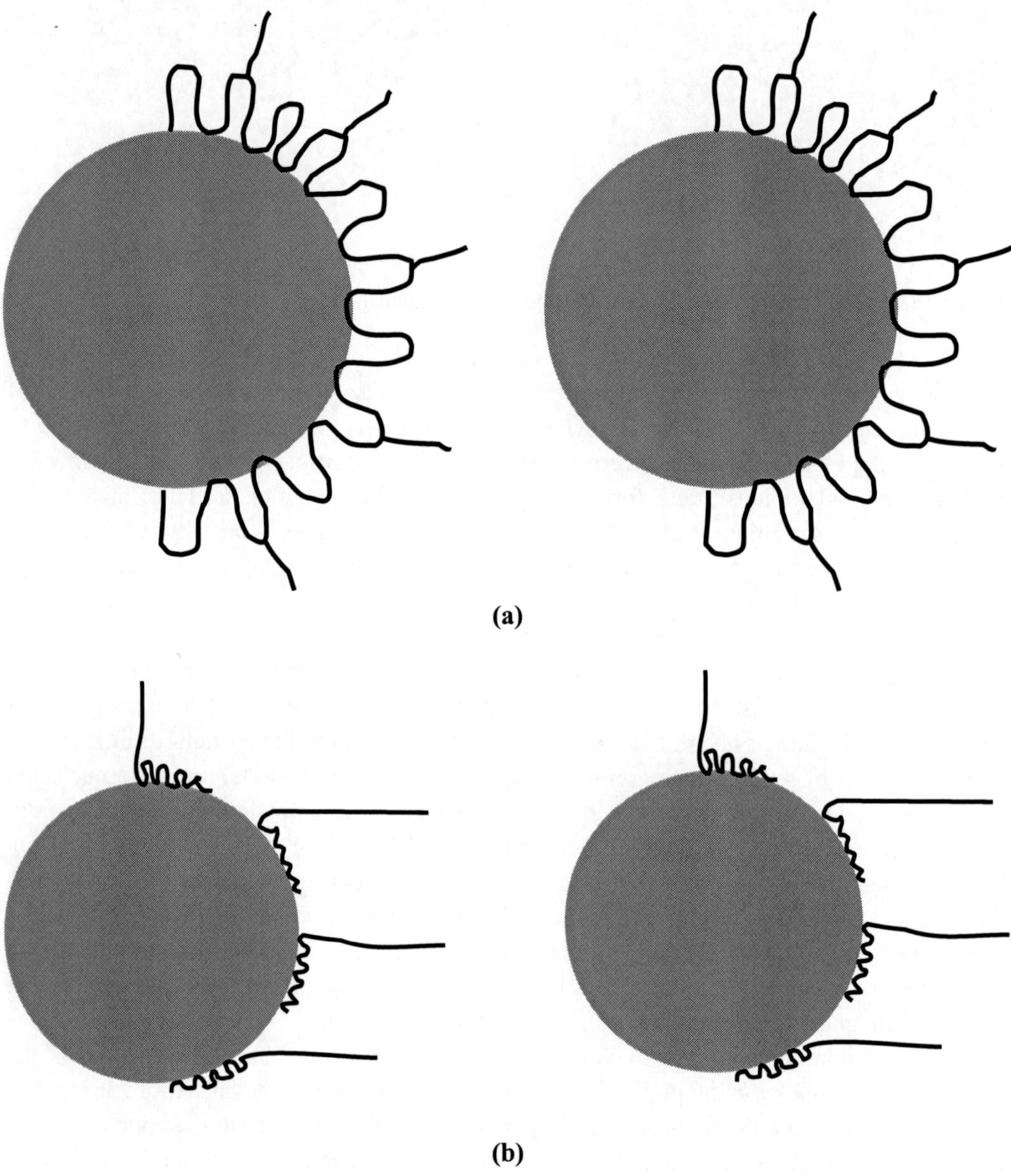

Figure 7. Steric stabilization of polymer particles using (a) graft copolymer chains and (b) block copolymer chains.

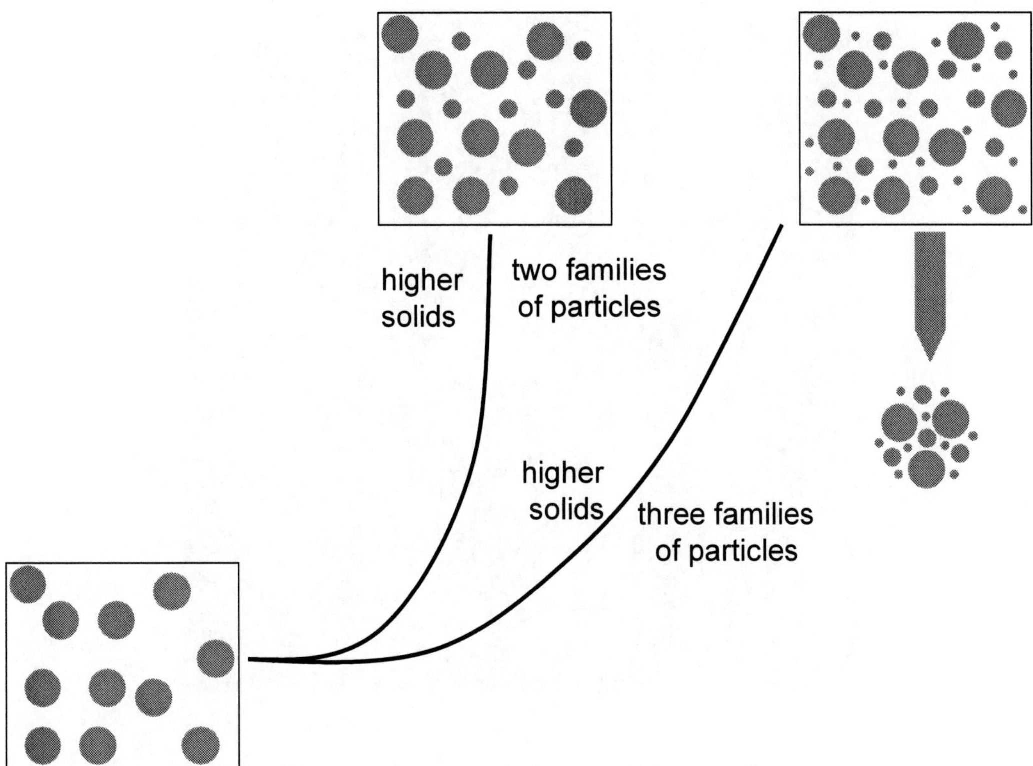

Figure 8. Generation of high solids in the latex using two or more families of particle sizes.

One interesting possibility is the achievement of reversible stability and destability of the latex particles when the latex particles are specifically functionalized [1-4]—for example, when the particles are grafted with a thermally reversible polymer, i.e., the polymer chains are hydrophobic at a temperature above the solution temperature of polymer and are hydrophilic at a temperature lower than the critical solution temperature. One such polymer to show reversible change in hydrophilicity and hydrophobicity as a function of temperature is poly(N-isopropylacrylamide). The polymer chains of poly(N-isopropylacrylamide) can be grafted onto the surface of polymer particles by carrying out either copolymerization of styrene with N-isopropylacrylamide or by grafting of the poly(N-isopropylacrylamide) chains on the preformed polystyrene seed particles by using controlled living polymerization techniques such atom transfer radical polymerization. These chains transform the nature of polymer particles and make them thermally responsive. When the temperature is lower than 32°C, which is the critical solution temperature, the chains form a hydrogen bond with the aqueous phase, thus, completely stabilizing the polymer particles. However, when the latex is heated above 32°, the chains collapse and become hydrophobic as they form the hydrogen bond with each other, thus, repelling the water molecules out. This generated hydrophobicity on the surface leads to a quick destabilization of the system, and the particles start to sediment. However, this destabilization is reversible as the particles become stable again as soon as the temperature is lowered below 32°C. Figure 5 represents this system of reversible stability and destability of the poly(N-isopropylacrylamide) functionalized latex particles below and above the critical solution temperature of 32°C.

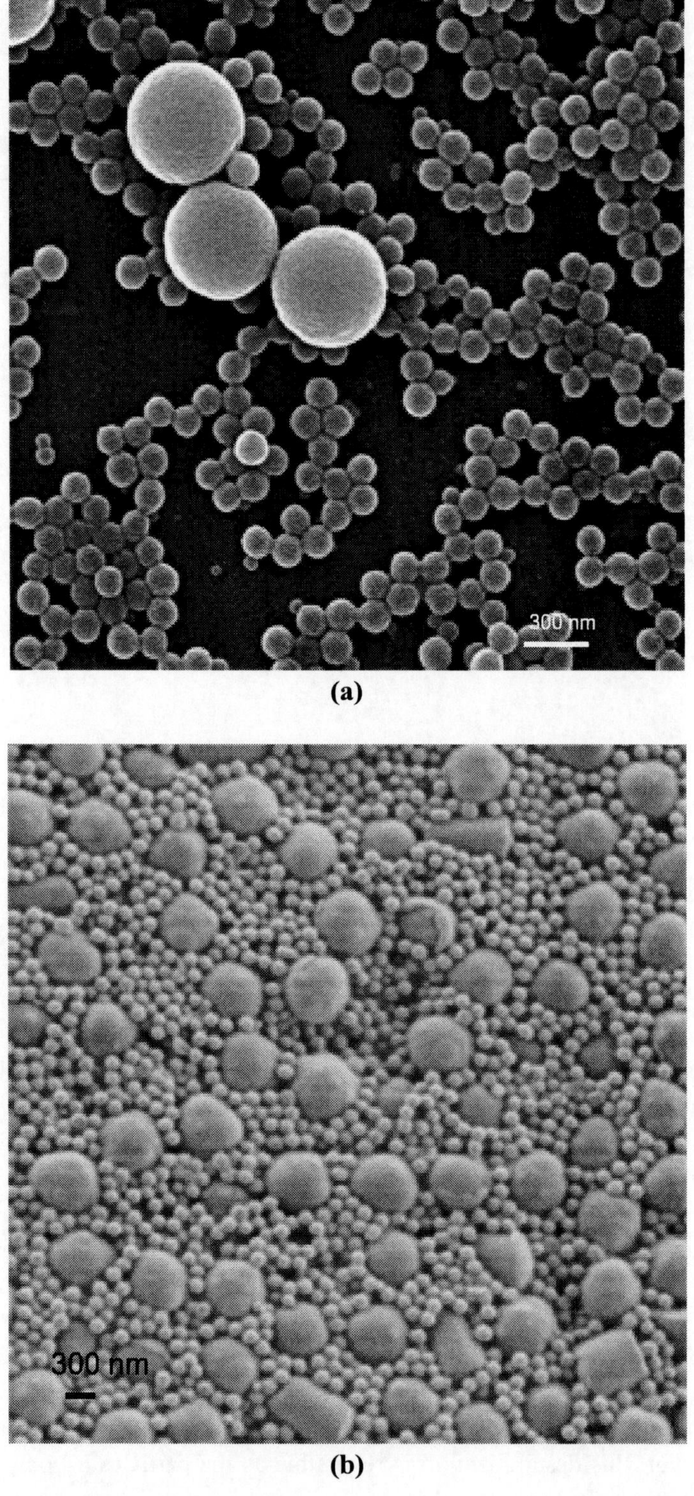

Figure 9. Examples of bimodal polymer latexes.

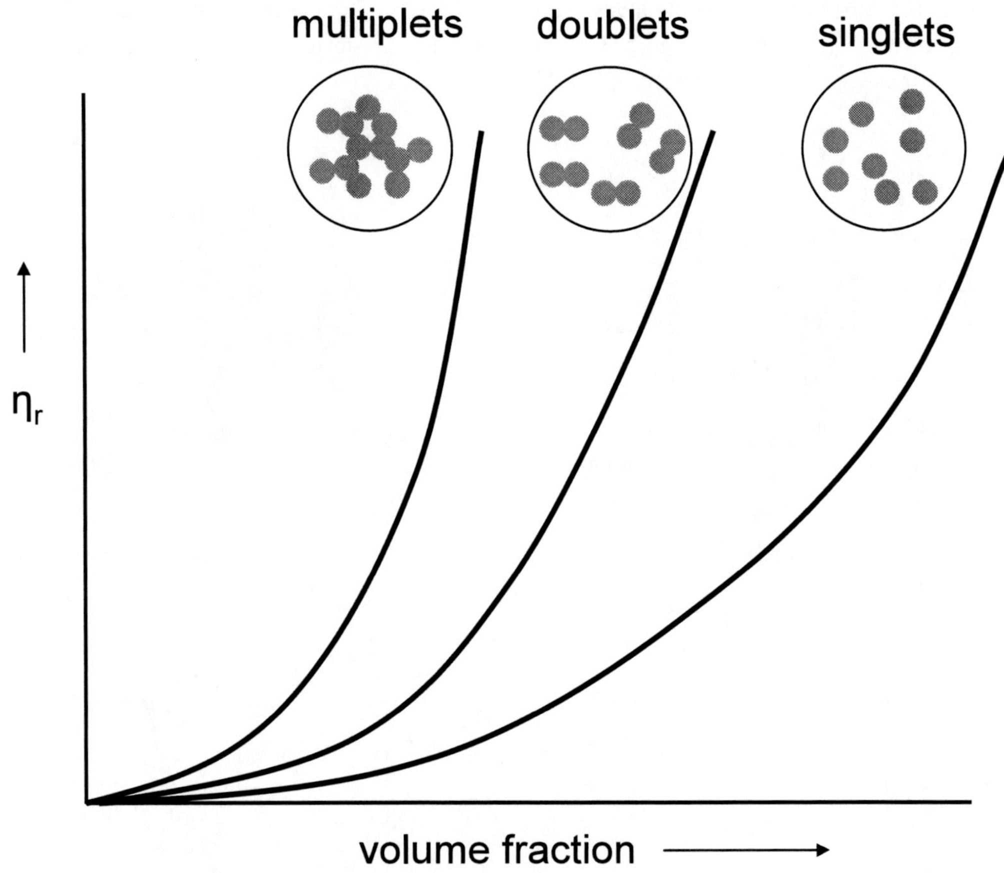

Figure 10. Dependence of latex viscosity as a function of coagulation or aggregation of the system.

Steric stabilization is another important means of achieving the stabilization of latex particles. In this case, long non-ionic polymer chains are adsorbed on the surface of the polymer particles. The systems stabilized by steric means are stable even at high salt concentrations, as the particles cannot be bridged by the additional amount of salt added to the system. The polymers used are hydrophilic and the loops of chains adsorbed on the surface form hydrogen bonds with the aqueous phase and help to stabilize the latex particles. Figure 6 is a representation of a sterically stabilized system. The adsorption of these polymer chains occurs on the surface of the particles only when energy of adsorption of these chains exceeds the loss of entropy of the polymer chains when they adsorb on the polymer particles. Specifically tailored adsorbents can be used to provide stabilization to the system. Generally, adsorbents with two different components are used: one of the components has a higher affinity for the polymer particles' surface and is insoluble in the aqueous phase, whereas the other component has a higher affinity for the aqueous phase or is soluble in water and has practically no interaction with the particles' surface. Graft and block copolymers find special applications in such stabilization processes. The main chain in the graft copolymer acts as a component that attaches itself to the particles, whereas the grafts on the main chain act as a second component that helps to stabilize the particle owing to its water solubility. In the case of block copolymers, one block of the chain attaches to the surface of polymer particles, whereas the other block radiates away from the surface of the particles because there is no interaction with the surface. Figure 7 shows examples of steric stabilization of particles by using graft as well as block copolymers.

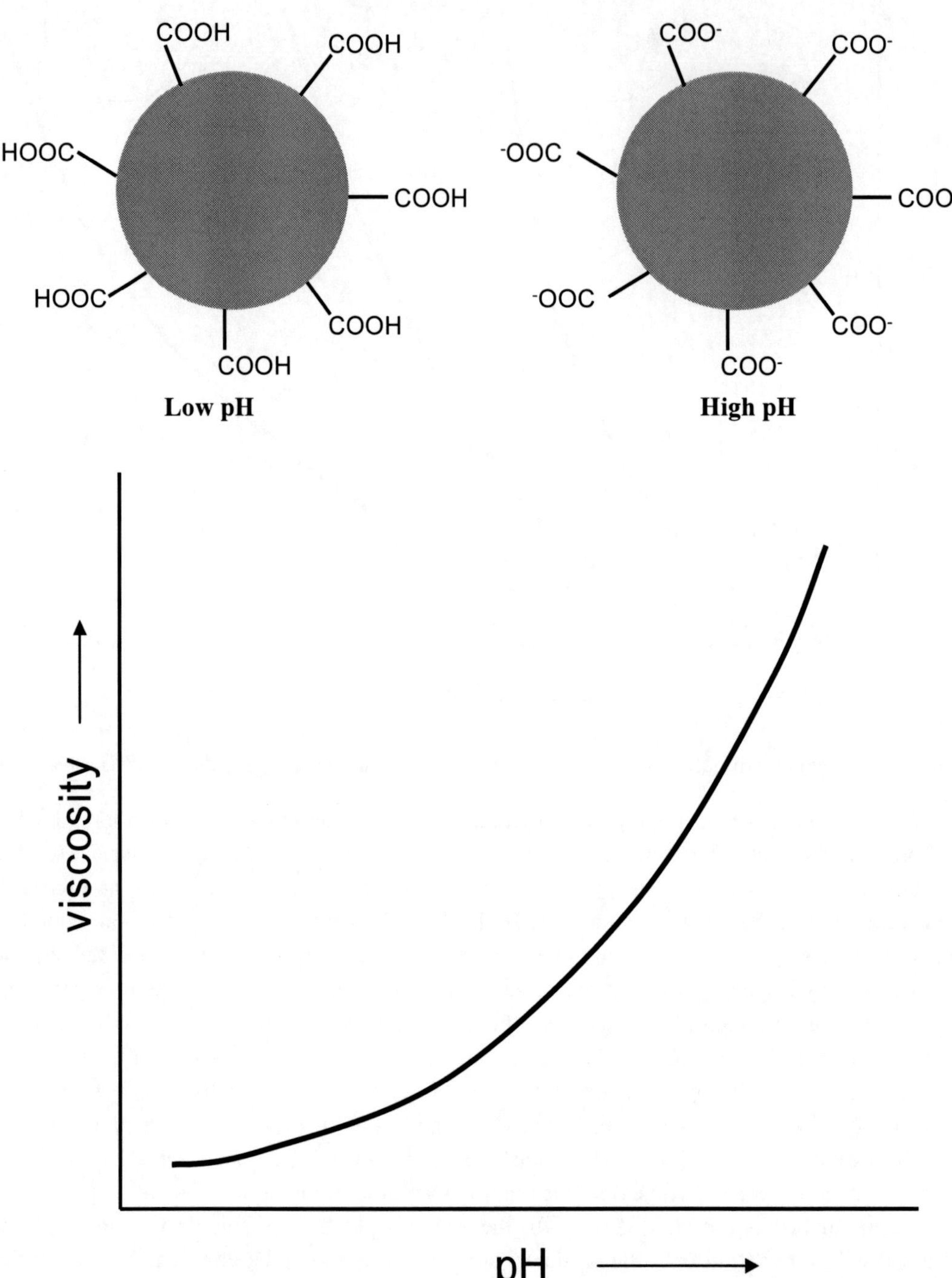

Figure 11. Effect of pH on the viscosity of the latex.

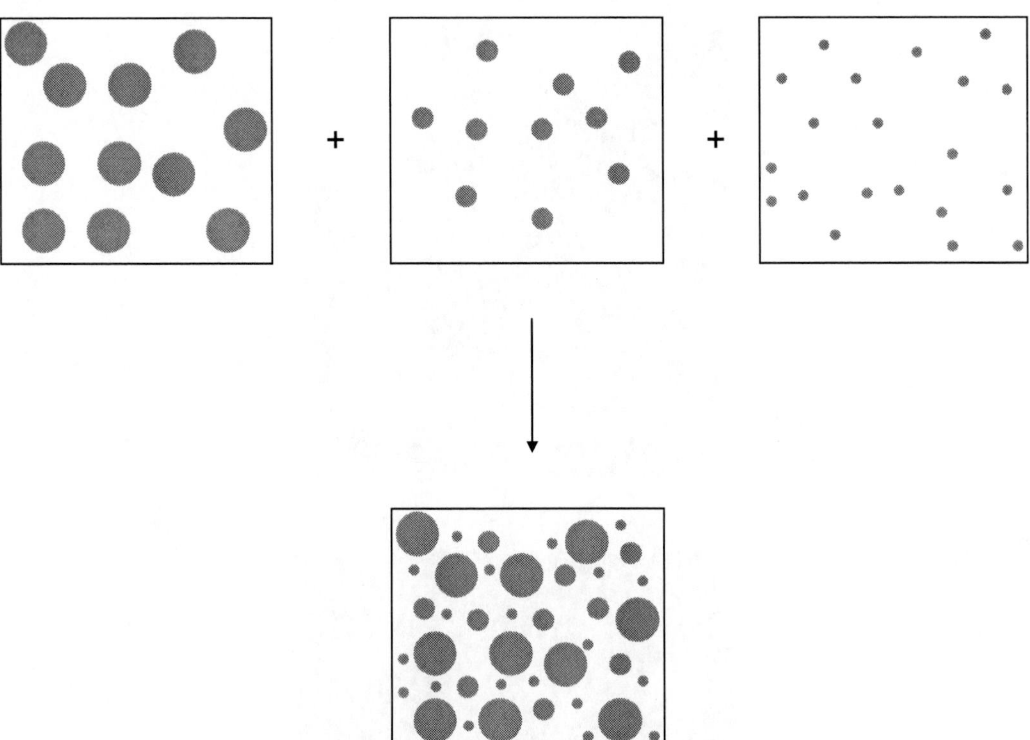

Figure 12. System of generation of multi-modal latexes (high solids) by the approach of latex mixing.

(a)

Figure 12. (Continues on next page.)

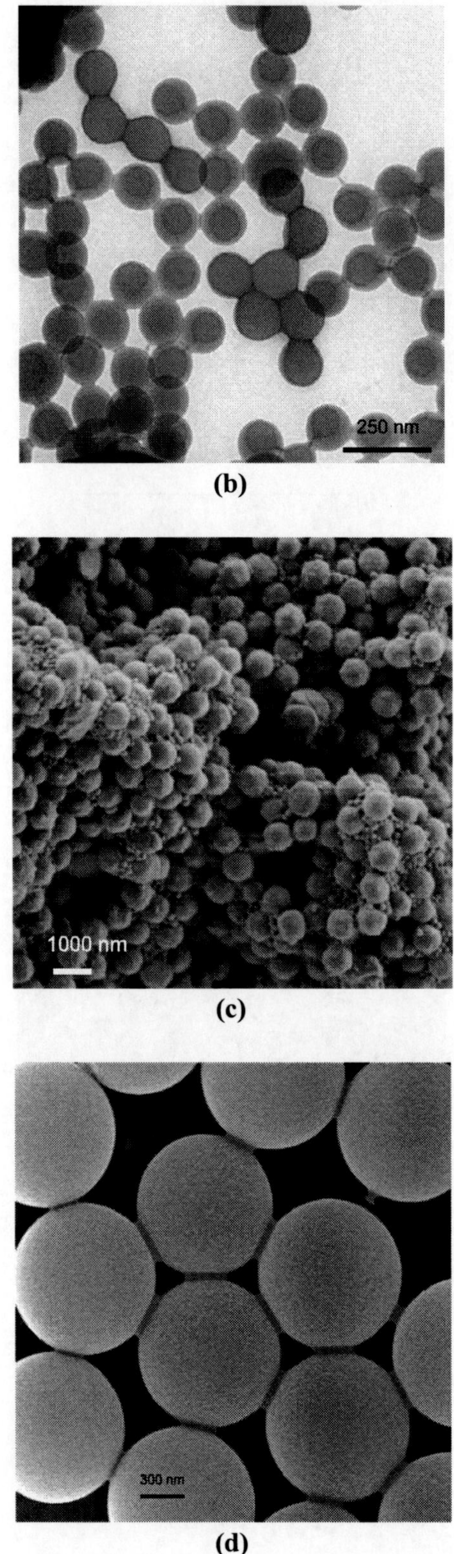

(b)

(c)

(d)

Figure 12. (Continues on next page.)

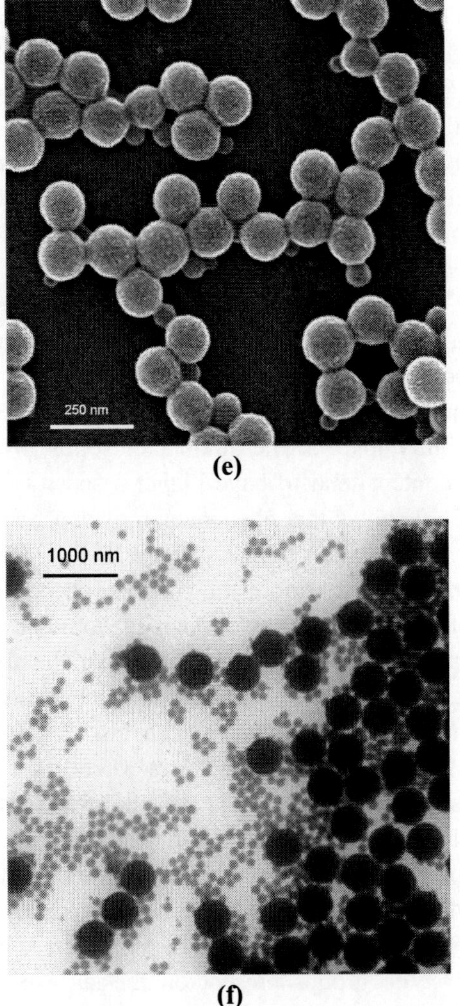

Figure 12. System of generation of multi-modal latexes (high solids) by the approach of latex mixing: (a,d) large particles, (b,e) small particles and (c,f) mixture of the two latexes.

C. Achievement of High Solids Latexes

As mentioned earlier, the achievement of high solids in the latexes has commercial importance, as it makes the polymerization process more efficient. Increasing the solid content in the latex leads to increasing the viscosity of the latexes and, after a certain solid content, the latexes generated can be very thick and the polymerization process can also be difficult to control. However, when particles of different sizes are either generated in the polymerization process or are simply mixed with each other, the high extent of solids can still be reached while keeping the viscosity low. Figure 8 is a depiction of such an effect, whereby the generation of higher solids with two or three families of particle sizes can be achieved to increase the workable amount of solids in the latexes. The smaller particles fit themselves into the interstitial spaces between the larger particles and hence do not require additional space. Similarly, in the case of three families of particle sizes, the middle-size particles sit in the interstitial spaces between the largest particles, whereas the smallest particles find places near the largest and middle-sized particles. Figure 9 is also an example of polystyrene particles copolymerized with a functional acrylic monomer where the bimodality of the particles is clearly seen and the smaller particles fit themselves in between the larger particles. The particles were achieved by first generating a seed of polymer particles by leading the full conversion of monomer in the seed particles. The particles were then again swollen with an additional batch of monomers and initiator, and, owing to the incompatibility of the forming copolymer chains with the surface of seed particles, new stable copolymer particles were formed in the system. By controlling the amount of the secondary batch of monomers, the size of these secondary particles can be controlled, and hence the rheology of the latex system can also be subsequently controlled.

The state of dispersion of polymer particles can also lead to significant influences on the viscosity and, thus, the ability to pack the amount of solids in the latex. The aggregated latexes at the same solid content tend to have higher viscosity compared to the uniformly dispersed particles. Figure 10 shows this phenomenon of dependence of latex viscosity as a function of aggregation or coagulation of the system. In fact, the extent of aggregation or coagulation also has an effect, as the more the coagulation occurs, the more the viscosity increases in the latex. In other words, the same viscosity would be reached in the latex system in the case of aggregated particles at lower solid content than the dispersed particles. To retain lower viscosities, thus, one needs to be sure to avoid any possibility of the aggregation of the particles. On the other hand, this phenomenon can also be used to advantage in many commercial applications, such as viscosity modifiers where the aggregated particles can be used to increase the viscosity of the system at lower solid fractions. Also, the effect of pH or other change in the conditions of the aqueous medium significantly affects the latex rheology. As shown in Figure 11, for latex particles with carboxylic groups on the surface, at low pH, the viscosity of the system is much lower. However, when the pH of the aqueous medium is increased, the viscosity becomes much higher at the same solid fraction, owing to the neutralization of the carboxylic groups present on the surface of the particles. Thus, by changing the interface of the latex particles, different properties can be achieved.

The importance of the generation of multimodal distributions in the latexes to control the increase in viscosity as a function of solid fraction was demonstrated in an interesting study by Chu et al. [5]. The viscosity of a mono-modal latex at 65% solid content was observed to be 1953 mPa.s, whereas the tri-modal latex with similar chemical composition of the particles as well as solid content had a viscosity of 84 mPa.s. The tri-modal latex was generated by synthesizing the mono-modal latexes first, and these latexes were subsequently mixed with each other and the aqueous phase was evaporated to concentrate the solid fraction in the resulting latex. The particle sizes of 75, 135 and 477 nm were used to form the tri-modal latex. The amount of particles in the solids corresponded to 80/10/10 percent of large, medium and small particles, and these values were observed to be optimum to achieve maximum packing of the particles in the latex. As mentioned earlier, multi-modal latexes were also generated by using the secondary nucleation approach. In this approach, a preformed seed was used, to which was added an additional amount of monomers, initiator and surfactant, and secondary particles with smaller particle sizes could be obtained. The system could be controlled according to need by adjusting the amount of seed particles, the amount of second batch of monomers as well as by adjusting the amount of surfactant added during the secondary nucleation. Figure 12 represents the system of generation of multi-modal latexes with the approach of latex mixing; micrographs are shown for polystyrene latexes to generate bimodal latex. Figure 13 shows the semi batch emulsion polymerization approach to generate multi-modal latexes. As shown in the micrographs 13a and 13c, the seed particles are first generated, which are then added with the second batch of monomers to generate the secondary particles, as shown in micrographs 13b and 13d. One can add the second batch of monomers either when the seed is fully polymerized, as shown in Figure 13d, or when the seed is partially polymerized, as shown in Figure 13b. In both cases, depending on the interaction of the generated copolymer chains with the fully or partially polymerized seed particles, different morphologies and sizes of secondary particles can be obtained. Figures 13e and 13f are the corresponding SEM images of the particles shown in Figures 13c and 13d. However, one must be careful about the stability of the secondary particles, as it is very important for them to reach a size quickly during the course of polymerization where they are colloidally stable. In the case that this condition is not reached, they would tend to collapse with each other and more often on the surface of seed particles, thus, losing the desired effectiveness of the latex, i.e., lower viscosity or high in solids. This phenomenon is explained in Figure 14. The polystyrene seed particles were used to generate a secondary population of particles of a copolymer of styrene and functional acrylate. However, the generated particles did not reach a stable state quickly and, due to collisions with the seed particles owing to agitation, collapsed on the surface of the seed particles. The system was difficult to stabilize because the polymerization was carried out without the use of surfactant. Caution is also called for in the case of surfactant free emulsion polymerization systems, in which the occasional generation of very large particles is caused by the coagulation of a large number of small particles, which may lead to incorrect interpretation of the results of size determination in characterization methods like light scattering. Figure 15 shows one such instance in which a very large particle was generated by the collapse of small particles.

D. Scale Up

Figure 16 is a schematic of scale up from the lab scale to pilot plant or production scale. As mentioned earlier, the reaction conditions that lead to the successful generation of the latexes and their surface functionalization on the lab scale may not lead to optimum results when synthesized on a larger scale. Because of changes in agitation, heat control, etc., the generated system morphology may be totally different. An example is shown in Figure 17, in which particles formed on a small lab scale and particles formed in pilot plant scale reactors are compared. Owing to the limited agitation, the particles formed on the lab scale were very rough in surface, which was a result of the collapse of secondary particles on the seed particles. However, in the pilot scale reactor, a great degree of efficient control over agitation was achieved, which led to the generation of particles with smoother surfaces. Therefore, one has to carefully optimize the reaction conditions that would lead to successful translation of the technology on different scales.

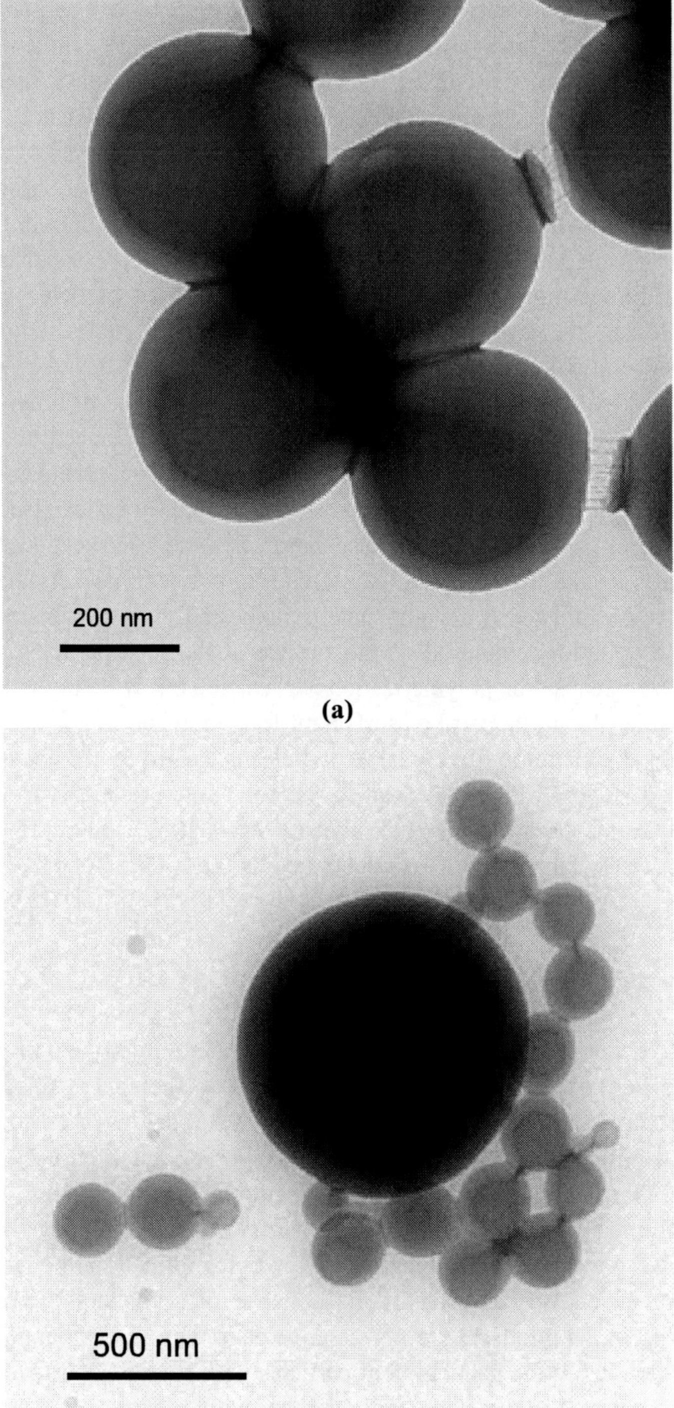

Figure 13. (Continues on next page.)

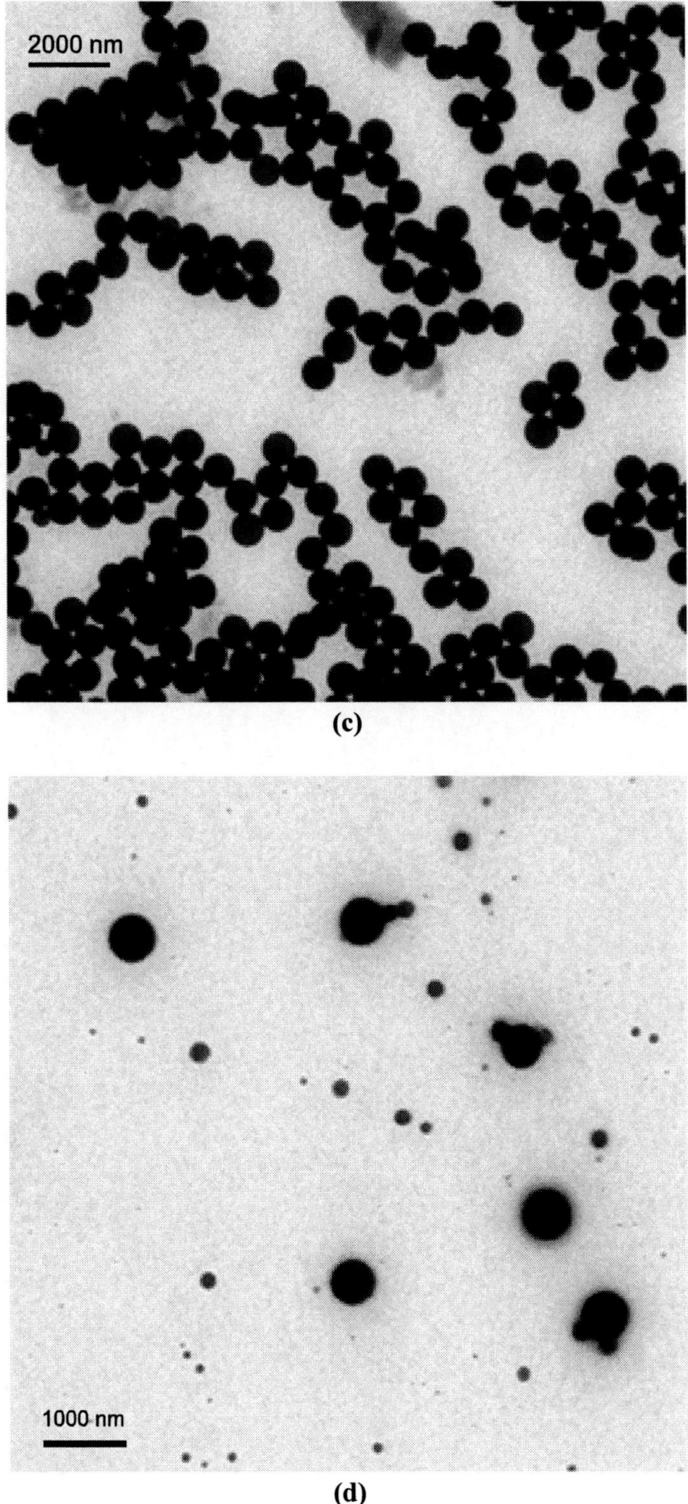

(c)

(d)

Figure 13. (Continues on next page.)

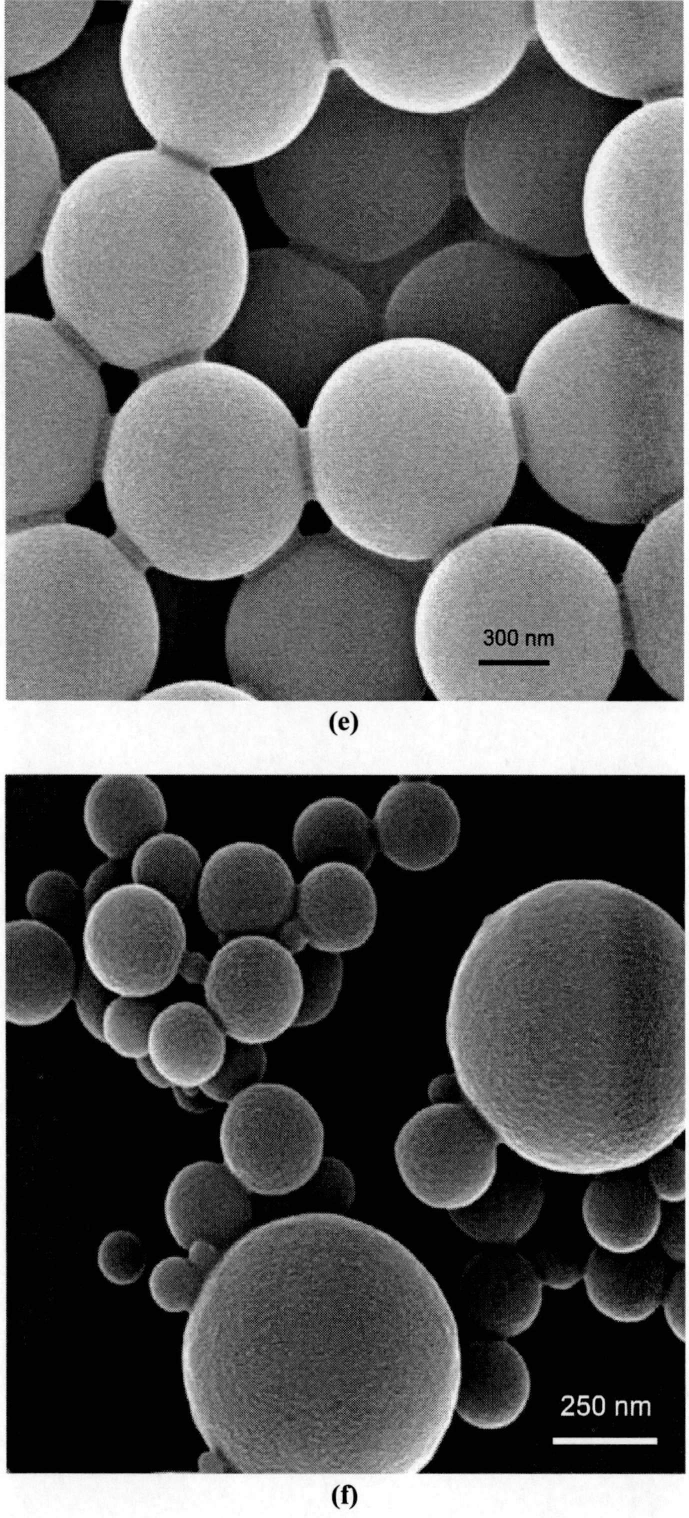

Figure 13. Semi batch emulsion polymerization approach to generate the multi-modal latexes: (a) seed, (b) resulting particles with two families of sizes, (c) seed, (d) resulting particles with two families of sizes, (e) seed, and (f) resulting particles with two families of sizes.

(a)

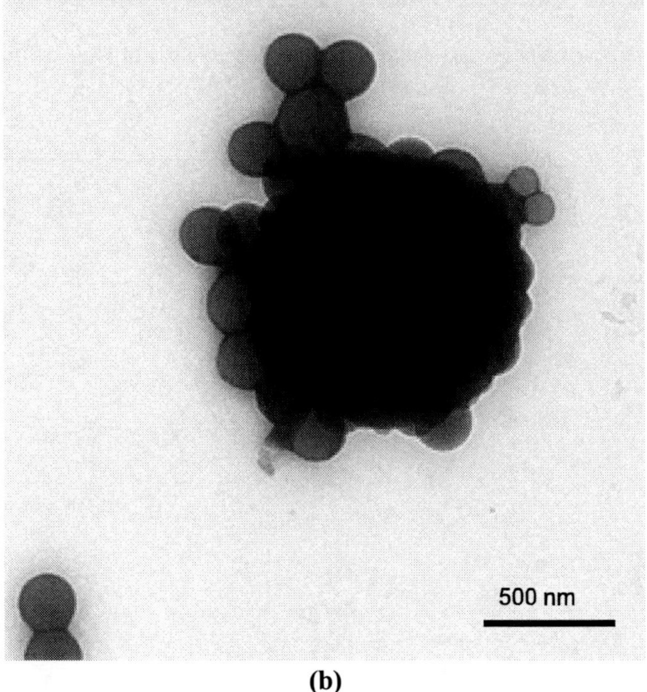

(b)

Figure 14 (a) and (b). Collapsed secondary particles on the seed particles.

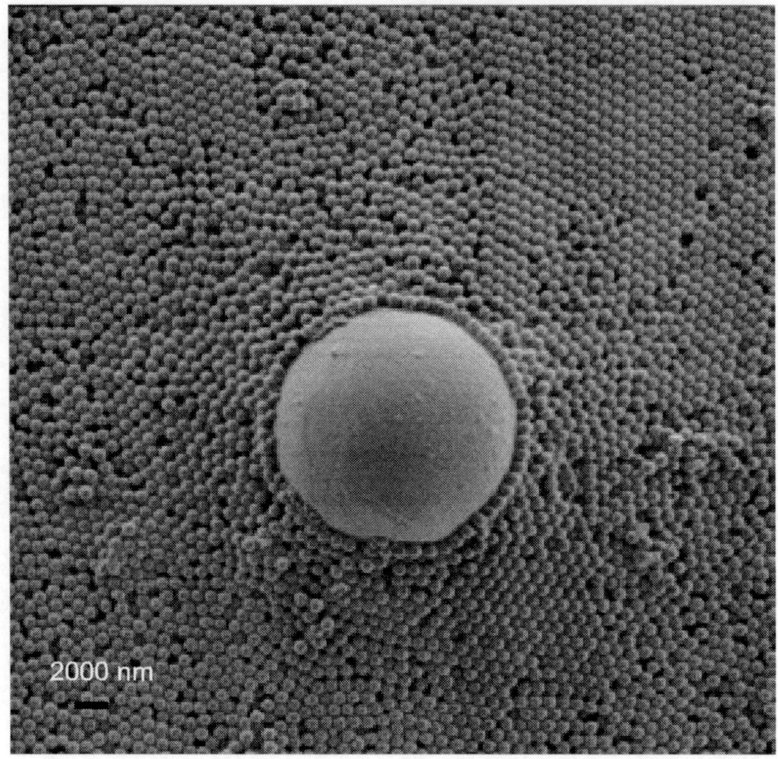

Figure 15. Demonstration of a large particle generated by the collapse of large number of small particles.

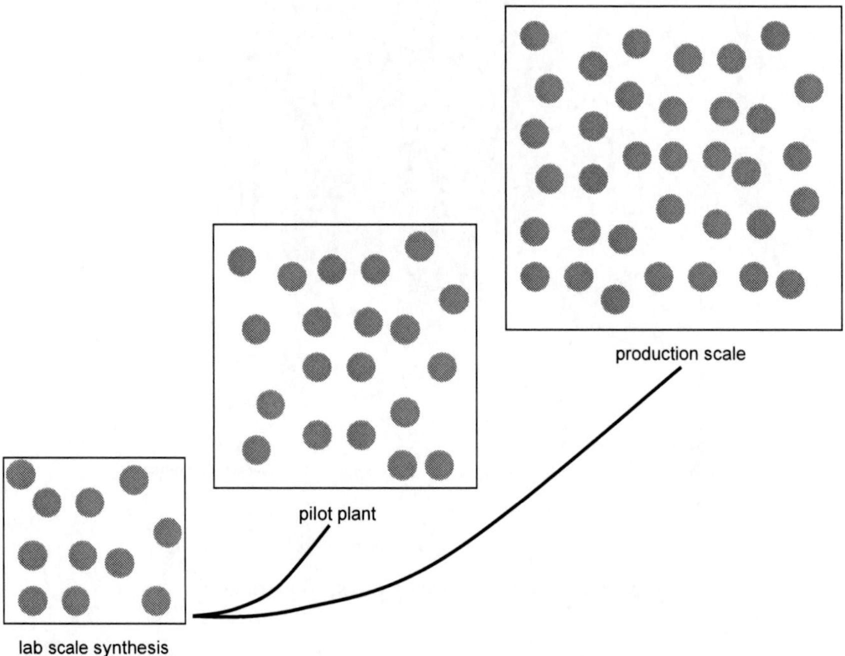

Figure 16. Schematic of scale up from laboratory scale to pilot plant and production scale.

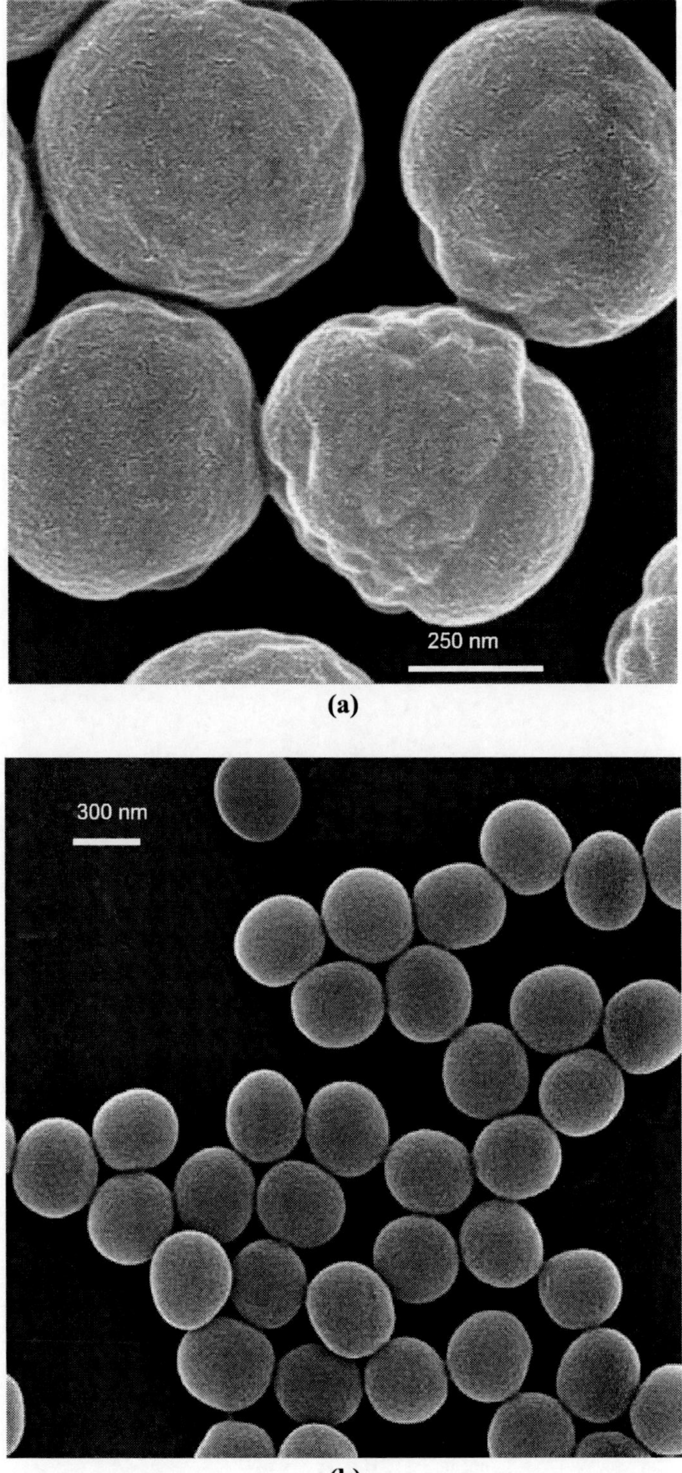

Figure 17. Comparison of the morphology of the particles generated in (a) a lab scale flask and (b) pilot plant scale reactor.

References

[1] V. Mittal, N. B. Matsko, A. Butté, and M. Morbidelli (2007). *Polymer,* 48, 2806-2817.

[2] V. Mittal, N. B. Matsko, A. Butté, and M. Morbidelli (2007). *European Polymer Journal*, 43, 4868-4881.

[3] V. Mittal, N. B. Matsko, A. Butté, and M. Morbidelli (2008). *Macromolecular Materials and Engineering,* 293, 491-502.

[4] V. Mittal, N. B. Matsko, A. Butté, and M. Morbidelli (2008). *Macromolecular Reaction Engineering*, 2, 215-221.

[5] F. Chu and A. Guyot (2001). *Colloid and Polymer Science*, 279, 361-367.

Chapter 10

THERMALLY SWITCHABLE POLYMER PARTICLES: A SPECIAL STUDY

A. INTRODUCTION

Surface properties of polymer particles have been among the most important contributors to their application, especially when the application involves the interaction of their surfaces with foreign media or entities. As an example, the surface properties are of tremendous importance when the chromatographic separations of proteins, polymers or any mixtures must be handled. There is always a requirement or desire to achieve extra functionalities on the surface by controlling the surface properties of these particles, which can be used for one application or another or a combination of a few [1,2]. These functionalized surfaces represent a novel class of materials, which at the forefront of technology can lead to revolutionary changes in the conventional processes. In one such approach, an environmentally responsive character has been generated on the surfaces. Generation of this property on the surface of the particles leads to functionalized materials in which their nature, e.g., hydrophobicity or hydrophilicity, can be finely controlled by environmental stimulants such as temperature, salt, pH, etc. This property translates the use of these particles in a number of applications, including temperature controlled adsorption and desorption of biomolecules, temperature controlled drug delivery processes, etc.

Different ways have been developed to generate the thermal responsive behavior on the surfaces, different materials have been used to generate thermal responsiveness and different substrates have been used on which thermal responsive character has been achieved. Many techniques to functionalize the surfaces have been reported in the literature, such as physical adsorption of oligomeric or polymeric chains on the surface [3-5], grafting of polymer chains "to" the surface [6,7] and grafting of polymer chains "from" the surface [8,9]. Although physical adsorption has the advantage of being able to lead to the adsorption of long polymer chains of well-defined length, molecular weight distribution and composition to the surface, this approach suffers from the limitations of steric hindrance posed by the long polymer chains, thus, seriously limiting the final surface density of the grafts. Apart from that, as the chains are only physically bound to the surfaces, the efficiency of these linkages for load transfer or resistance towards washing or cleaving off may not be sufficient. Grafting of polymer chains "to" the surface approach focuses on chemical immobilization of end-functionalized polymer chains to the surface by reacting the ends with the reactive groups already immobilized on the surface. It can also be achieved either by attaching a monomer on the surfaces and subsequently polymerizing the monomer in the presence of externally added monomer and initiator. In both the cases, as the preformed long chains approach the surfaces, therefore, the steric hindrance concerns many still hinder the generation of high density grafts on the surface. The generated grafts may also not be very uniform in terms of molecular weight and chain lengths, etc., owing to the uncontrolled nature of the process, which is generally achieved by non-living conventional free radical polymerization.

Another important and extensively studied approach is the grafting of polymer chains "from" the surface. In this approach, an initiator is covalently bound to the surface which is subsequently polymerized with the external monomer without the external addition of any further initiator. This approach has enjoyed more success because of its ability to generate densely packed polymer functionalities in the absence of any kinetic hindrance, along with the advantage of covalently bound polymer chains on the surface [8-11]. Such structures are referred to as polymer brushes, when the distance between the grafted chains is less than twice the radius of gyration of the unperturbed polymer chain [12]. A number of different polymerization modes have been used to graft polymer chains on the surfaces, like non-living free radical polymerization [13], controlled polymerization techniques such as ionic [14,15], nitroxide mediated polymerization [13], atom transfer radical polymerization (ATRP) [11,16-18] and reversible addition fragmentation transfer (RAFT) [19-21]. Conventional polymer systems suffer from unwanted bimolecular termination reactions, thereby leading the grafting process to an abrupt end. However, the growth of living polymer chains from the surfaces generated by controlled living polymerization techniques ensures better control over the molecular weight distribution and the amount of grafted polymer. ATRP has been particularly useful techniques because of ease of operation, good control over the properties of the grafts, and the possibility of polymerization in aqueous phase [3-10, 16,17,19,22-39].

As mentioned earlier, environmental responsiveness means reversible changes in the hydrophobicity and hydrophilicity, dimension and density of the polymer chains, owing to changes in the intramolecular interactions stimulated by changes in stimulants like temperature, salt concentration, pH, electrical potential, etc. Out of a number of known polymers known to exhibit this behavior, water soluble poly(N-isopropylacrylamide) (PNIPAAM) is a very common and extensively studied material. It has a lower critical solution temperature (LCST) of about 32°C [40-45], and below this temperature the chains exhibit chain extended conformations and random coil structure. The intermolecular hydrogen bonding with the water molecules due to the chain extended morphology generates the hydrophilic nature of the chains. The chains transform into more collapsed globular form above the lower critical solution temperature owing to the domination of intramolecular bonding between the CO and NH groups over the external hydrogen bonding. The highly reversible responses of PNIPAAM chains to temperature as well as other stimulants have led to its extensive use in the synthesis of stimulus sensitive surfaces. Due to these characteristics, PNIPAAM chains grown on various surfaces have been used for protein adsorption, responsive gels, biological separations, etc. [46-56].

B. GENERATION OF THERMALLY REVERSIBLE BEHAVIOR ON THE SURFACES

In an interesting study to generate flat surfaces with thermally responsive characteristics, PNIPAAM was generated from silicon substrates and the effect of surface roughness on the behavior of PNIPAAM grafts was studied [18]. To generate these surfaces, as shown in Figure 1, a clean silicon substrate was immersed in an aqueous NaOH solution followed by the addition of HNO_3 to generate surface hydroxyl groups. The surface was heated in reflux with toluene solution of aminopropyl trimethoxysilane (ATMS) to obtain chemically bonded NH_2 groups on the surface. The surface was then immersed in pyridine solution followed by the addition of polymerization initiator bromoisobutyryl bromide. Grafting of PNIPAAM brushes from the surface could be achieved by immersing the ATRP initiator immobilized silicon substrate in a solution of N-isopropylacrylamide containing CuBr and pentamethyl diethylenetriamine (PMDETA). It was pointed out that the surface characteristics of the material on which the polymer chains are grafted affect the functionalization with PNIPAAM. When grown on rough surfaces, the PNIPAAM chains transformed the materials to become superhydrophobous and superhydrophilic [20]. Contact angles with water of ~0° and ~150° were reported at 25°C and 40°C, respectively, while values of ~64° and ~93° were found for the same temperatures when PNIPAAM was grown on unroughened surfaces. As the polymer particles can be expected to be rough on the surface, it is possible therefore to translate the same performance on the surface of the polymer particles when these particles are covered with this thermally responsive polymer system.

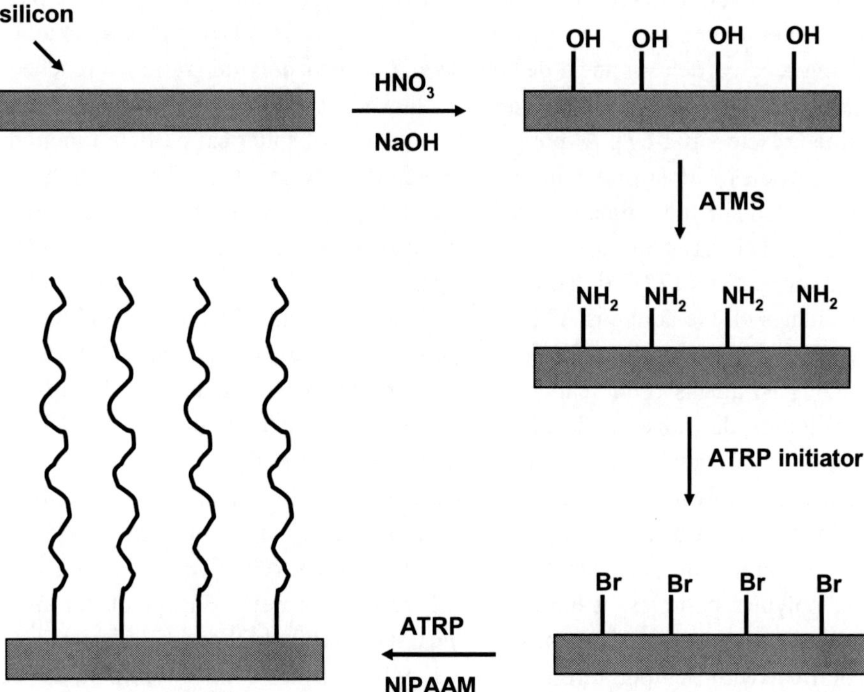

Figure 1. Generation of thermally responsive behavior on flat surfaces: use of ATRP methods on silicon wafers functionalized with ATMS and ATRP initiator to graft PNIPAAM chains [18].

It is clear from the literature that many of the reported studies on grafting thermally responsive polymers on surfaces employ the use of flat substrates [57-60]. On the other hand, spherical particles are also important materials of choice for a number of applications and, owing to their higher surface to volume ratio, are interesting substrates to study. As the polymer latex particle surfaces in water have been reported to be very different from the flat silica surfaces commonly used, because of surface charges, charge concentration, wetting properties, etc. [61], their behavior in the above-mentioned functionalization processes can also be very different and can only be quantized when these systems are practically studied. Also, as the surface roughness was observed to enhance the thermally responsive behavior of the surfaces, one would expect that polymeric porous materials, owing to their surface roughness, could exhibit similarly enhanced reversible hydrophobic to hydrophilic transitions.

Emulsion polymerization is the commonly employed method to generate surface functionalized spherical polymer particles. Different methods can be used to achieve such a functionalized particle. In a representative example, styrene and NIPAAM could be copolymerized and, owing to the different nature of monomers, the hydrophilic monomer forms the shell in the polymerized particle. As an example, to generate more ordered particles, the functionalizing moiety, e.g., ATRP or RAFT initiator, can be covalently immobilized onto seed polymer latex particles, and the particles can be subsequently grafted with functional polymers—PNIPAAM, for example—for thermal responsiveness. Other polymers with functional groups on the backbone that can be used for specific interactions with foreign media can also be similarly grafted. The process, thus, requires first the formation of polymer seed with the desired characteristics followed by the immobilization of the initiator on the surface as a thin shell. This process forms a well-defined core shell morphology; however, a mixed morphology can also be generated by a single step process in which the monomer carrying the initiator (e.g., for NMP or ATRP process) is added to the polymerizing seed particles, once a desired degree of their polymerization is reached. In both cases, various surface morphologies can be achieved depending on various thermodynamic and the kinetic factors [62-67]. Although these methods of spherical particle functionalization can lead to particles with strong commercial potential for a number of applications, very little has been reported for their functionalization [68-72]. Also, when used to generate polymer brushes from flat surfaces and particles, ATRP is also quite efficient, although not as much as in solution applications [73-76]. To this regard, Kizhakkedathu et al. have recently reported extensive studies on the aqueous ATRP for the growth of substituted acrylamide layers from polystyrene latex particles [68,69,77]. Matyjaszewski et al. have reported the synthesis of versatile ATRP agents end capped with acrylic and methacrylic groups for easy polymerization on the substrate surfaces [78]. Figure 2 illustrates the recent advances in the technology of functionalization of spherical particles with PNIPAAM.

Although the polymer particles are not exactly same as other spherical substrates like silica particles in terms of their colloidal interactions, many reported surface modifications of silica or other inorganic particles with poly(N-isopropylacrylamide), however, can also be applied to polymer particles. Kanazawa et al. reported a new concept of chromatographic separations using temperature-responsive silica stationary surfaces [47]. The silica surfaces were modified with temperature-responsive polymers to exhibit temperature-controlled hydrophilic/hydrophobic changes. To achieve this functionalization, poly(*N*-isopropyl-acrylamide) (PNIPAAM) was grafted onto (aminopropyl) silica using an activated esteramine coupling method. The results showed that the temperature-dependant hydrophilic and hydrophobic character of PNIPAAM chains could be observed on the silica surfaces, as grafted silica surfaces showed hydrophilic properties at lower temperatures and were transformed to hydrophobic surfaces once the temperature was raised above the lower critical solution temperature of PNIPAAM. It could be successfully shown that the steroids separation could be controlled by the temperature of the aqueous phase in high performance chromatography.

Kikuchi et al. also reported a review on the techniques used to generate thermally responsive surfaces [48]. Porous glass beads were grafted with PNIPAAM by Gewehr et al. with an application as functional column materials in chromatographic separation processes [79]. PNIPAAM chains end functionalized with mercaptopropionic acid were prepared through telomerization polymerization and were reacted with the functional amino groups present on the glass beads. These amine groups on the glass beads were immobilized by treatment with the silane coupling agents. Adsorption and desorption of dextran with different molecular weights were studied, and it was observed that, owing to the conformational changes in the polymer chains grafted on the surface with temperature, the effective pore size also changed. It was also stressed that the pore size of base materials as well as molecular weight of grafted chains were important criteria to be optimized.

In another interesting study, Hosoya et al. used porous polystyrene beads of roughly 1 μm diameter as base material, and PNIPAAM was polymerized and selectively grafted on the surface of the beads using porogens [80-81]. By using cyclohexanol as porogen, it was possible to graft the PNIPAAM chains inside and outside of the porous polymer particles owing to the solubility of propagating PNIPAAM radical chains in cyclohexanol. On the other hand, it was only possible to graft PNIPAAM on the outer surface of porous particles when toluene was used as porogen, as the propagating radical chains could not enter the inner voids of porous particles due to their insolubility in toluene. The functionalized particles could be successfully shown to be applicable for a number of chromatographic separations that could be controlled by the use of temperature, indicating that the PNIPAAM functionalization was useful in avoiding the harsh separation conditions of salt and pH used conventionally.

Go et al. reported the generation of functional chromatographic separation media based on PNIPAAM grafting [82]. In this approach, the polymerizable functional groups were immobilized on the surfaces of silica particles. These polymerizable methacryloylpropyl groups were generated by treatment of silica beads with silane coupling agents. NIPAAM was then copolymerized with the monomer units by the addition of initiator. The PNIPAAM chains start to form away from the surface of the beads; however, during polymerization, the growing radical attacks the double bond immobilized on the silica surfaces and transfers the radical to the surface. Subsequently, the chains start to grow from the surface. This way, the whole surface is grafted with PNIPAAM chains covalently bound to the surface. It also produces a large amount of free polymer in the solution, which is required to be properly cleaned if any useful application from the grafted particles is to be expected. The above functionalized particles were very efficient in separation of dextrans with different molecular weights. As a result of PNIPAAM functionalization, the elution times were also observed to shorten in general.

PNIPAAM brushes on a polystyrene particle were also synthesized by copolymerization of NIPAAM with styrene [83-85]. According to this process, initiation of hydrophilic (NIPAAM) monomer in the beginning is followed by polymerization of hydrophobic (styrene) monomer, which typically takes place inside the forming particles. Thick layers of PNIPAAM on the particles' surface as well as rough surface morphology can be achieved in this case. These particles were generated by initiating first the batch polymerization of styrene and N-isopropylacrylamide (NIPAAM) using 2,2′-azobis(2-amidinopropane) dihydrochloride as a cationic initiator. After 70% polymerization was achieved, a second batch of monomers containing NIPAAM, N,N′-methylenebisacrylamide (MBA) and aminoethyl methacrylate hydrochloride (AEMH) along with the initiator were added to form a shell containing a majority of PNIPAAM. The adsorption-desorption behavior of bovine serum albumin (BSA) protein was studied on the PNIPAAM functionalized particles. Adsorption of BSA protein onto such cationic and thermosensitive particles was found negligible below the LCST of PNIPAAM, whereas it was much higher above it, indicating that the thermally responsive behavior of PNIPAAM could be translated onto the particles. Mittal et al. also reported similar PS-PNIPAAM copolymer articles [85] and compared the swelling-deswelling behavior of these particles with PNIPAAM grafted PS particles by using the atom transfer radical polymerization approach. PNIPAAM grafted particles were observed to show a swelling degree of more than 3 times, whereas the copolymer particles were swollen to less than 100% at the same conditions. The accurately defined brush morphology was suggested to be responsible for the much better response of the grafted particles. In particular, it was believed that the copolymer structure did not allow PNIPAAM to fully exhibit its potential in switching between hydrophilic and hydrophobic behavior, nor did the presence of crosslinking. This also indicated that much better control over the process can only be achieved if a grafting route involving controlled polymerization is selected to functionalize the particles. Thus, it was confirmed by these studies that, in spite of the simplicity of the copolymerization process, more tunable properties of the functionalized material can be achieved using grafting techniques, as they allow the independent tuning of the support characteristics and of the grafting points. Moreover, the control over the density of the grafted brushes can also be achieved by tuning the initial density of the initiator on the grafting surface.

Physical adsorption of PNIPAAM on surfactant-free polystyrene nanoparticles was also reported by Gao et al. [44], and methods like static and dynamic laser light scattering were employed to quantify the adsorption. The adsorption was achieved by mixing polystyrene latex particles with preformed PNIPAAM chains with controlled molecular weights. By using these light scattering techniques, the temperature dependence of the hydrodynamic radius of the nanoparticles adsorbed with PNIPAAM was monitored to reveal the "coil-to-globule" transition of PNIPAAM on the particle surface. It was also reported that the adsorbed chains had a lower LCST of 29°C than the free chains in water (32°C), indicating slight changes in their physical characteristics after adsorption. It was also observed by the authors that, for a given temperature below the LCST, on increasing the amount of the added PNIPAAM, the thickness of the adsorbed PNIPAAM layer increased, but the average density of the adsorbed PNIPAAM layer decreased, suggesting an extension of the adsorbed chains.

Kizhakkedathu et al. reported a series of studies on the grafting of thermally responsive polymers from the latex particles using controlled ATRP methods [56,68-69]. A polystyrene shell latex (PSL) was synthesized, which was then modified by forming a shell of 2-(methyl-2′-chloropropionato) ethyl acrylate (HEA-Cl) on the polystyrene particles. HEA-Cl contained two functionalities in the molecule. By one terminal double bond, it could be polymerized along with styrene to form a thin shell on the seed particles, whereas by the other terminal

ATRP initiator moiety, it could be used to subsequently graft the poly(N,N′-dimethylacrylamide) chains from the surface of the spherical particles. Different densities of the ATRP initiator could be achieved on the surface of the particles by changing the feed ratio of monomers in the shell-forming batch. It was observed by the authors that molecular weight of the grafted poly(N,N′-dimethylacrylamide) chains varied linearly with monomer concentration, and grafting density was roughly independent of monomer concentration except at the highest initiator concentration. Grafting density was observed to vary as (initiator surface concentration). In another study from the authors, PNIPAAM brushes could be grafted from the surface of polymer particles by using similar ATRP methods, and the grafted brushes showed a second-order temperature-dependent collapse at a lower temperature than free polymer. Molecular weights of the grafted PNIPAAM chains were found to depend on the concentration of copper (II) complex and the presence of external initiator in the reaction medium along with the monomer concentration. A comparison of the ligands was also reported for their effect on the molecular weight of the grafted PNIPAAM chains, and HMTETA/CuCl catalyst was observed to produce higher molecular weight chains than PMDETA/CuCl. The authors also studied the block copolymerization of N,N′-dimethylacrylamide from PNIPAAM-grafted latex. It confirmed the successful application of ATRP methods on the spherical particles. The hydrodynamic thickness of generated brushes could also be calculated by dynamic light scattering, and these values were found to be sensitive to temperature and salt concentration.

Control of surface morphology of the particles has always been a challenge, as the homogenous surface distribution of the functionalizing entities (e.g., ATRP initiator) is of utmost importance in order to achieve subsequently homogenous polymer brushes. Reaction conditions such as the weight ratio of the functionalizing molecule (with terminal double bond and ATRP initiator moieties) to the seed polymer, preswelling of the seed particles with the functionalizing molecule, mode of addition of the monomer to the seed, etc. dramatically influence the resulting particles' size, shape and morphology. Depending on the nature of the initiator and its ability to immobilize on preformed polymer latex particles, the resulting morphology in terms of initiator surface concentration and distribution can dramatically change [86]. Based on the kinetic and thermodynamic factors affecting the course of the polymerization, it is often challenging to find the optimum conditions leading to the uniform distribution of ATRP initiator on the latex particles. In order to characterize the quality of the functionalization, the characterization of particle size—along with its distribution before and after brush growth as well as changes in brush dimension because of thermal response—is essential. This is mostly carried out by laser light scattering. However, small amounts of aggregation can sometimes be observed, especially in the particles synthesized with surfactant-free emulsion polymerizations, thus making the use of light scattering difficult. In such cases, other qualitative or quantitative investigation tools are required. Mittal et al. [86] studied the conditions to optimize the surface functionalization of the polymer particles with ATRP initiator 2-(2-bromopropionyloxy) ethyl acrylate (BPOEA) by extensive use of microscopy. It was observed that a number of operating conditions led to the generation of secondary nucleation of the shell-forming monomers. Apart from that, the surface morphology of the ATRP initiator-modified particles was always different, depending upon the synthesis technique. It was reported that by using the one-step process, i.e., by addition of the shell-forming particles to the seed particles when the seed particles have 70% monomer conversion, led to the elimination of the secondary nucleation. It was also observed that the addition of a small amount of crosslinker divinylbenzene (DVB) led to the total elimination of secondary nucleation, and particles could be successfully functionalized with ATRP initiator by following any method of synthesis. In the study it was concluded, therefore, that to achieve crosslinked material, the functionalization procedure should be a one-step process with starved addition of the shell-forming monomers at the end stages of polymerization of the seed. If non-crosslinked material is required, then good results are obtained by the shot addition of shell-forming monomers to the 70% polymerized polystyrene seed particles. The control of the density of functional groups could also be achieved by changing the styrene/BPOEA mole ratio in the shell-generating monomer mixture added to the polystyrene seed.

Similar ATRP methods to graft PNIPAAM particles from the surface of the particles were recently applied, and different morphologies of the particles below and above the LCST were characterized by using transmission electron microscopy [87]. The atom transfer radical polymerization of NIPAAM was carried out with an HMTETA/CuBr/$CuBr_2$ system along with additionally added Cu powder to latex particles pre-functionalized with ATRP initiator. The presence of a PNIPAAM layer was also confirmed by electron energy loss spectroscopy (EELS), which indicates the presence of a layer containing oxygen and nitrogen around the particles. The particles were observed to be thermally responsive, and their thermally responsive character was quantified by measuring the amount of the swelling of the grafted chains below and above the LCST. A potential use of these materials as stationary phases for bioseparations was suggested, and it was reported that the adsorption behavior of tobacco mosaic virus could successfully be achieved in these particles. The process was also observed to be fully reversible with temperature and fast in response.

Polymer particles are extensively used to generate chromatographic supports in which, based on the pore size of the supports, the separations of various molecules could be achieved. As seen in the earlier references, PNIPAAM chains could be grafted within the pores of the particles by free radical polymerization, and then these particles could be used to separate the mixture by making use of the thermally responsive character of PNIPAAM chains. However, these particles were packed to form the chromatographic column. Recently, a new technique named "reactive gelation" was reported [88], by which macroporous monolithic materials were generated by controlled aggregation of emulsion latexes which were initially swollen with monomers. The aggregated latexes were then re-polymerized to fix the so-obtained macroporous particle network. As an additional functionalization, these particles could first be functionalized with ATRP initiator and could then be formed onto macroporous monoliths. PNIPAAM chains could subsequently be grafted from these particles bound into monolithic structures [89]. The resulting structures showed similar behavior to free particles once they were functionalized with PNIPAAM brushes. The reduced swelling capabilities and the slower kinetics in swelling and deswelling were, however, observed, probably due to the constrained environment of the monolith, as opposed to free particles. Figures 2 j–m show the monolithic structures obtained by the networking of particles. One can also control the extent of functionalization of these particles as well as the porosity of the generated networks. Figure 2n shows the comparison of the PNIPAAM grafted particles with the grafted monoliths in terms of the swelling-deswelling characteristics. As mentioned above, the PNIPAAM chains in monoliths have more constraints; therefore, their extent of swelling is a bit smaller than the free particles. The swelling and deswelling as a function of time is also affected.

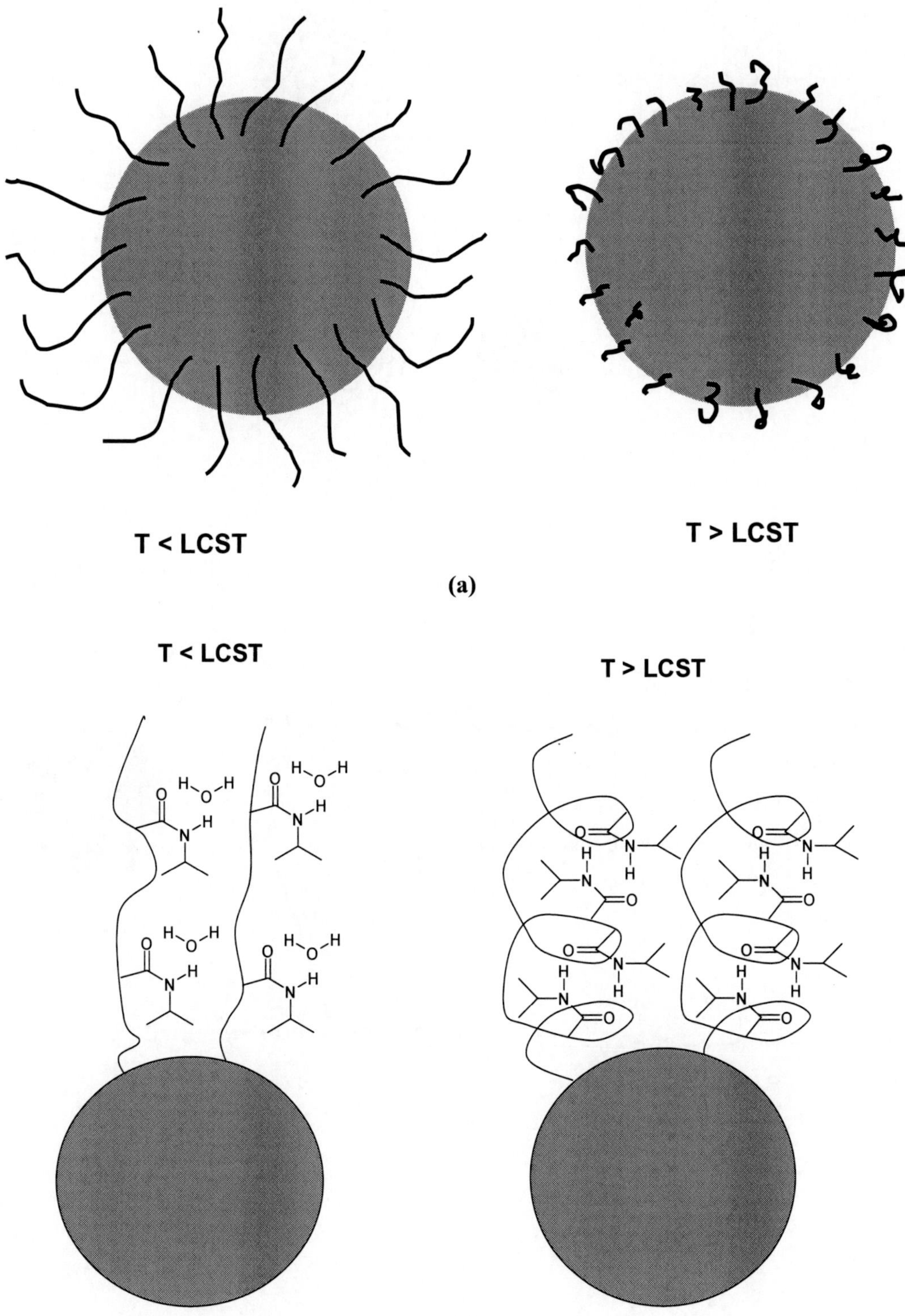

Figure 2. (Continues on next page.)

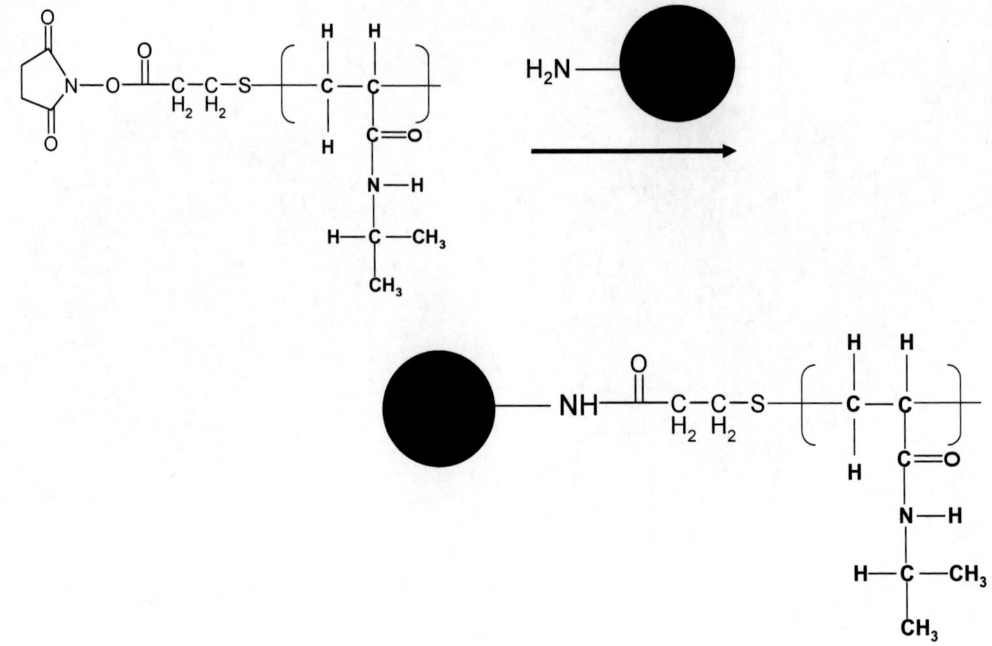

(c)

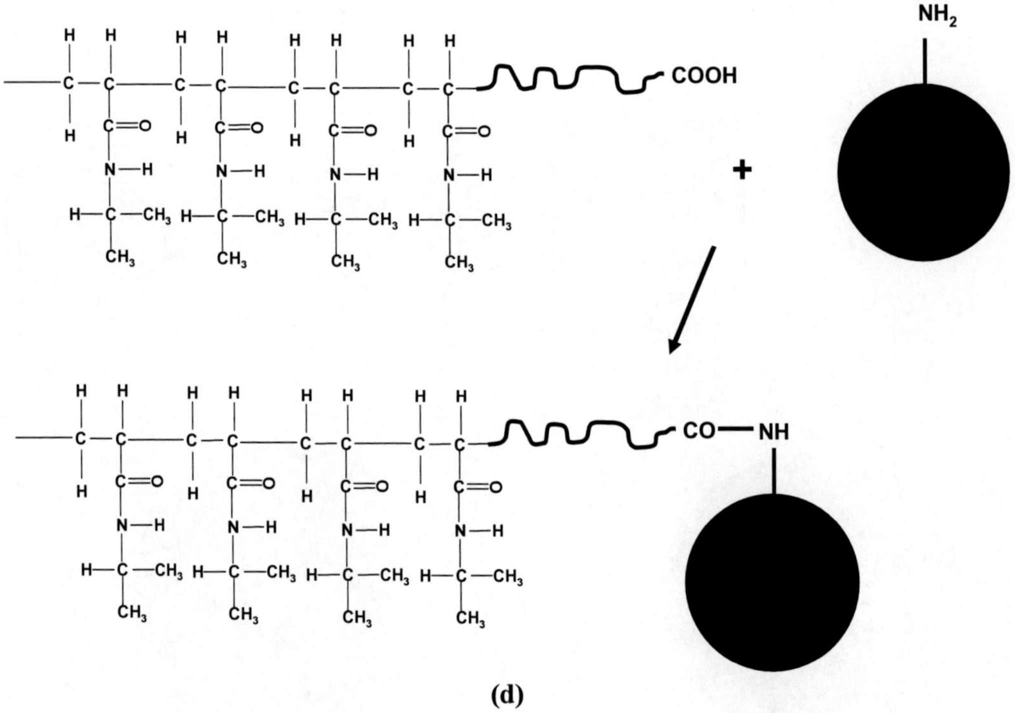

(d)

Figure 2. (Continues on next page.)

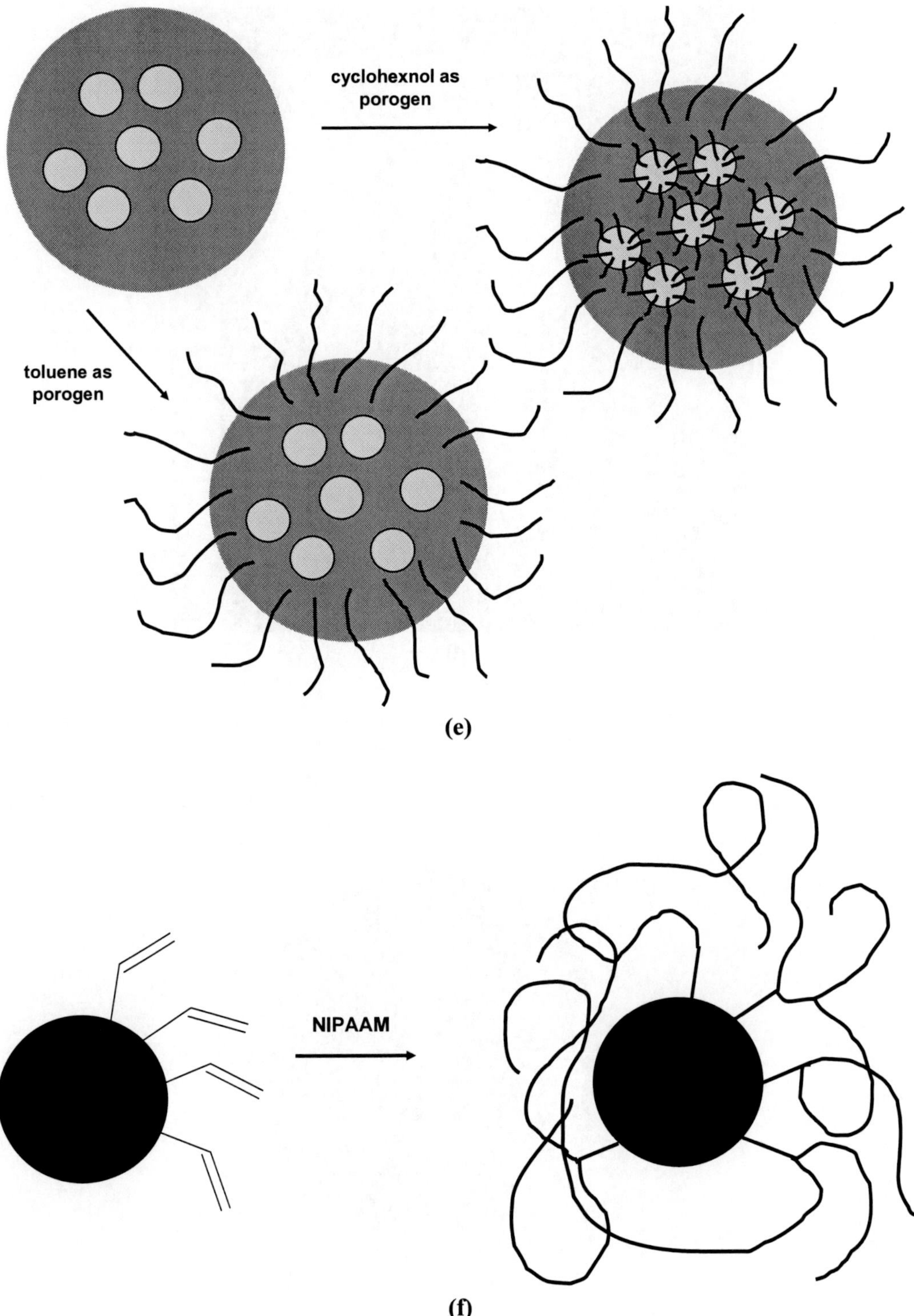

Figure 2. (Continues on next page.)

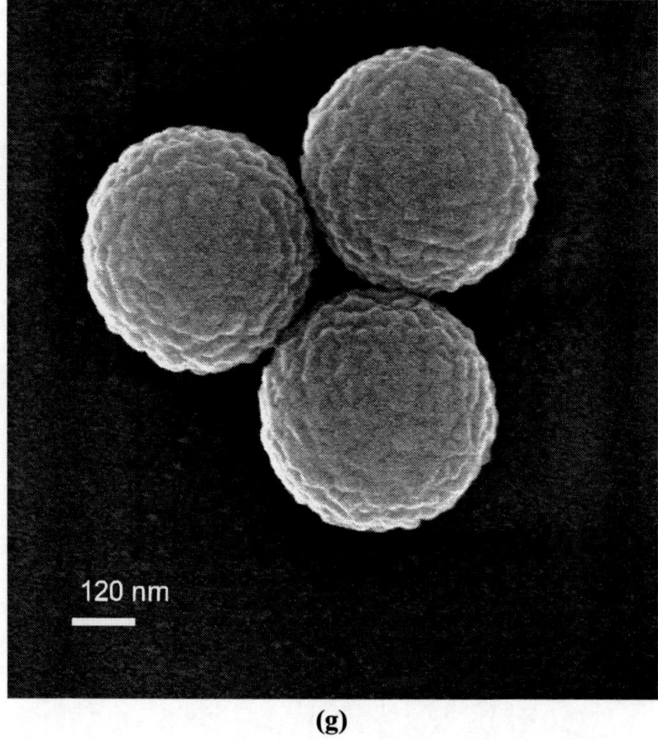

(g)

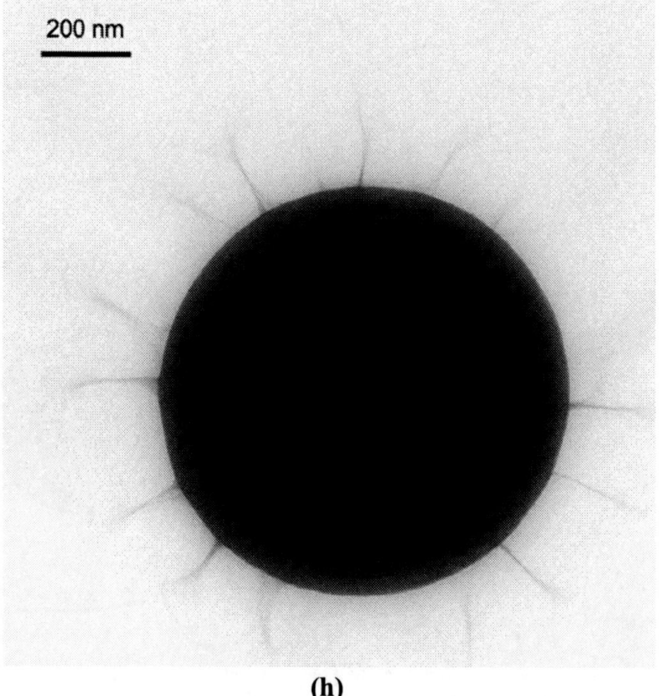

(h)

Figure 2. (Continues on next page.)

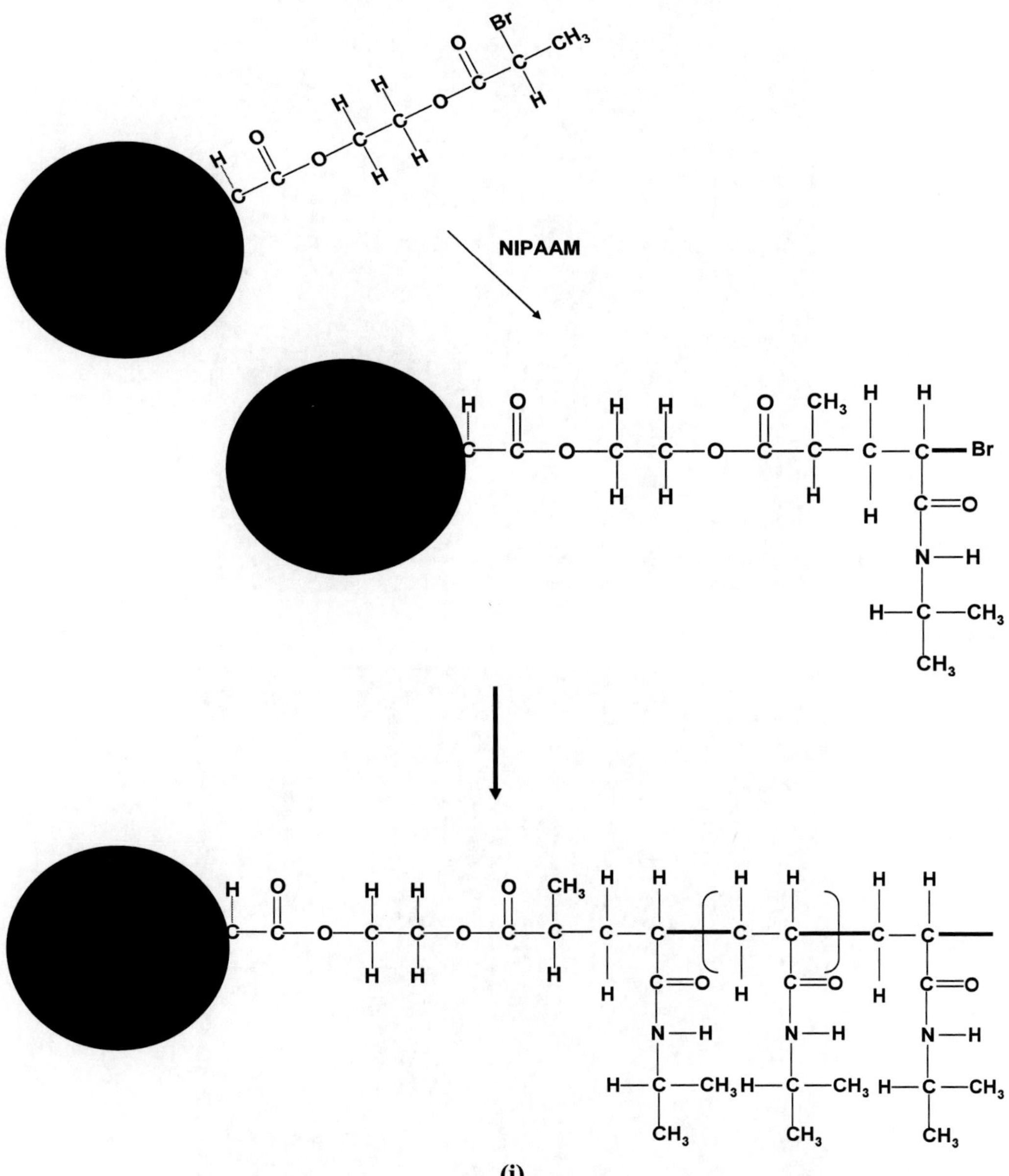

Figure 2. Generation of thermally responsive behavior on spherical surfaces: (a) and (b), schemes showing the generation of hydrophobic and hydrophilic surfaces of the particles by PNIPAAM modifications [47, 87]; (c) amine treated silica reacted by esteramine reactions with PNIPAAM chains [47]; (d) modification of amine functionalized glass beads with carboxyl PNIPAAM chains [79]; (e) PNIPAAM functionalization by using porogens [80-81]; (f) modification of particles by copolymerization of NIPAAM with methacryloyl-propyl groups immobilized on the surface [82]; (g) generation of thermally responsive particles by copolymerization of styrene and NIPAAM [85], (h) generations of PNIPAAM grafts from the surface of the particles by using ATRP method [87], (i) scheme describing the modification of particles surfaces with ATRP initiator followed by ATRP of NIPAAM to graft brushes from the surface [87].

Figure 2 (Continues)

(j)

(k)

Figure 2 (continued). Generation of thermally responsive behavior on spherical surfaces. (j) and (k): SEM micrographs of PNIPAAM functionalized polymer monoliths synthesized by networking latex particles [85, 89].

Figure 2 (continued). Generation of thermally responsive behavior on spherical surfaces. (l) and (m): SEM micrographs of PNIPAAM functionalized polymer monoliths synthesized by networking the latex particles [85, 89].

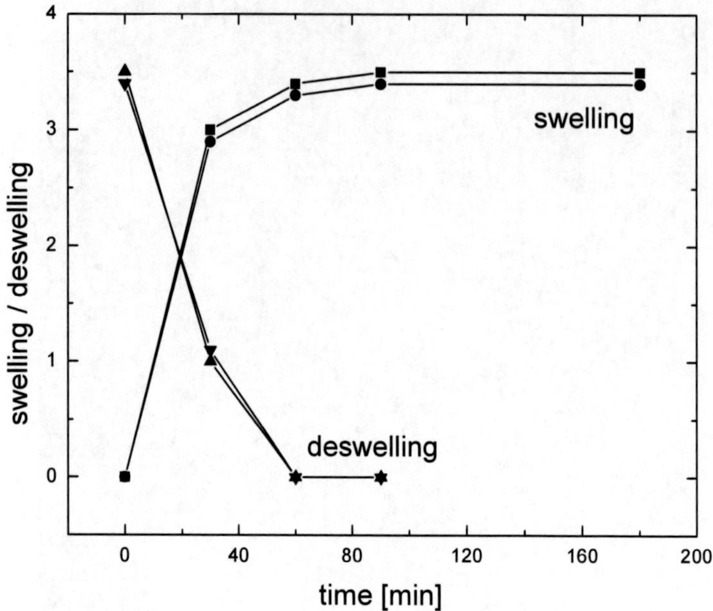

FREE PARTICLES

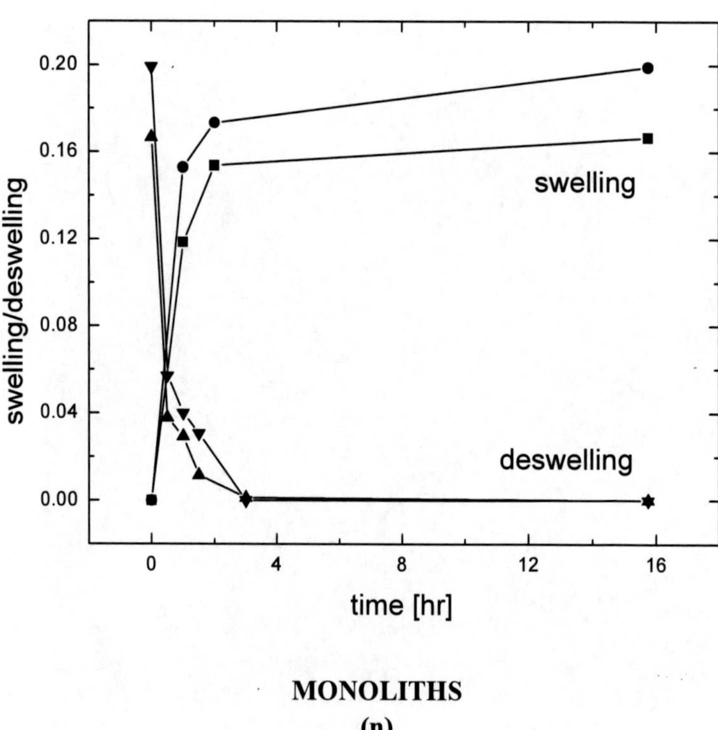

MONOLITHS

(n)

Figure 2 (continued). (n) Comparison of the swelling deswelling characteristics of the free particles with that of networked particles in monoliths.

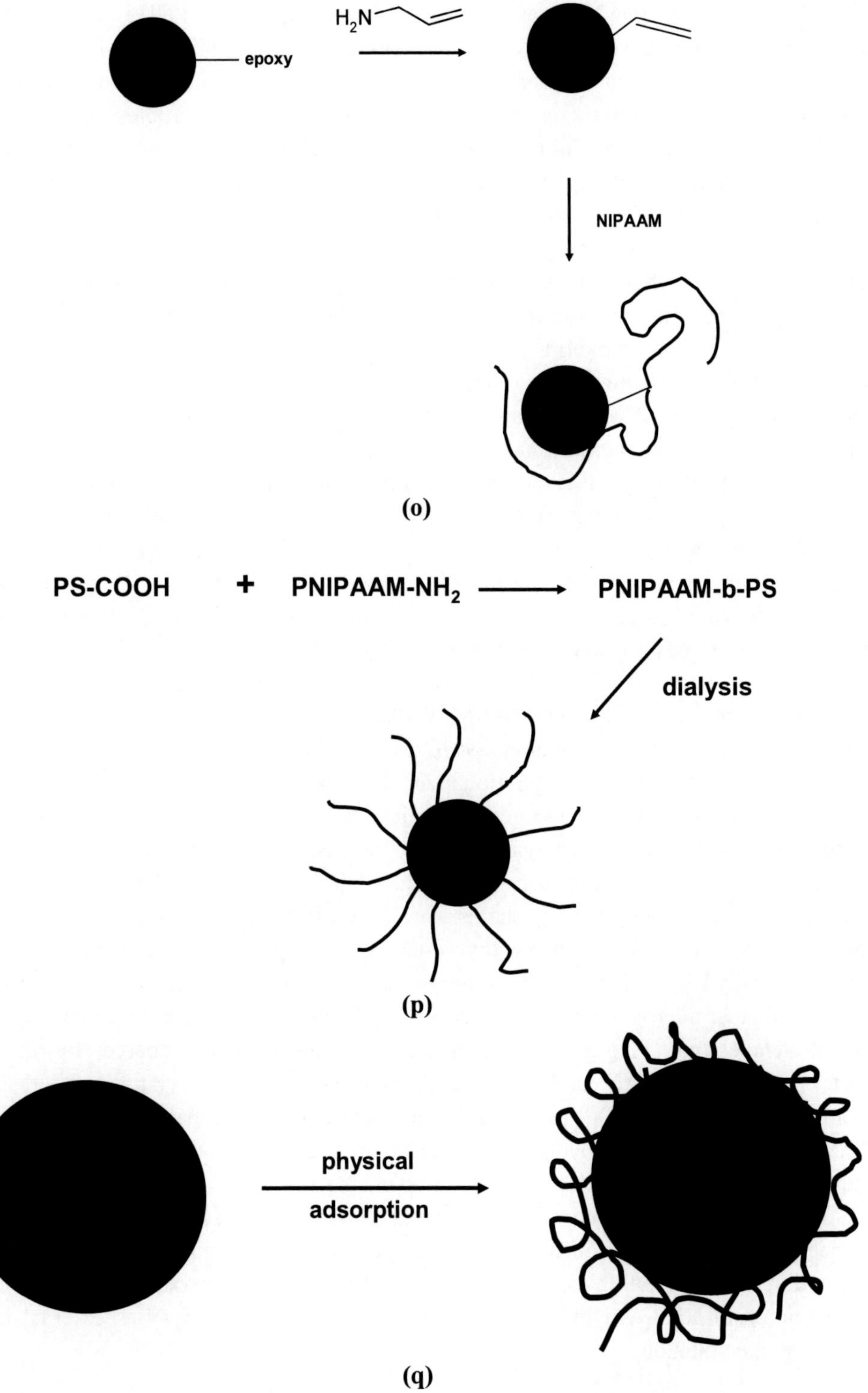

Figure 2 (continued). Generation of thermally responsive behavior on spherical surfaces: (o) use of epoxide functional groups to generate monomer moieties on the surface which can then be copolymerized with NIPAAM [90], (p) generation of hydrophobic and hydrophilic components of the chains formed into micelles [91] and (q) sketch showing the physical adsorption process [44].

An "aggregate grafting" process was also used to graft the PNIPAAM chains from particles in the reported study [85]. In this method, which is slightly different from the reactive gelation method mentioned above, the latex particles functionalized with ATRP initiator were first shear mixed to form physical networks. These particles were then mixed with NIPAAM solution containing a small amount of crosslinker and the catalysts. The crosslinked material was highly stable and porous. The same behavior of thermo-reversible hydrophobicity and hydrophilicity of the PNIPAAM chains was observed in the monoliths, though the extent and rate of swelling is hindered owing to the large degree of crosslinking among PNIPAAM brushes and space constraints on the chains. The solid fractions were also altered in these monoliths in order to estimate the effect of availability of free volume to the PNIPAAM chains in the monolith on their swelling deswelling properties. It was observed that a higher extent of swelling deswelling character occured in the monolith with higher free volume, confirming the effect of the above-mentioned space constraints.

In another study to generate functional monoliths, a method to modify the internal pore surface of rigid poly(glycidyl methacrylate-co-ethylene dimethacrylate) (GMA-EDMA) monoliths with poly(N-isopropylacrylamide) (PNIPAAM) chains [90] was reported. The functionalization was achieved in two steps: vinylization of the pore surface was achieved by the reaction of the epoxide moiety on the surface with allyl amine followed by subsequent in situ radical polymerization of NIPAAM within these pores. The monolith structures so formed were observed to change their hydrophobic and hydrophilic character by the use of temperature.

In an interesting study, PNIPAAM-block-PS copolymers were used to generate thermoresponsive polymeric micelles [91]. Carboxyl functionalized polystyrene polymer chains were reacted with amine-functionalized PNIPAAM chains, and the resulting block copolymer chains were dialyzed to generate micelles. These micelles had hydrophilic shells containing PNIPAAM chains, whereas the core consisted of hydrophobic polystyrene chains.

In a study reporting the use of reversible addition fragmentation chain transfer (RAFT), linear poly(N-isopropylacrylamide) chains were successfully grafted onto spherical poly(N-isopropylacrylamide) and 2-hydroxyethyl ester (PNIPAAM/HEA) copolymer microgel to achieve thermally responsive core-shell morphology [92]. Both the tethered PNIPAAM shell and the copolymer microgel core were observed to shrink as the temperature increased, but in different temperature ranges. PNIPAAM/HEA microgel was prepared by dispersion polymerization of NIPAAM and HEA using N,N′-methylenebisacrylamide as crosslinker, sodium dodecyl sulphate (SDS) as emulsifier and potassium persulphate (KPS) as initiator. At 70°C, PNIPAAM became hydrophobic and collapsed so that hydrophilic HEA was presumably located on the periphery of the resultant copolymer microgels. RAFT initiator was subsequently immobilized on the surfaces of particles first by reacting them in the presence of butyl acid dithiobenzoate, and dicyclohexycarbodiimide in THF followed by precipitation of the particles. The grafting polymerization was carried by RAFT polymerization of NIPAAM in THF and using small amount of AIBN as external initiator followed by precipitation.

C. Use of PNIPAAM Functionalized Latex Particles

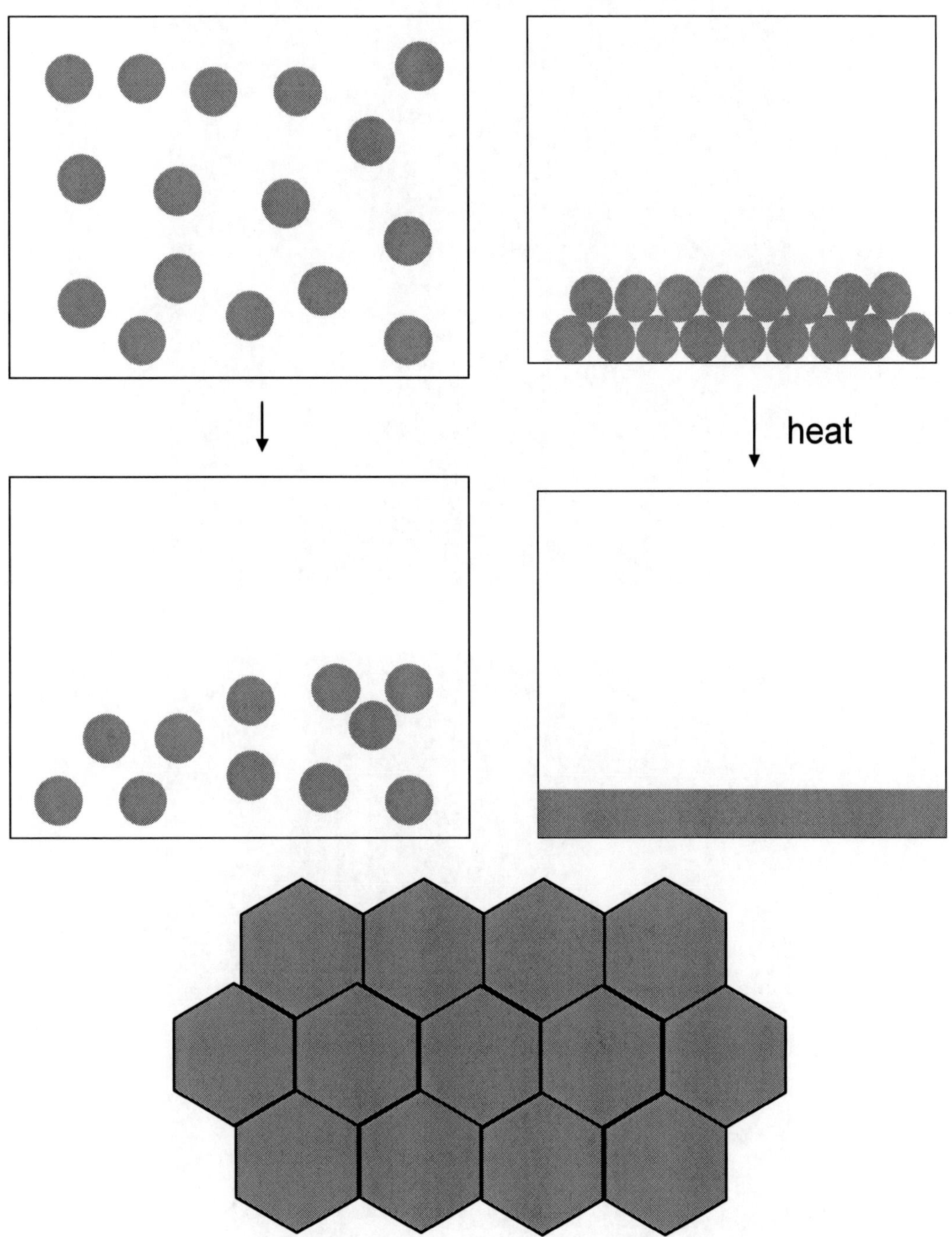

Figure 3. Process of film formation of polymer particles on a flat substrate and top view of the uniform film.

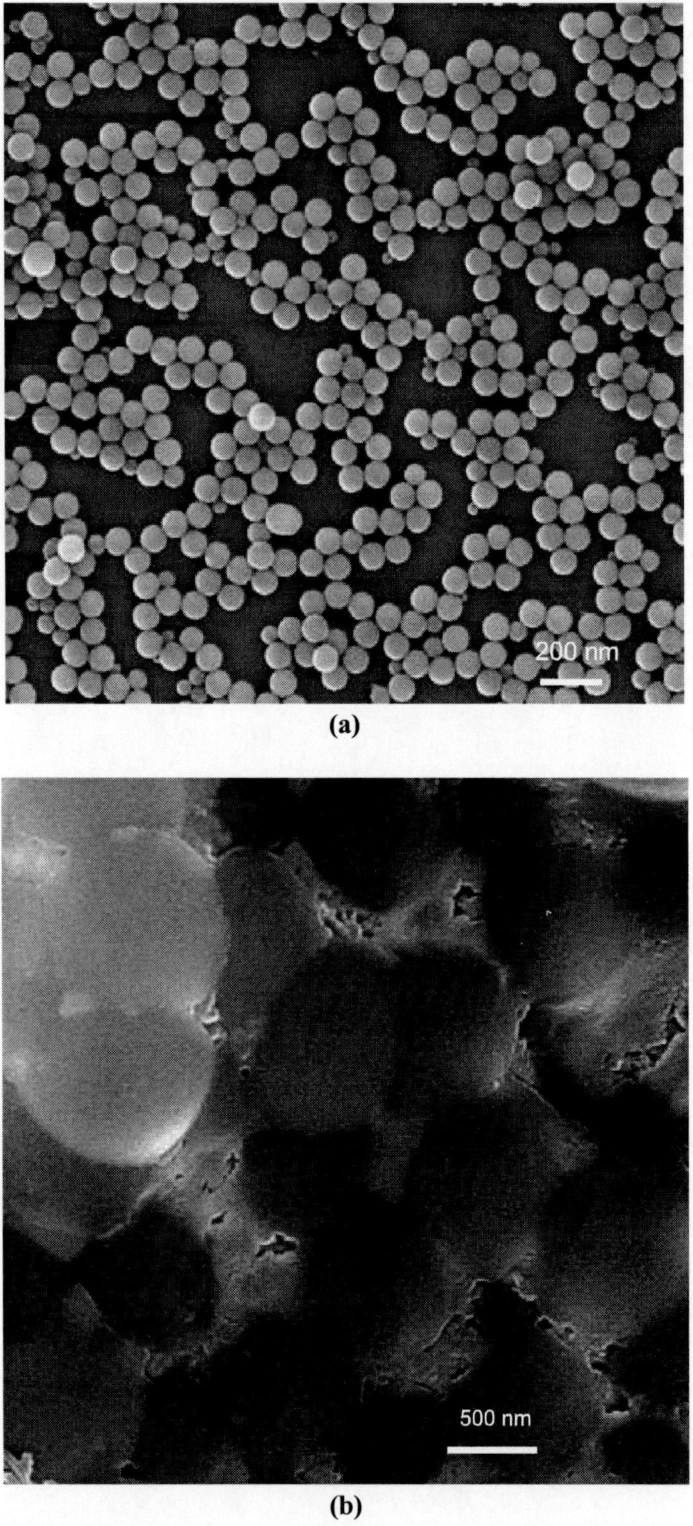

Figure 4. Comparison of the film formation at (a) room temperature and (b) at 70°C.

PNIPAAM functionalized particles can find applications in many different directions. As a first example, these particles could be demonstrated to generate thermally switchable macro surfaces apart from the generation of thermally reversible monoliths. These surfaces were first coated with a thin film of particles carrying the ATRP initiator on the surface. The particles form a uniform layer on the surface, and these particles were then subjected to PNIPAAM grafting on the surface. In this way, the thermally responsive behavior could be translated on the macro surfaces. One must be careful to achieve a uniform layer before the PNIPAAM grating, as the quality of the grafted brushes is dependant on the quality of the coated film. Figure 3 is a scheme of the forming of a uniform layer of particles on the surface.

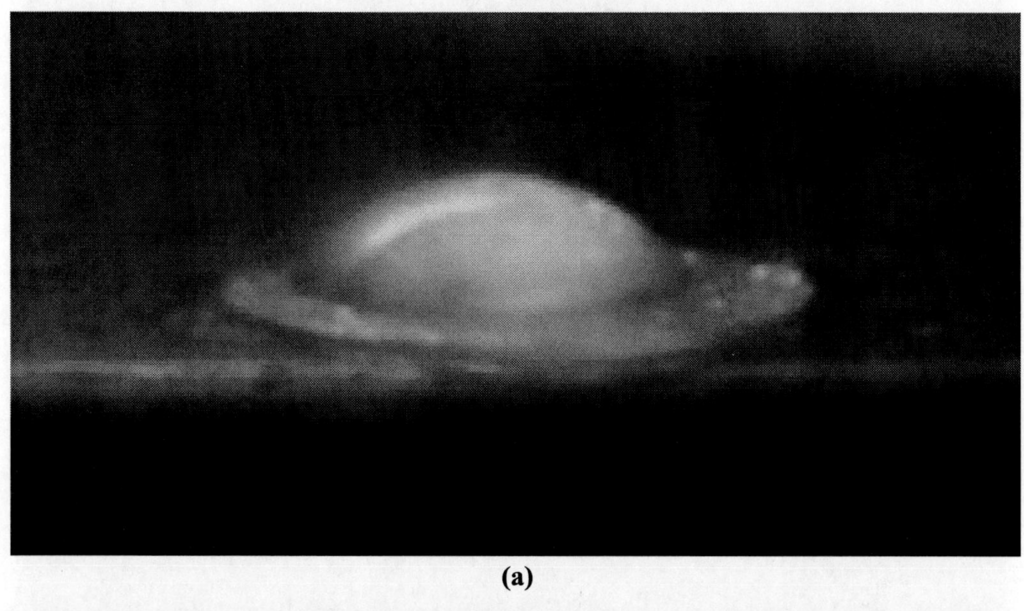

(a)

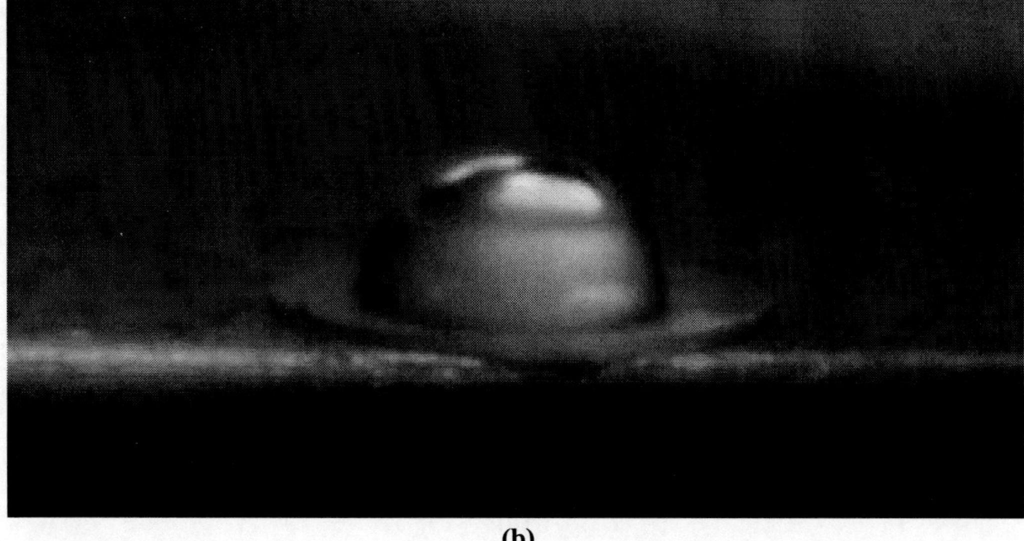

(b)

Figure 5. Behavior of the drops of water on the flat surfaces functionalized by PNIPAAM at (a) room temperature and (b) at 40°C.

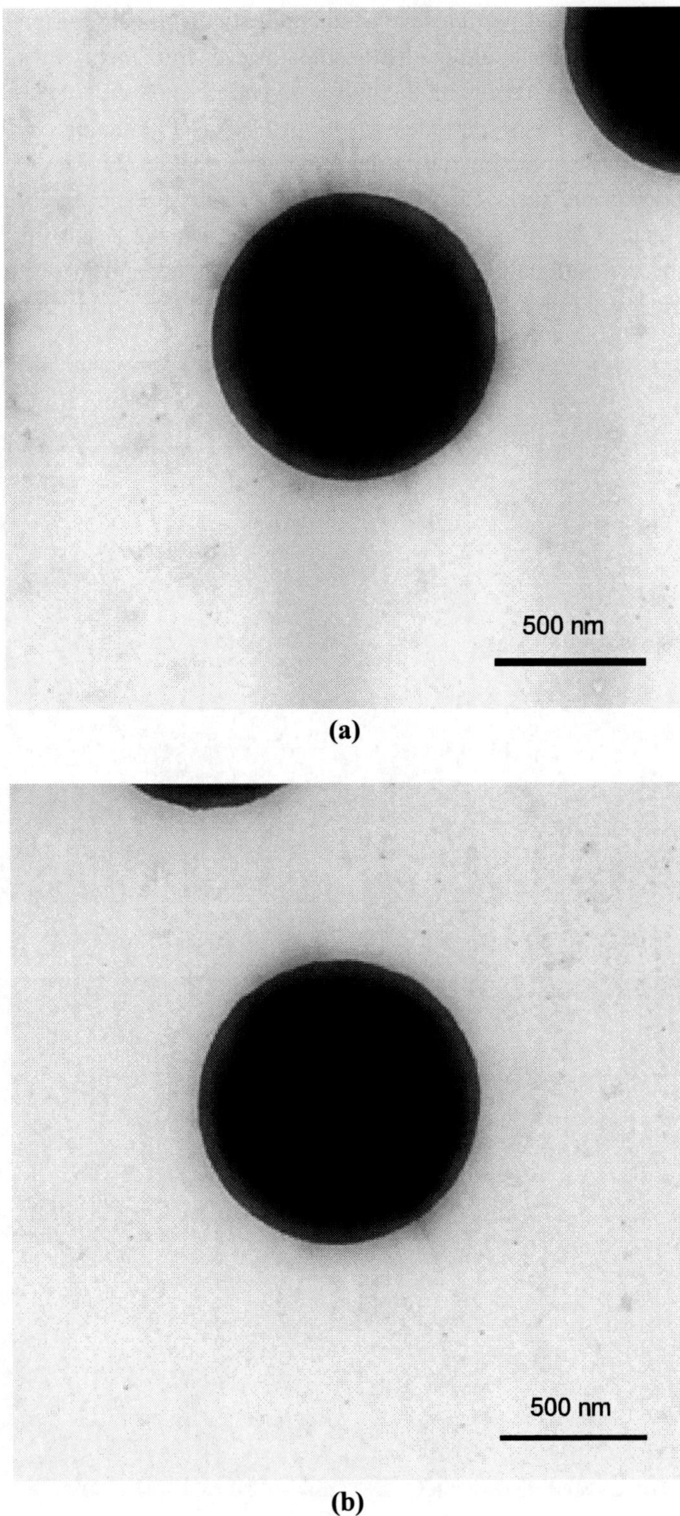

Figure 6 (a) and (b). The desorption of protein from the polymer particle surfaces when the PNIPAAM modified particles were cooled to 10°C.

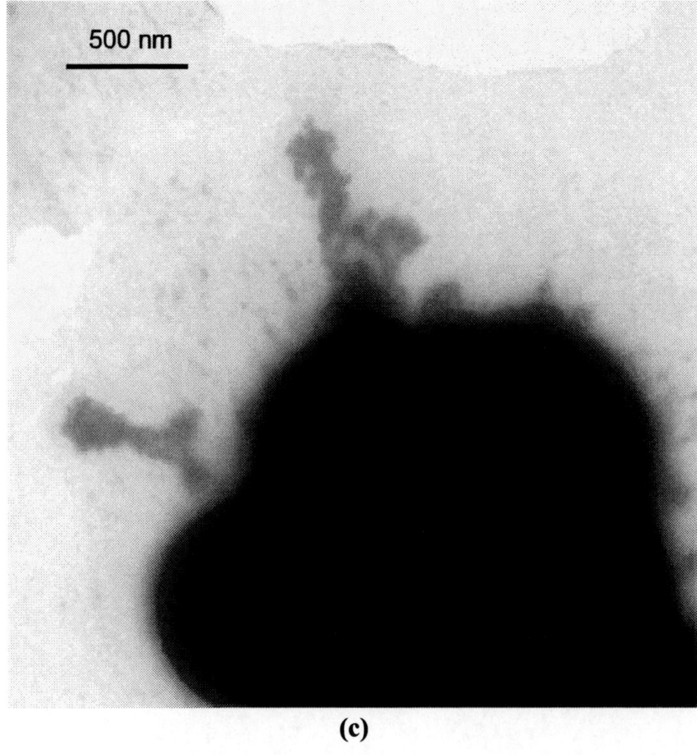

(c)

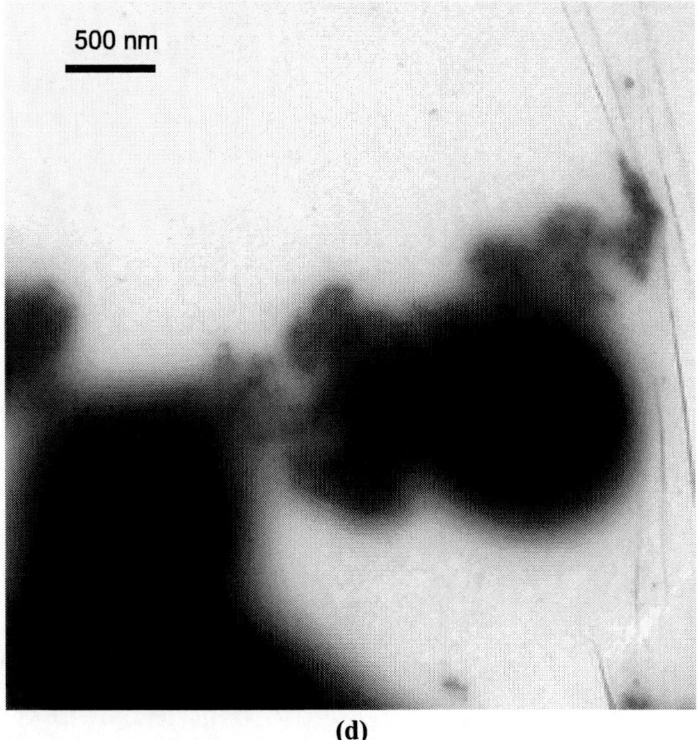

(d)

Figure 6 (c) and (d). The adsorption of protein from the polymer particle surfaces when the PNIPAAM modified particles were heated to 37°C.

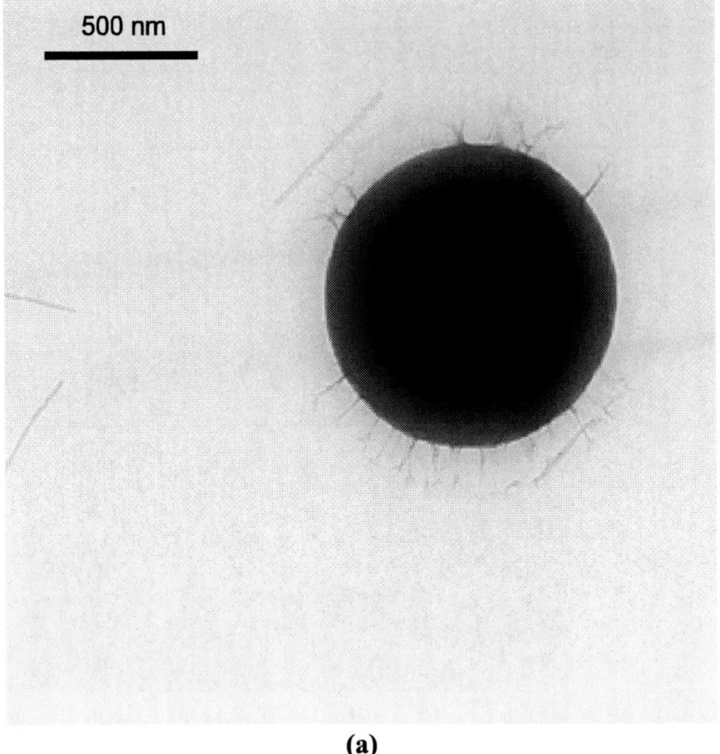

(a)

(b)

Figure 7 (a) and (b). The desoprtion of tobacco mosaic virus from the polymer particle surfaces when the PNIPAAM modified particles were cooled to 10°C.

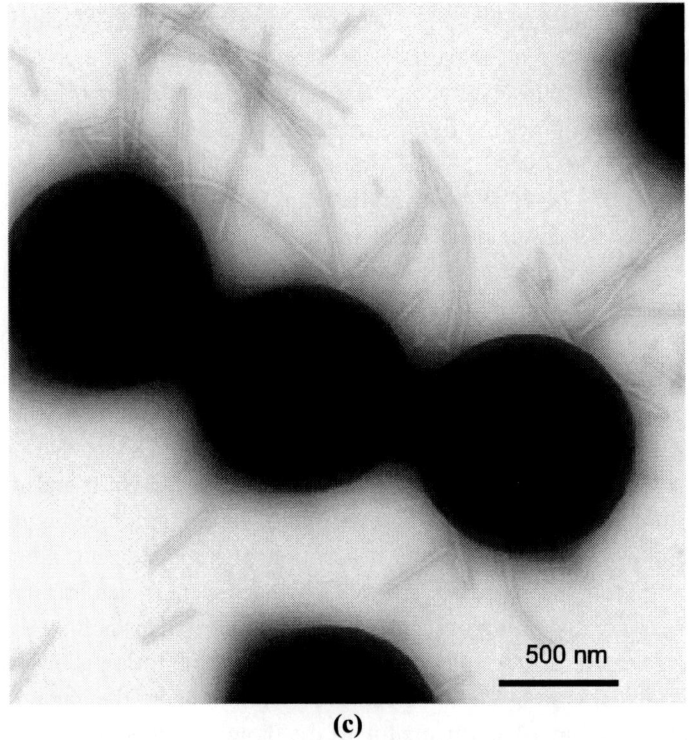

(c)

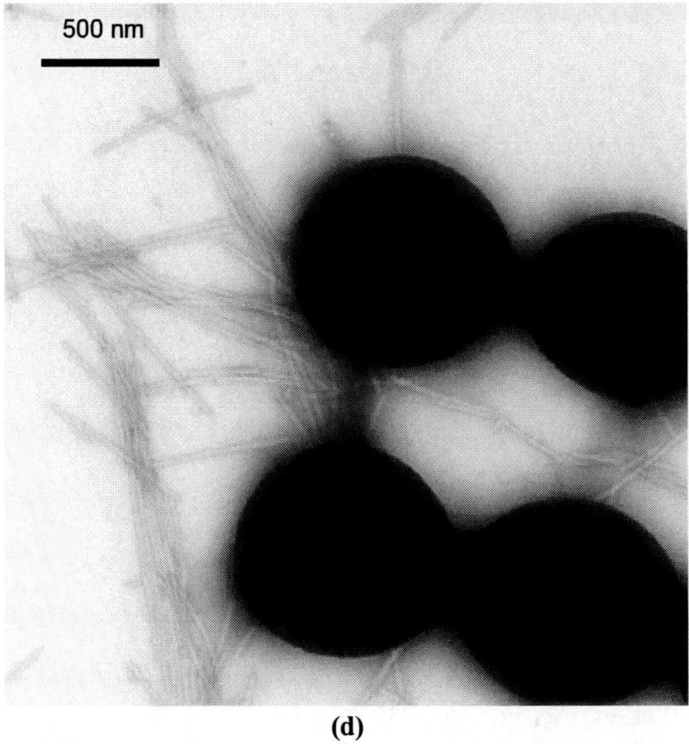

(d)

Figure 7 (c) and (d). The adsorption of tobacco mosaic virus from the polymer particle surfaces when the PNIPAAM-modified particles were heated to 37°C.

The particles must be heated to roughly 70°C for a few seconds to seal the interstitial gaps between the particles. However, the particles should not be heated at high temperatures or for too long a time as they would lose the surface initiator, which may diffuse deep inside the molten particles. Figure 4 shows the film generated from the particle at room temperature and after heating for a few seconds at 70°C, indicating the better quality of film when heat treatment was provided. The PNIPAAM-grafted surfaces were truly thermally reversible, as confirmed by the behavior of the droplets of water placed on them, as shown in Figure 5. At 10°C, when the surfaces are hydrophilic, the water droplet spreads out, whereas at 37°C, when the surfaces are hydrophobic in nature, the water droplet does not spread out and remains as it is. Although the generated surfaces are not super hydrophobic or super hydrophilic, such a reversible change in nature has been achieved just by the adsorption of particles and does not require costly techniques such as laser etching, which are generally required to treat the surfaces to generate super hydrophilicity and super hydrophobicity.

The PNIPAAM-grafted particles could also be shown to adsorb and desorb biological entities just as a function of a change of a few degrees in temperature. Figures 6 and 7 show the temperature-controlled adsorption and desorption of human serum albumin protein and tobacco mosaic virus on the particles. At 10°C, when the particles are hydrophilic on the surface, there is seen no adsorption or the adsorbed protein or virus is fully desorbed, whereas when the system is heated to 37°C, a clear protein or virus adsorption is observed. Thus, these separations of biological entities could be achieved without using the harsh conditions of salt or pH, which sometimes can affect the quality of the biological media.

REFERENCES

[1] Lahann, J., Mitragotri, S., Tran, T.N., Kaido, H., Sundram, J., Choi, I.S., Hoffer, S., Somarjai, G.A., and Langer, R. (2003). *Science*, 299, 371.

[2] Sidorenko, A., Minko, S., Schenk-Meuser, K., Duschner, H., and Stamm, M. (1999). *Langmuir*, 15, 8349.

[3] Chaudhury, M.K. and Whitesides, G.M. (1992). *Science*, 256, 1539.

[4] Nagasaki, Y. and Kataoka, K. (1996). *Trends in Polymer Science*, 4, 59.

[5] Fytas, G., Anastasizdis, S.H., Seghrouchni, R., Vlassopoulos, D., Li, J., Factor, B.J., Theobald, W., and Toprakcioglu, C. (1996). *Science*, 274, 2041.

[6] Mansky, P., Liu, Y., Huang, E., Russell, T.P., and Hawker, C.J. (1997). *Science*, 275, 1458.

[7] Mir, Y., Auroy, P. and Auvray, L. (1995). *Physical Review Letters*, 75, 2863.

[8] Prucker, O. and Ruhe, J. (1998). *Macromolecules*, 31, 592.

[9] Prucker, O. and Ruhe, J. (1998). *Macromolecules*, 31, 602.

[10] Zhao, B. and Brittain, W.J. (1999). *Journal of the American Chemical Society*, 121, 3557.

[11] Wu, T., Efimenko, K. and Genzer, J. (2001). *Macromolecules*, 34, 684.

[12] Milner, S.T. (1991), *Science*, 251, 905.

[13] Mittal, V. (2007). *Journal of Colloid and Interface Science*, 314, 141.

[14] Jordan, R., Ulman, A., Kang, J.F., Rafailovich, M.H., and Sokolov, J. (1999). *Journal of the American Chemical Society*, 121, 1016.

[15] Ingall, M.D.K., Honeyman, C.H., Mercure, J.V., Bianconi, P.A., and Kunz, R.R. (1999). *Journal of the American Chemical Society*, 121, 3607.
[16] Zhao, B. and Brittain, W.J. (2000). *Macromolecules*, 33, 8813.
[17] Kim, J., Bruening, M.L. and Baker, G.L. (2000). *Journal of the American Chemical Society*, 122, 7616.
[18] Sun, T., Wang, G., Feng, L., Liu, B., Ma, Y., Jiang, L., and Zhu, D. (2004). *Angewandte Chemie International Edition*, 43, 357.
[19] You, Y.Z., Hong, C.Y., Pan, C.Y., and Wang, P.H. (2004). *Advanced Materials*, 16, 1953.
[20] D'Agosto, F., Charreyre, M.-T., Pichot, C., and Gilbert, R.G. (2003). *Journal of Polymer Science, Part A: Polymer Chemistry*, 41, 1188.
[21] Lamb, D., Anstey, J.F., Fellows, C.M., Monteiro, J.M., and Gilbert, R.G. (2001). *Biomacromolecules*, 2, 518.
[22] Wang, X.S., Jackson, R.A. and Armes, S.P. (2000). *Macromolecules*, 33, 255.
[23] Nishikawa, T., Ando, T., Kamigaito, M., and Sawamoto, M. (1997). *Macromolecules*, 30, 2244.
[24] Coca, S., Jasieczek, C., Beers, K.L., and Matyajaszewski, K. (1998). *Journal of Polymer Science, Part A: Polymer Chemistry*, 36, 1417.
[25] Gaynor, S.G., Qiu, J. and Matyajaszewski, K. (1998). *Macromolecules*, 31, 5951.
[26] Qiu, J., Gaynor, S.G. and Matyajaszewski, K. (1999). *Macromolecules*, 32, 2872.
[27] Matyjaszewski, K. (Ed.) (1998). Controlled Radical Polymerization, ACS Symposium Series, 685, American Chemical Society, Washington, DC.
[28] Wang, J.S. and Matyjaszewski, K. (1995). *Journal of the American Chemical Society*, 117, 5614.
[29] Granel, C., Dubois, P., Jerome, R., and Teyssie, P. (1996). *Macromolecules*, 29, 8576.
[30] Kato, M., Kamigaito, M., Sawamoto, M., and Higashimura, T. (1995). *Macromolecules*, 28, 1721.
[31] Nishikawa, T., Ando, T., Kamigaito, M., and Sawamoto, M. (1997). *Macromolecules*, 30, 2244.
[32] Ashford, E.J., Naldi, V., O'Dell, R., Billingham, N.C., and Armes, S.P. (1999). *Polymer Preprints*, 40, 405.
[33] Makino, T., Tokunaga, E. and Hogen-Esch, T.E. (1998). *Polymer Preprints*, 39, 288.
[34] Teodorescu, M. and Matyjaszewski, K. (1999). *Macromolecules*, 32, 4826.
[35] Teodorescu, M. and Matyjaszewski, K. (2000). *Macromolecular Rapid Communications*, 21, 190.
[36] Rademacher, J.T., Baum, M., Pallack, M.E., Brittain, W.J., and Simonsick, W. J. J. (2000). *Macromolecules*, 33, 284.
[37] Jewrajka, S.K. and Mandal, B.M. (2003). *Macromolecules*, 36, 311.
[38] Perruchot, C., Khan, M.A., Kamitsi, A., Armes, S.P., von Werne, T., and Patten, T.E. (2001). *Langmuir*, 17, 4479.
[39] Ejaz, M., Yamamoto, S., Ohno, K., Tsujii, Y., and Fukuda, T. (1998). *Macromolecules*, 31, 5934.
[40] Ringsdorf, H., Sackmann, E., Simon, J., and Winnik, F.M. (1993). *Biochimica Biophysica Acta*, 1153, 335.
[41] Bae, Y.H., Okano, T. and Kim, S.W. (1990). *Journal of Polymer Science, Part B: Polymer Physics*, 28, 923.

[42] Heskins, M. and Guillet, J.E. (1968). *Journal of Macromolecular Science: Chemistry*, A2, 1441.
[43] Yoshida, R., Uchida, K., Kaneko, Y., Sakai, K., Kikuchi, A., Sakurai, Y., and Okano, T. (1995). *Nature*, 374, 240.
[44] Gao, J. and Wu, C. (1997). *Macromolecules*, 30, 6873.
[45] Park, T.G. and Hoffman, A.S. (1993). *Macromolecules*, 26, 5045.
[46] Janzen, J., Le, Y., Kizhakkedathu, J.N., and Brooks, D.E. (2004). *Journal of Biomaterials Science*, Polymer Edition, 15, 1121.
[47] Kanazawa, H., Yamamoto, K., Matsushima, Y., Takai, N., Kikuchi, A., Sakurai, Y., Okano, T. (1996). *Analytical Chemistry*, 68, 100.
[48] Kikuchi, A., Okano, T. (2002). *Progress in Polymer Science*, 27, 1165.
[49] Cunliffe, D., Heras Alarcon, C., Peters, V., Smith, J.R., and Alexander, C. (2003). *Langmuir*, 19, 2888.
[50] Okano, T., Yamada, N., Okuhara, M., Sakai, H., and Sakurai, Y. (1995). *Biomaterials*, 16, 297.
[51] Chen, G., Ito, Y. and Imanishi, Y. (1997). *Biotechnology and Bioengineeringg*, 53, 339.
[52] Okajima, S., Yamaguchi, T., Sakai, Y., and Nakao, S. (2005). *Biotechnology and Bioengineering*, 91, 237.
[53] Hosoya, K., Watabe, Y., Kubo, T., Hoshino, N., Tanaka, N., Sano, T., and Kaya, K. (2004). *Journal of Chromatography A*, 1030, 237.
[54] Collier, T.O., Anderson, J.M., Kikuchi, A., and Okano, T. (2002). *Journal of Biomedical Materials Research*, 59, 136.
[55] Jones, D.M., Smith, R.R., Huck, W.T.S., and Alexander, C. (2002). *Advanced Materials*, 14, 1130.
[56] Kizhakkedathu, J.N., Norris-Jones, R., and Brooks, D.E. (2004). *Macromolecules*, 37, 734.
[57] Takei, Y.G., Aoki, T., Sanui, K., Ogata, N., Sakurai, Y., and Okano, T. (1994). *Macromolecules*, 27, 6163.
[58] Morra, M., and Cassinelli, C. (1995). *Polymer Preprints*, 36, 55.
[59] Volpe, C.D., Cassinelli, C. and Morra, M. (1998). *Langmuir*, 14, 4650.
[60] Liang, L., Feng, X., Liu, J., Rieke, P.C., and Fryxell, G.E. (1998). *Macromolecules*, 31, 7845.
[61] Kizhakkedathu, J.N. and Brooks, D.E. (2003). *Macromolecules*, 36, 591.
[62] Du, Y.Z., Ma, G.H., Ni, H.M., Nagai, M., and Omi, S. (2002). *Journal of Applied Polymer Science*, 84, 1737.
[63] Cho, I. and Lee, K.W. (1985). *Journal of Applied Polymer Science*, 30, 1903.
[64] Sundberg, D.C., Casassa, A.P., Pantazopoulous, J., Muscato, M.R., Kronberg, B., and Berg, J. (1990). *Journal of Applied Polymer Science*, 41, 1425.
[65] Juang, M.S. and Krieger, I.M. (1976). *Journal of Polymer Science*, *Polymer Chemistry Edition*, 14, 2089.
[66] Zukoski, C.F. and Saville, D.A. (1985). *Journal of Colloid and Interface Science*, 104, 583.
[67] van Herk, A.M. and Gilbert, R.G. In: van Herk, A.M. (Ed.) (2005). Chemistry and technology of emulsion polymerization, Blackwell Publishing, Oxford.
[68] Kizhakkedathu, J.N., Norris-Jones, R., and Brooks, D.E. (2004), *Macromolecules*, 37, 734.

[69] Kizhakkedathu, J.N., Takacs-Cox, A., and Brooks, D.E. (2002). *Macromolecules*, 35, 4247.

[70] Guo, X., Weiss, A. and Ballauff, M. (1999). *Macromolecules*, 32, 6043.

[71] Guerrini, M.M., Charleux, B. and Varion, J.P. (2000). *Macromolecular Rapid Communications*, 21, 669.

[72] Kim, D.J., Heo, J.Y., Kim, K.S., and Choi, I.S. (2003). *Macromolecular Rapid Communications*, 24, 517.

[73] Matyajaszewski, K., Miller, P.J., Shukla, N., Immaraporn, B., Gelamn, A., Luokala, B.B., Siclovan, T.M., Kickelbick, G., Vallant, T., Hoffmann, H., and Pakula, T. (1999). *Macromolecules*, 32, 8716.

[74] von Werne, T. and Patten, T.E. (2001). *Journal of the American Chemical Society*, 123, 7497.

[75] Husseman, M., Malmstrom, E.E., McNamara, M., Mate, M., Mecerreyes, D., Benoit, D.G., Herdrick, J.L., Mansky, P., Huang, E., Russell, T.P., and Hawker, C.J. 1999, *Macromolecules*, 32, 1424.

[76] Huang, W.X., Kim, J.B., Bruening, M.L., and Baker, G.L. (2002). *Macromolecules*, 35, 1175.

[77] Kizhakkedathu, J.N., Goodman, D., and Brooks, D.E. In: Matyajaszewski, K. (Ed.), 2003, Advances in Controlled/Living Polymerization, ACS Symposium Series 854, American Chemical Society, Washington, DC.

[78] Matyajaszewski, K., Gaynor, S.G., Kulfan, A., and Podwika, M. (1997). *Macromolecules*, 30, 5192.

[79] Gewehr, M., Nakamura, K., Ise, N., and Kitani, H. (1992). *Makromolekular Chemie,* 193, 249.

[80] Hosoya, K., Sawada, E., Kimata, K., Araki, T., and Tanaka, N. (1994). *Macromolecules*, 27, 3973.

[81] Hosoya, K., Kimata, K., Araki, T., Tanaka, N., and Frechet, J. M. J. (1995). *Analytical Chemistry*, 67, 1907.

[82] Go, H., Sudo, Y., Hosoya, K., Ikegami, T., and Tanaka, N. (1998). *Analytical Chemistry*, 70, 4086.

[83] Duracher, D., Veyret, R., Elaissari, A., and Pichot, C. (2004). *Polymer International*, 53, 618.

[84] Duracher, D., Sauzedde, F., Elaissari, A., Perrin, A., and Pichot, C. (1998) *Colloid and Polymer Science*, 276, 219.

[85] Mittal, V., Matsko, N.B., Butte, A., and Morbidelli, M. (2008). *Macromolecular Materials and Engineering*, 293, 491.

[86] Mittal, V., Matsko, N.B., Butte, A., and Morbidelli, M. (2007). *Polymer*, 48, 2806.

[87] Mittal, V., Matsko, N.B., Butte, A., and Morbidelli, M. (2007). *European Polymer Journal*, 43, 4868.

[88] Marti, N., Quattrini, F., Butte, A., and Morbidelli, M. (2005). *Macromolecular Materials and Engineering*, 290, 221.

[89] Mittal, V., Matsko, N.B., Butte, A., and Morbidelli, M. (2008). *Macromolecular Reaction and Engineering*, 2, 215.

[90] Peters, E.C., Svec, F., and Frechet, J. M. J. (1997). *Advanced Materials*, 9, 630.

[91] Cammas, S., Suzuki, K., Sone, C., Sakurai, Y., Kataoka, K., and Okano, T. (1997). *Journal of Controlled Release*, 48, 157.

[92] Hu, T., You, Y., Pan, C., and Wu, C. (2002). *Journal of Physical Chemistry B*, 106, 6659.

INDEX

C

D

E

F

G

H

I

K

L

M

N

O

P

R

S

T

U

V

W

X

Y